Handbook of Wind Resource Assessment

Handbook of Wind Resource Assessment

Simon Watson
Delft University of Technology
Netherlands

WILEY

This edition first published 2023

Registered Offices
John Wiley & Sons, Inc., 111 River Street, Hoboken, NJ 07030, USA
John Wiley & Sons Ltd, The Atrium, Southern Gate, Chichester, West Sussex, PO19 8SQ, UK

For details of our global editorial offices, customer services, and more information about Wiley products visit us at www.wiley.com.

Wiley also publishes its books in a variety of electronic formats and by print-on-demand. Some content that appears in standard print versions of this book may not be available in other formats.

A catalogue record for this book is available from the Library of Congress

Hardback ISBN: 9781119055297; ePub ISBN: 9781119055396; ePDF ISBN: 9781119055303; oBook ISBN: 9781119055402

Cover Image: © Andrew Zarivny/Shutterstock
Cover Design: Wiley

Set in 9.5/12.5pt STIXTwoText by Integra Software Services Pvt. Ltd, Pondicherry, India

Contents

Preface *ix*
Acknowledgements *xi*
About the Author *xiii*

1 Introduction *1*
1.1 Early Wind Measurements *1*
1.2 The Need for Wind Resource Assessment *6*
1.3 A Brief Overview of the Wind Resource Assessment Process *7*
1.4 Layout of this Book *8*
References *9*

2 The Atmospheric Boundary Layer *11*
2.1 The Structure of the Atmospheric Boundary Layer *11*
2.2 The Surface Layer Wind Speed Profile *13*
2.3 The Gradient and Geostrophic Wind *22*
2.4 The Ekman Layer *27*
References *30*

3 Measurement *33*
3.1 History of Wind Speed Measurement *33*
3.2 Types of Measurement *34*
3.3 Terrestrial Measurements *35*
3.4 Remote Sensing *53*
3.5 Meteorological Measurement Stations *75*
3.6 The Effect of Climatic Conditions on Wind Speed Measurements *78*
References *80*

4 Wind Speed Variability and Distributions *85*
4.1 Power Spectrum *85*
4.2 The Spectral Gap *87*
4.3 Long-term Variation *88*

4.4 Synoptic Scale Weibull Wind Speed Distributions *90*
4.5 Weibull Parameter Variation with Height *94*
4.6 Synoptic Scale Wind Direction Distributions *95*
4.7 Turbulent Scale Distributions *99*
4.8 Extreme Value Statistics *101*
References *107*

5 Numerical Modelling *111*
5.1 The Governing Equations *111*
5.2 Microscale Models *114*
5.3 Mesoscale Models *136*
5.4 Global Models *142*
5.5 Model Chains *142*
5.6 The Importance of Atmospheric Stability *146*
References *150*

6 Wind Resource Estimation in Complex Terrain, Offshore, and in Urban Areas *157*
6.1 Coastal and Offshore *157*
6.2 Forested Terrain *163*
6.3 Urban Areas *169*
References *175*

7 Orographic Test Cases *179*
7.1 Askervein Hill *179*
7.2 Bolund Hill *184*
7.3 Big Southern Butte (BSB) and Salmon River Canyon (SRC) *187*
7.4 Perdigão *190*
7.5 Alaiz *193*
References *197*

8 Statistical Methods *199*
8.1 Spatial Coherence *199*
8.2 Measure–Correlate–Predict: An Overview *202*
8.3 Analogue Ensembles *210*
8.4 Wind Indices *213*
8.5 The Robust Coefficient of Variation *214*
8.6 Spatial Interpolation *216*
References *224*

9 Atmospheric Reanalyses and Wind Atlases *227*
9.1 What Is a Reanalysis? *227*
9.2 Global Reanalyses *229*
9.3 Regional Reanalyses *239*
9.4 Comparing Reanalyses *244*
9.5 Wind Atlases *250*
References *254*

10 Mesoscale Phenomena *263*
10.1 Thermally Driven Winds *263*
10.2 Internal Gravity Waves *266*
10.3 Low-level Jets *269*
10.4 Convective Cells *273*
10.5 Roll Vortices *275*
References *277*

11 Long-term Wind Climate Trends *281*
11.1 The Historical Evidence for Long-term Trends in Wind Speeds *281*
11.2 Climate Models *286*
11.3 Possible Future Changes and Uncertainties *287*
11.4 Impact on Resource Assessment *288*
References *289*

Index *291*

Preface

When Wiley first approached me to write this book, I naively thought it would be a process that would take about a year or so. Fast-forward five years and, owing to the pressures of work and a change of job (and country), progress had been painfully slow. Then came the infamous global pandemic. Finally, the constraints of lockdown focused my mind and gave me the required impetus to (electronically) put pen to paper and deliver the completed manuscript to a very patient publisher. I say 'complete', though I still consider it a work in progress in many ways. Nonetheless, I hope it now contains sufficient useful information to practitioners in the field of wind resource assessment.

Wind energy has come a long way since I started working in the field back in 1990. It promises to make a major contribution to the energy transition as the world attempts to move away from the climate-damaging effects of burning fossil fuels. Where to site wind turbines and, just as importantly, where not to site them is a very practical and economic consideration. An accurate prediction of the expected wind resource at a potential wind farm site is, therefore, vital. Several books have been written about wind resource assessment, but I wanted something that was a practical guide with enough underlying theory for students and researchers to understand the measurement techniques that are used and the physical and statistical processes that are being modelled in trying to predict the behaviour of the wind. I also wanted to give some impression of the limitations of the measurements and models and their accuracy. In so doing, I have drawn on my experience of over 30 years of research in the field of wind energy and wind resource and the material that I have taught my students. I have also tried to draw on a wide range of literature to cover the different aspects of wind resource assessment in the many environments where wind turbines and wind farms are sited. I have deliberately stayed away from translating wind resource into energy yield (apart from a couple of very brief forays) as to cover the topic of wind farm yield, including wakes, would warrant a book on its own. Maybe this will be my next book when I have had enough time to recover from this venture.

I hope you will find this book a useful reference in understanding the science and engineering of wind resource assessment. Comments and feedback are welcome as I work on the next edition of what is an evolving field of research.

Simon Watson
The Hague

Acknowledgements

I would like to thank Dries Allaerts for taking the time to proofread the manuscript and give helpful advice and comments. I would also like to thank the many people and companies for permission to use their figures and photographs, especially Adolf Thies, AQSystem, AWI, Campbell Scientific, Cian Desmond, DTU Wind Energy, Srinidhi Gadde, Gill Instruments, Nergica, Radiometrics, ReSoft, RPG, Scintech, Setra, Richard Stevens, Vaisala, Vector Instruments, WindSensor, and ZX Lidars. Finally, I would like to thank my family for all their support, in particular my wife, Christiane, for her love and support and for the numerous cups of tea provided during the long evenings drafting the manuscript.

About the Author

Simon Watson is Professor of Wind Energy Systems at Delft University of Technology. He has worked in the field of wind energy for over 30 years having graduated with a degree in physics from Imperial College London and a PhD from Edinburgh University. He began his career at the Energy Research Unit of the Rutherford Appleton Laboratory, near Oxford, UK before moving to a small start-up company which later became the green electricity supply company Good Energy. In 2001, he moved to the Centre for Renewable Energy Systems Technology at Loughborough University, Leicestershire, UK, becoming a full professor in 2010. In 2017, he moved to the Netherlands to take up his present position. Professor Watson has published widely in a range of wind energy topics, including resource assessment, wind turbine condition monitoring, and wind energy integration.

1

Introduction

1.1 Early Wind Measurements

Wind power has come a long way since the earliest windmills were built to grind corn for flour or wind pumps installed to drain the marshes for growing crops. The siting of these early machines was driven more by the practical necessity to be close to the local populace than for maximum economic benefit. Estimates of the local wind climate could mostly be obtained from local informed opinion or visual clues such as the angle of growth of trees, or indeed the complete absence of trees at extremely windy sites. Measurement of wind direction, as opposed to wind speed, has been made for far longer. One of the earliest-known examples of this is a weathervane in the shape of the god Triton which stood atop the Tower of the Winds in Athens when it was first built around 50 BCE or possibly earlier (see Figure 1.1). Weathervanes on buildings such as churches in the Western world have also been a common sight for centuries. Although giving an indication of the local wind direction at the time of observation, such early devices were not generally used in the recording of historical wind direction distribution. Before the development of instruments to measure the magnitude of the wind speed, human perception or slightly more objective visual indicators were used to provide a wind speed scale. For example, the famous English writer Daniel Defoe, following the Great Storm of 1703 which caused significant destruction in southern England, proposed an 11-point scale based on common phrases, as detailed in Table 1.1 (The Weather Window, n.d.). Scale point 6, describing 'a top sail gale', gives a clue to the importance of such information for the maritime community, and several such empirical wind speed scales devised by sailors were known to exist in the seventeenth century and probably much earlier. Possibly the most famous such scale was developed in 1805 by the

Handbook of Wind Resource Assessment, First Edition. Simon Watson.

Figure 1.1 The Tower of Winds in Athens. *Source:* Joanbanjo / Wikipedia Commons / CC BY-SA 3.0.

Irish hydrographer Francis Beaufort, a Royal Navy officer who later became a rear admiral and trained Robert FitzRoy who, in turn, founded what later became the UK Meteorological Office. The so-called Beaufort scale was designed as a 13-point scale based on the impact the wind speed had on the sails of a ship. These initial visual descriptors would be later converted into

Table 1.1 Daniel Defoe's proposed verbal wind speed scale from around 1704.

Scale point	Description
0	Stark calm
1	Calm weather
2	Little wind
3	A fine breeze
4	A small gale
5	A fresh gale
6	A top sail gale
7	Blows fresh
8	A hard gale of wind
9	A fret of wind
10	A storm
11	A tempest

an actual measurement scale, which is still in use today for some applications (e.g. shipping broadcasts) and is detailed in Table 1.2.

Table 1.2 describes three units of measurement including the knot (often abbreviated to 'kt', or 'kts' if plural) which may be less familiar. This is a unit of speed derived from the days of sailing ships when sailors would use a long length of rope to which pieces of wood would be tied using knots at regular intervals which would be paid out from the stern (rear) of a ship as it travelled for a defined period of time determined using an hourglass. The number of knotted pieces of wood paid out in this time period would be used to calculate the speed of the ship. The modern knot is defined as one nautical mile (1.852 km) per hour which is equivalent to 0.514 ms^{-1}. This unit is still used to describe the speed of ships or aircraft and is sometimes used by meteorological agencies to measure wind speed, though the unit of ms^{-1} is more widely used nowadays.

These early wind measurements were primarily of interest for mariners where the wind was an important source of motive power. Today, it is generally only leisure shipping which still uses the wind to provide propulsion. However, now our attention has turned to use of the wind for generating electrical power and the quest to decarbonise the energy system. In this book, we look at the science of wind resource assessment from the point of view of measurement and modelling.

Table 1.2 The Beaufort scale with modern equivalent units of wind speed.

Wind force	Description	Wind speed			Specifications (*italics refer to conditions at sea*)	Wave height		Sea state
		km/h	mph	knots		Probable (m)	Max. (m)	
0	Calm	<1	<1	<1	Smoke rises vertically *Sea like a mirror*	—	—	0
1	Light air	1–5	1–3	1–3	Direction shown by smoke drift but not by wind vanes *Sea rippled*	0.1	0.1	1
2	Light breeze	6–11	4–7	4–6	Wind felt on face; leaves rustle; wind vane moved by wind *Small wavelets on sea*	0.2	0.3	2
3	Gentle breeze	12–19	8–12	7–10	Leaves and small twigs in constant motion; light flags extended *Large wavelets on sea*	0.6	1.0	3
4	Moderate breeze	20–28	13–18	11–16	Raises dust and loose paper; small branches moved *Small waves, fairly frequent white horses*	1.0	1.5	3–4
5	Fresh breeze	29–38	19–24	17–21	Small trees in leaf begin to sway; crested wavelets form on inland waters *Moderate waves, many white horses*	2.0	2.5	4

6	Strong breeze	38–49	25–31	22–27	Large branches in motion; whistling heard in telegraph wires; umbrellas used with difficulty *Large waves, extensive foam crests*	3.0	4.0	5
7	Near gale	50–61	32–38	28–33	Whole trees in motion; inconvenience felt when walking against the wind *Foam blown in streaks across the sea*	4.0	5.5	5–6
8	Gale	62–74	39–46	34–40	Twigs break off trees; generally impedes progress *Wave crests begin to break into spindrift*	5.5	7.5	6–7
9	Strong gale	75–88	47–54	41–47	Slight structural damage (chimney pots and slates removed) *Wave crests topple over, spray affects visibility*	7.0	10.0	7
10	Storm	89–102	55–63	48–55	Seldom experienced inland; trees uprooted; considerable structural damage *Sea surface largely white*	9.0	12.5	8
11	Violent storm	103–117	64–72	56–63	Very rarely experienced; accompanied by widespread damage *Medium-sized ships lost to view behind waves; sea covered in white foam; visibility seriously affected*	11.5	16.0	8
12	Hurricane	≥118	≥73	≥64	Devastation *Air filled with foam and spray; very poor visibility*	≥14	—	9

Source: taken from The Royal Meteorological Society (n.d.).

1.2 The Need for Wind Resource Assessment

The need for accurate assessment of the wind conditions at a site has been driven by the rapid expansion of wind power worldwide. By the end of 2020, a total capacity of 743 GW had been installed around the globe (Lee and Zhao, 2021). The global weighted-average installed cost for onshore wind energy in 2020 was \$1355 kW^{-1} (IRENA, 2020). For a 100 MW wind farm this equates to a total investment of \$140 million. The equivalent cost offshore is somewhat higher at around \$3185 kW^{-1} (though costs have fallen significantly) and a 'typical' offshore wind farm may have a combined capacity approaching 1000 MW. This equates to an investment cost of just over \$3 billion. In terms of the levelised cost of energy, the most recent figures are stated to be \$0.041 kW^{-1} onshore and \$0.084 kW^{-1} offshore (IRENA, 2020). A well-used measure of the operational efficiency of a wind farm is that of the capacity factor which is the ratio of the long-term power output of a farm to its total rated power capacity. A typical onshore wind farm may have a capacity factor of 25–35%, and an offshore wind farm may exceed 50%. Figure 1.2 shows the dependence of capacity factor on average wind speed (Watson et al., 2015) based on a Rayleigh distribution of wind speeds (wind speed distributions are discussed in Chapter 4). Assuming an onshore average hub height wind speed of 7 ms^{-1}, and a value of 10 ms^{-1} offshore, a 5% uncertainty in the long-term wind speed estimate represents an uncertainty of 9% in the capacity factor for the onshore farm and 5% for the offshore site in this example.

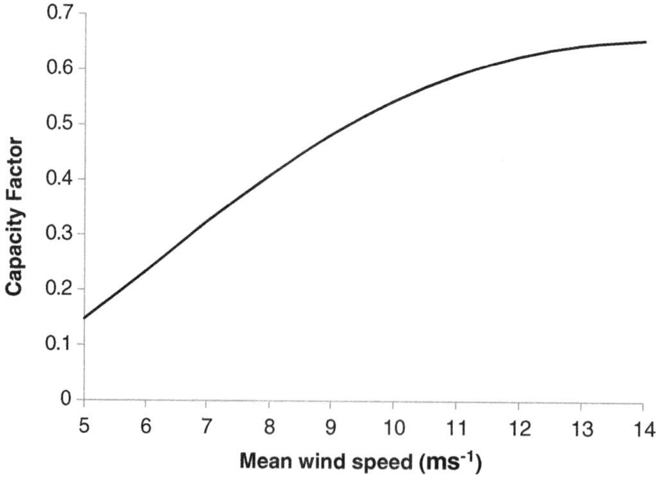

Figure 1.2 Capacity factor of a Vestas V80 2MW wind turbine as a function of wind speed. *Source:* adapted from Watson et al. (2015).

This uncertainty equates to $24 million for a 100 MW onshore wind farm and $460 million for a 1000 MW offshore wind farm. It is clear from these numbers that a modest investment to reduce the uncertainty in a long-term mean wind speed estimate for a site can have a large impact in terms of the viability of a wind farm, where frequently the lowest projected energy yield given the expected uncertainties is used in a financial investment decision.

1.3 A Brief Overview of the Wind Resource Assessment Process

To assess the long-term energy yield of a wind farm is a multistep process and is shown schematically in Figure 1.3. Initially, a wind atlas may be used to assess the approximate wind resource over a wide area to identify a candidate wind farm site. This site will then be the subject of a more comprehensive measurement campaign based on measurements made onsite at one or more masts covering the area of potential interest. A measurement campaign, for practical reasons, may last only 1–2 years, which is insufficient to capture the climatological variability that would be expected over the lifetime of a typical wind farm (20–25 years). To account for this, some form of long-term correction is required. This, in turn, will only give detailed information about the long-term resource at

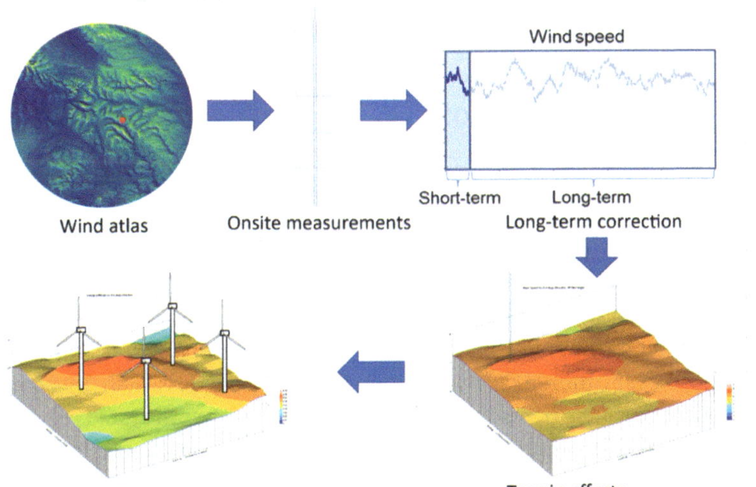

Figure 1.3 The wind resource assessment process. The scope of this book is shown within the dashed box. *Source:* the 3D surface plots were reproduced from ReSoft WindFarm software (resoft.co.uk) / with permission of Alan Harris.

the mast (or masts). The next step is to extrapolate the wind speed conditions from the mast (or masts) over the wider area within which a wind farm is expected to be built. This area will depend on several factors, including the area available; the expected total capacity of the wind farm, which will be driven by factors that may be economically related or linked to the capacity of the nearby grid; and the size of the turbines. The spatial extrapolation of the wind speed, both horizontally and vertically, to take account of different potential turbine sizes, is generally carried out using a physical or statistical model which can be of various levels of sophistication depending on the complexity of the terrain and the size of the wind farm, and thus the size of the investment. The final step is to determine the optimum siting of the turbines considering the spatial variation in the wind speed and the aerodynamic interactions between the turbines (i.e. array effects, including wakes). This last step may require an iterative process, so it may be that a simpler, faster numerical, or physically based empirical model is used first that may be slightly less accurate than a more sophisticated model but is accurate enough to determine an optimum layout. A final stage of fine-tuning the layout and determining the expected long-term energy yield may then be carried out using a more sophisticated but computationally intensive numerical model. It should be noted that the emphasis of this book is on assessing the long-term *wind speed* rather than the energy yield and so only the steps within the dashed box in Figure 1.3 are covered in the remaining chapters. Array effects are relatively complex and warrant a detailed book in themselves. It is for this reason that they are considered outside the scope of this book.

1.4 Layout of this Book

The remainder of this book describes the important elements in the wind resource assessment process. Chapter 2 introduces the basic atmospheric properties of the wind and the structure of the atmospheric boundary layer. Chapter 3 looks at the measurement of the key variables in the process of assessing wind resource, including a description of the most commonly used instruments. Chapter 4 considers the temporal variability in wind speed over different scales, ranging from years to less than a second, and their importance in resource assessment and reviews the most popular mathematical distributions which are used to characterise the statistical properties of the wind for the long and short term. Chapter 5 describes how the spatial variation of the wind is modelled, starting with the governing equations for fluid flow and the models which are based on these.

Chapter 6 looks at empirical and semi-empirical modelling approaches which can be used to predict spatial variation in the wind in specific cases. Chapter 7 reviews some of the well-known orographic test cases which have been used to validate and test different wind resource model predictions. Chapter 8 considers statistical approaches which are used to extrapolate short-term wind speed measurements to the long-term and techniques to interpolate spatial data that do not require complex physical models. Chapter 9 describes the technique of reanalysis to reconstruct long-term wind speed time series. It also looks at some of the more popular reanalysis datasets, and their use in wind resource assessment, including the production of wind atlases. Chapter 10 reviews mesoscale phenomena and how they might potentially affect the variability of the local wind resource. Chapter 11 reviews the evidence for potential future trends in global wind speed patterns and how this may affect wind resource assessment.

References

IRENA, 2020. *Renewable Power Generation Costs in 2020*. International Renewable Energy Agency.

Lee, J., Zhao, F., 2021. *GWEC global wind report*. Global Wind Energy Council. https://gwec.net/global-wind-report-2021/ (accessed 3 October 2022).

The Royal Meteorological Society, n.d. *The Beaufort scale*. https://www.rmets.org/metmatters/beaufort-scale (accessed 29 April 2022).

Watson, S.J., Kritharas, P., Hodgson, G.J., 2015. Wind speed variability across the UK between 1957 and 2011. *Wind Energy* 18: 21–42.

The Weather Window, n.d. *Historical and contemporary versions of the Beaufort scale*. https://weather.mailasail.com/Franks-Weather/Historical-And-Contemporay-Versions-Of-Beaufort-Scales (accessed 29 April 2022).

2

The Atmospheric Boundary Layer

In wind resource assessment, the spatial and temporal characteristics of wind are most relevant close to the ground, though with the advent of airborne wind energy (AWE) the resource higher up is of some relevance. Nonetheless, the conditions within the atmospheric boundary layer (ABL) are of most importance for modern wind energy generation systems. In this chapter, the structure of the ABL and how it may be parameterised for the purposes of wind resource assessment are described.

2.1 The Structure of the Atmospheric Boundary Layer

The ABL defines that region close to the earth's surface where the influence of the surface has an impact on the wind profile, above which is the free atmosphere where the wind is assumed to be geostrophic, which is described in Section 2.3. Typically, the height of the ABL, h_{ABL}, is around 1–2 km during daytime when the air is well mixed but can be much lower during calm night-time periods. The ABL is divided into two main layers: the Ekman layer and the surface layer as shown in Figure 2.1.

In the Ekman layer, the behaviour of the wind is little influenced by the surface but is strongly influenced by the Coriolis force and there is a significant turning of the wind direction with height. The surface layer extends to around 100–200 m above the ground, and this is the layer of most significance for wind energy. In this layer, Coriolis effects are somewhat less important, but the wind profile is strongly influenced by the characteristics of the underlying surface. This layer consists of an upper inertial sublayer where the wind profile shows a strong shear with height which depends on the orography and the aerodynamic surface roughness length z_0. Below z_0, in the roughness sublayer, the wind flow is stagnant and molecular

Handbook of Wind Resource Assessment, First Edition. Simon Watson.

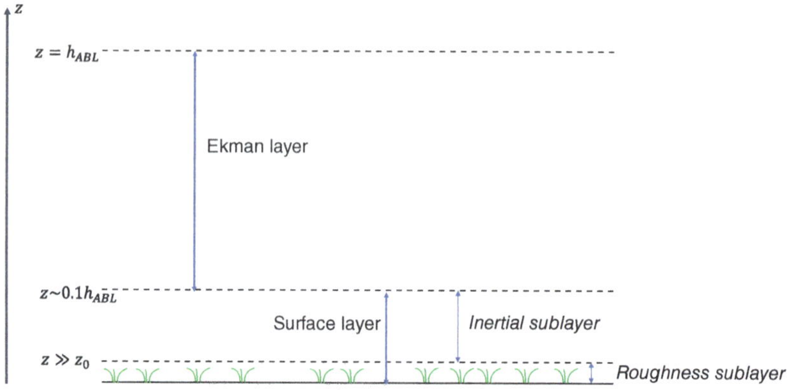

Figure 2.1 The structure of the atmospheric boundary layer (ABL): h_{ABL} is the height of the ABL, z is height above the surface, and z_0 is the aerodynamic roughness length due to roughness elements on the surface. *Source:* adapted from Garratt (1994).

diffusion processes are important. From observations, a classification scheme has been developed relating terrain type to roughness length. This was originally developed by Davenport and was later updated by Wieringa (Davenport, 1960; Wieringa, 1992). Table 2.1 describes this scheme and also lists the approximate equivalent shear exponent factor for flat terrain explained in Section 2.2.

Over water surfaces, the roughness length is not constant and depends on the wind speed above the surface. A commonly used expression to calculate the roughness length over a water surface which is applicable for open sea is the *Charnock relationship* (Charnock, 1955)

$$z_0 = \alpha_c \frac{u_*^2}{g} \tag{2.1}$$

where α_c is a constant determined from empirical observations and has a value of ≈ 0.014 (although it has been seen to have a higher value for coastal sites), $g = 9.81\ \mathrm{ms^{-2}}$ is the acceleration due to gravity, and u_* is the friction velocity, which, as we will see in Section 2.2, is related to shear stress at the surface. Other more sophisticated relationships have been formulated to take account of sea fetch and wave age, but this equation remains the most used in wind resource assessment.

Table 2.1 Wieringa's updated version of Davenport's original roughness classification scheme (Wieringa, 1992). Approximately equivalent flat terrain shear exponent factors (α) are also shown (see Section 2.2).

z_0 (m)	α	Terrain description
0.0002 (Sea)	0.09	Open sea or lake (irrespective of the wave size), tidal flat, snow-covered flat plain, featureless desert, tarmac, and concrete, with a free *fetch* (length of sea over which the wind blows) of several kilometres
0.005 (Smooth)	0.12	Featureless land surface without any noticeable obstacles and with negligible vegetation (e.g. beaches, pack ice without large ridges, morass, and snow-covered or fallow open country)
0.03 (Open)	0.16	Level country with low vegetation (e.g. grass) and isolated obstacles with separations of at least 50 obstacle heights (e.g. grazing land without windbreaks, heather, moor, and tundra, runway area of airports)
0.10 (Roughly open)	0.20	Cultivated area with regular cover of low crops, or moderately open country with occasional obstacles (e.g. low hedges, single rows of trees, isolated farms) at relative horizontal distances of at least 20 obstacle heights
0.25 (Rough)	0.24	Recently developed 'young' landscape with high crops or crops of varying height, and scattered obstacles (e.g. dense shelterbelts, vineyards) at relative distances of about 15 obstacle heights
0.5 (Very rough)	0.29	'Old' cultivated landscape with many rather large obstacle groups (large farms, clumps of forest) separated by open spaces of about 10 obstacle heights Also, low large vegetation with small interspaces, such as bushland, orchards, young densely planted forest
1.0 (Closed)	0.37	Landscape totally and quite regularly covered with similar-size large obstacles, with open spaces comparable to the obstacle heights (e.g. mature regular forests, homogeneous cities or villages)
2.0 (Chaotic)	$\geq$ 0.49	Centres of large towns with mixture of low-rise and high-rise buildings Also, irregular large forests with many clearings

2.2 The Surface Layer Wind Speed Profile

When estimating the mean wind speed at the hub height of a wind turbine, it is important to have an accurate model of the variation of wind speed with height in the surface layer. An engineering approximation to calculate the wind speed u_2 at a height z_2 based on the wind speed u_1 at a height z_1 is the *power law*

$$u_2 = u_1 \left(\frac{z_2}{z_1} \right)^{\alpha} \tag{2.2}$$

where α is the shear exponent factor, which depends on the terrain orography, surface coverage, and surface layer stability. Table 2.1 gives approximate values of α for different terrain types, assuming flat terrain and a neutral surface layer.

The most widely used parameterisation of the vertical variation in wind speed which has a more physically based justification is the *logarithmic profile*, the stability dependent (*diabatic*) form of which is given by the integration of the surface layer relationship

$$\frac{\partial u}{\partial z} = \frac{u_*}{\kappa z} \phi_m \left(\frac{z}{L} \right) \tag{2.3}$$

to give

$$u(z) = \frac{u_*}{\kappa} \left[\ln \left(\frac{z}{z_0} \right) - \psi_m \left(\frac{z}{L} \right) \right] \tag{2.4}$$

where κ is the von Kármán constant (≈ 0.4), ϕ_m is a stability function which depends on height z, and *Monin–Obukhov length* L and ψ_m is given by

$$\psi_m = \int_0^{z/L} \frac{(1 - \phi_m)}{z'} dz' \tag{2.5}$$

Strictly, the logarithmic profile is only applicable in flat homogeneous terrain. The friction velocity u_* relates to the shear stress τ_x at the ground and the air density ρ

$$u_* = \sqrt{\frac{\tau_x}{\rho}} \tag{2.6}$$

The shear stress for a Newtonian fluid is given by the expression

$$\tau_x = \mu \frac{\partial u}{\partial z} \tag{2.7}$$

where μ is the dynamic viscosity. This equation is normally applied on the microscopic scale, but by analogy it can be applied at the macroscopic scale for wind flow where the microscopic viscosity is replaced by an eddy viscosity K_M.

An alternative expression for τ_x relates the stress to the rate at which the horizontal momentum of the air is transferred vertically to the surface by turbulence

$$\tau_x = -\rho\overline{u'w'} \tag{2.8}$$

where u' and w' are turbulent fluctuations in the streamwise and vertical wind speeds, respectively. Combining Eqs. (2.6) and (2.8) gives

$$u_*^2 = -\overline{u'w'} \tag{2.9}$$

So u_* is a measure of the momentum flux close to the ground and is assumed to be constant in the surface layer. This means that Eq. (2.4) can be used to extrapolate a measured wind speed from one height to another in a similar way to the power law.

At this point, it is useful to introduce several variables:

- T : the absolute temperature
- T_v : the virtual absolute temperature
- θ : the potential temperature
- θ_v : the virtual potential temperature.

The virtual absolute temperature T_v of a moist parcel of air is the equivalent absolute temperature T at which a dry parcel of air would have a total pressure and density the same as the moist parcel of air and is given by

$$T_v = T\,(1 + 0.61q) \tag{2.10}$$

where q is the specific humidity which defines the mass of water vapour per unit mass of moist air.

The potential temperature θ is the absolute temperature that would result if a parcel of dry air were brought adiabatically[1] to a reference pressure ($p_{ref} = 1000\,\text{hPa}$)

$$\theta = T\left(\frac{p}{p_{ref}}\right)^{-R_d/C_p} \tag{2.11}$$

where R_d is the specific gas constant of dry air ($= 287.05\,\text{J kg}^{-1}\cdot\text{K}^{-1}$) and C_p is the specific heat capacity at constant pressure ($= 1005.7\,\text{J kg}^{-1}\cdot\text{K}^{-1}$ at 273 K for dry air).

The virtual potential temperature θ_v is the equivalent of θ for a moist parcel of air based on the virtual absolute temperature T_v, i.e.

$$\theta_v = T_v\left(\frac{p}{p_{ref}}\right)^{-R_d/C_p} \tag{2.12}$$

1 Without any heat transfer to or from the parcel of air.

In a neutral surface layer, the potential temperature (or virtual potential temperature if the air is moist) is constant with height. This situation arises when there is no net heating or cooling of the underlying surface. If the surface undergoes heating during the day, it will be warmer than the air above. A parcel of air that starts to move upwards will be warmer than its surroundings and thus will gain buoyancy, which increases with height as the potential temperature reduces with height. This creates what is called an *unstable* or *convective boundary layer* (CBL) where turbulent mixing of the air is promoted throughout the ABL. The third case arises when the ground cools, for example at night. In this case, the ground is cooler than the overlying air and the potential temperature increases with height. A parcel of air perturbed upwards will thus be cooler than its surroundings and will sink back down to its equilibrium position. This creates a stable surface layer where turbulent mixing is suppressed. This variation in turbulent mixing with a surface stability condition is important as it affects the wind speed variation with height or the wind shear. The shear is increased during stable conditions and decreased during unstable conditions relative to a neutral surface layer. This is parameterised by the ϕ_m function in Eq. 2.3.

There are a number of empirical formulations of the ϕ_m function including those of Businger and Dyer (Businger et al., 1971; Dyer, 1974). Under stable conditions ($0 \leq z/L \leq 1$)

$$\phi_m = 1 + \frac{5z}{L} \tag{2.13}$$

giving

$$\psi_m = -\frac{5z}{L} \tag{2.14}$$

Under unstable conditions ($-2 \leq z/L \leq 0$)

$$\phi_m = \left(1 + 16\left|\frac{z}{L}\right|\right)^{-\frac{1}{4}} \tag{2.15}$$

giving

$$\psi_m = 2\ln\left[\frac{(1+x)}{2}\right] + \ln\left[\frac{(1+x^2)}{2}\right] - 2\tan^{-1}(x) + \frac{\pi}{2} \tag{2.16}$$

where

$$x=\left(1-16\frac{z}{L}\right)^{\frac{1}{4}} \tag{2.17}$$

In the limit where $L \rightarrow \infty$, then $\phi_m \rightarrow 1, \psi_m \rightarrow 0$ and the logarithmic profile corresponds to that of a neutral surface layer. This is an approximation that is frequently made in the absence of stability information, in which case the logarithmic profile is said to be *adiabatic*.

Figure 2.2 shows an example of three logarithmic profiles calculated using Eq. (2.4) and Eqs. (2.14)–(2.17) with a roughness length $z_0 = 0.0003$ m (corresponding to conditions offshore), a common wind speed of 6 ms^{-1} at 10 m above the surface and three stability conditions corresponding to stable, neutral, and unstable conditions. The wind shear differs markedly for three different stability classes, highlighting the importance of taking stability conditions into account when extrapolating a wind profile to calculate the wind resource at turbine hub height, particularly if the measurement height is significantly different from the hub height.

The Monin–Obukhov length, or often just the Obukhov length L, was defined by Monin and Obukhov when they developed their similarity relationships, which determine the effects of buoyancy on turbulent flows (Monin and Obukhov, 1954)

$$L=-\frac{u_*^3\overline{\theta}_v}{\kappa g\left(\overline{w'\theta_v'}\right)_0} \tag{2.18}$$

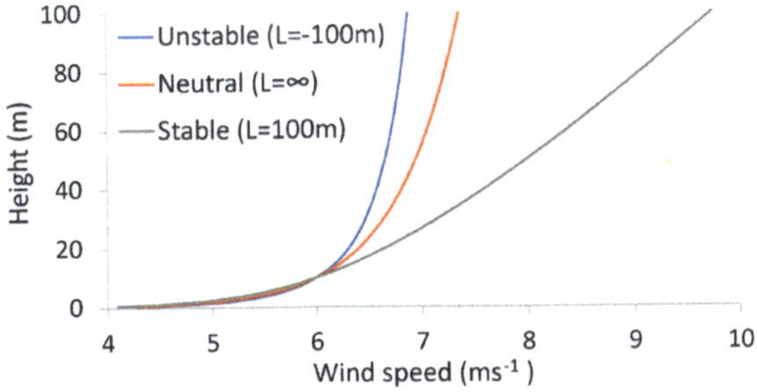

Figure 2.2 Logarithmic profiles for three different stability values, z_0 = 0.0003 m and a common reference wind speed of 6 ms^{-1} at 10 m above the surface.

where $\overline{\theta}_v$ is the mean surface layer virtual potential temperature, and θ_v' is the turbulent fluctuation in virtual potential temperature. The quantity $\left(\overline{w'\theta_v'}\right)_0$ represents the virtual potential temperature flux at the surface. Using a 3D sonic anemometer, it is possible to measure L directly. However, a lower cost solution relies on the use of an alternative metric which can be converted into L, namely the *gradient Richardson number*

$$Ri_g = \frac{\dfrac{g}{\overline{\theta}_v}\dfrac{\partial \overline{\theta}_v}{\partial z}}{\left(\dfrac{\partial \overline{u}}{\partial z}\right)^2 + \left(\dfrac{\partial \overline{v}}{\partial z}\right)^2} \tag{2.19}$$

This expression relies on knowledge of the instantaneous vertical gradient of the virtual potential temperature and the wind horizontal speed components $\bar{u}$ and $\bar{v}$. However, a practical approximation to this is to infer the gradient Richardson number over a finite layer of the atmosphere

$$Ri_g = \frac{\dfrac{g}{\overline{\theta}_v}\dfrac{\Delta \overline{\theta}_v}{\Delta z}}{\left(\dfrac{\Delta \overline{u}}{\Delta z}\right)^2 + \left(\dfrac{\Delta \overline{v}}{\Delta z}\right)^2} = \frac{g \Delta \overline{\theta}_v \Delta z}{\overline{\theta}_v \left((\Delta \overline{u})^2 + (\Delta \overline{v})^2\right)} \tag{2.20}$$

This merely requires the measurement of the wind speed and virtual potential temperature at two heights. A further approximation that is made is to estimate the gradient Richardson number taking the lower height to be the ground or sea surface where the wind speed is zero. This is known as the *bulk Richardson number*

$$Ri_b = \frac{g\left(\overline{\theta}_v - \overline{\theta}_{v0}\right)z}{\overline{\theta}_v \left((\overline{u})^2 + (\overline{v})^2\right)} \tag{2.21}$$

where $\overline{\theta}_v$, $\overline{u}$, and $\overline{v}$ are measured at height z, and $\overline{\theta}_{v0}$ is measured at the ground or sea surface. When calculating the bulk Richardson number, care is required when determining the surface layer temperature to ensure it is truly the interfacial temperature between air and ground (or sea). This is especially true offshore, when temperature sensors can be located up to a few metres under the water. In this case, a skin cooling effect correction is required (Fairall et al., 1996).

If the specific humidity does not vary significantly over the two heights, which is a reasonable approximation for a height difference < 100 m (except at sea where the difference can be more significant), $\overline{\theta}_v$ can be substituted by $\overline{\theta}$. Moreover, by assuming hydrostatic balance

$$\frac{\partial p}{\partial z} = -\rho g \tag{2.22}$$

and using the equation of state

$$p = \rho R_d T \tag{2.23}$$

it can be shown that

$$\frac{1}{\theta}\frac{\partial \theta}{\partial z} = \frac{1}{T}\left(\frac{\partial T}{\partial z} + \frac{g}{C_p}\right) \tag{2.24}$$

gives the linear approximation, which can be used in the surface layer

$$\theta = T + \frac{g}{C_p}\Delta z \tag{2.25}$$

The quantity $g/C_p = 9.8\,\mathrm{K\,km^{-1}}$ is known as the *dry adiabatic lapse rate.*

Thus, it is possibly to estimate Ri_g using a standard meteorological mast with cup anemometers and temperature sensors at two heights separated by Δz. The value of Ri_g can be converted to L using the empirical formulae (Kaimal and Finnigan, 1994)

$$\frac{z}{L} = Ri_g \quad \text{for } -2 < Ri_g < 0 \text{ (unstable conditions)} \tag{2.26}$$

$$\frac{z}{L} = \frac{Ri_g}{1 - 5Ri_g} \quad \text{for } 0 < Ri_g < 0.2 \text{ (stable conditions)} \tag{2.27}$$

The conversion of Ri_b to Ri_g (and thus L) is not so straightforward and usually involves the use of a nomogram. Nonetheless, there are equations of a similar form to Eqs. (2.26) and (2.27) which have been proposed and validated against measurements (Grachev and Fairall, 1997)

$$\frac{z}{L} = \frac{10Ri_b}{1 - Ri_b/4.5} \quad \text{for } -2 < Ri_b < 0 \text{ (unstable conditions)} \tag{2.28}$$

$$\frac{z}{L} = \frac{10Ri_b}{1 - 5Ri_b} \quad \text{for } 0 < Ri_b < 0.2 \text{ (stable conditions)} \tag{2.29}$$

(N.B. It should be noted that Eq. (2.28) has been validated against offshore data only (Grachev and Fairall, 1997), and although Eq. (2.29) is theoretically valid for Ri_b up to 0.2, it has only been validated up to around 0.04 (Högström, 1988)).

The stability parameters Ri_g, Ri_b, and L reflect the relative balance between buoyancy and inertial forces. At higher wind speeds, the inertial forces tend to dominate and surface layer stability becomes more neutral. Moreover, due to the different levels of irradiation and thus surface heating, surface layer stability shows strong diurnal and seasonal cycles, especially on land. An example of the stability conditions at Høvsøre, Denmark is shown in Figure 2.3. Høvsøre is on the west coast of Denmark and the stability conditions differ significantly by direction showing far less variation with time of day from the seaward facing direction. This can be understood

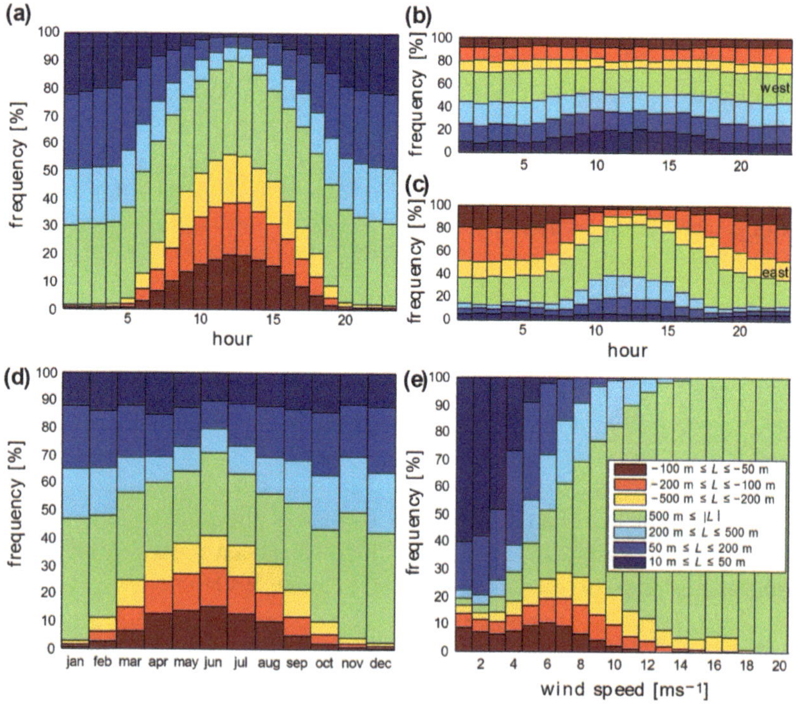

Figure 2.3 Stability conditions at Høvsøre, Denmark by: (a, b, c) hour of the day, (d) month, and (e) wind speed. Plots (a), (d), and (e) are based on a sonic anemometer on the main meteorological mast at 10 m above ground level. Plots (b) and (c) are based on a 160 m sonic anemometer on a nearby lighting tower for selected directions. *Source:* taken from Peña et al. (2016) / Springer Nature / CC BY 4.0.

considering the large thermal mass of the sea whose temperature changes little during the day. The fetch (length of sea over which the wind blows) is largely from the west, so the airmass coming from that direction can be assumed to have reached thermal equilibrium with the surface and so its temperature will also change little during the day (see Figure 2.4 as air moves from sea to land). This means that the diurnal variation in stability is likely to be small as observed.

An approximation made in extrapolating wind speed with height using Eq. 2.4 is that conditions are mainly neutral above around $6\,\mathrm{ms}^{-1}$. This is broadly justified by Figure 2.3(e), where the green band represents neutral conditions. For comparison, the stability conditions by wind speed are shown for two offshore sites, Egmond aan Zee and Horns Rev, in Figure 2.5. The class labelled N is comparable with the neutral class '$500\,\mathrm{m} \leq |L|$' in Figure 2.3. Neutral conditions do not dominate until much higher wind speeds ($\gtrsim 15\,\mathrm{ms}^{-1}$). Stability conditions at sea for sites relatively close to the coast often show significant periods of non-neutral stability. Air masses on land undergo significant diurnal variation in temperature. As they move (advect) from land to sea with the wind, they may have a very different temperature to the sea surface thus creating a stratified layer with a strong temperature gradient (see Figure 2.4 as air moves from land to sea). This highlights the important of taking stability into account when using the logarithmic profile, particularly for offshore sites less than 100 km from

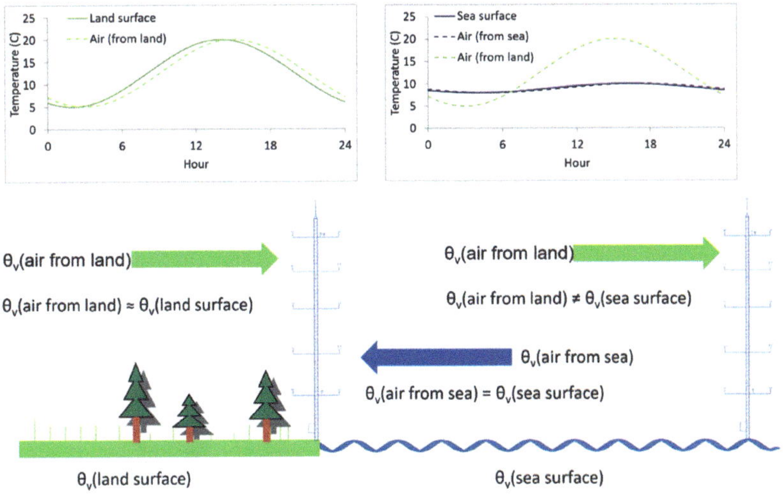

Figure 2.4 Temperature stratification at a coastal mast with an onshore wind and at offshore mast with an offshore wind.

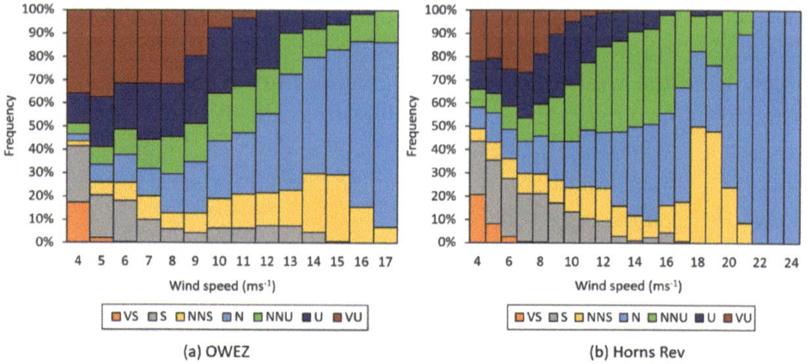

Figure 2.5 Stability conditions by wind speed at (a) Egmond aan Zee and (b) Horns Rev. *Source:* reproduced from Sathe et al. (2011) / with permission of John Wiley & Sons.

land. Even for those offshore sites far from land, the movement of warm and cold air masses associated with weather fronts can lead to stratified conditions and frequent periods of non-neutral stability.

2.3 The Gradient and Geostrophic Wind

As the earth rotates, different parts of its surface are exposed to the sun, and the radiation flux varies with latitude being highest near the equator and lowest at the poles. This sets up temperature and pressure gradients as heat flows from the equator to the poles. The earth's surface is far from uniform, and this complicates things further as the thermal properties of the different surface types (e.g. sea, plains, forests, deserts, etc.) vary significantly. The atmosphere is thus a patchwork quilt of pressure gradients and air masses of different temperatures. It is this which drives the weather and, of course, the winds we experience. A pressure gradient will drive a tendency for air to flow from high to low pressure. However, this is not the full picture; the earth is rotating once per day on its axis, so as an air mass moves, it experiences both a centrifugal and a Coriolis force within the rotating frame of reference of the earth. The second of these forces is the most significant and causes a significant deflection as a large air mass moves from high to low pressure. This can be visualised by looking at Figure 2.6. In this figure, an air mass is initially moving from point A to point B in the northern hemisphere under the influence of a pressure gradient. If the earth rotates with angular velocity Ω, points at the higher latitude have a lower tangential velocity (Ωr_1) than points at the lower latitude $\left(\Omega r_2\right)$ as $r_2 > r_1$. This means

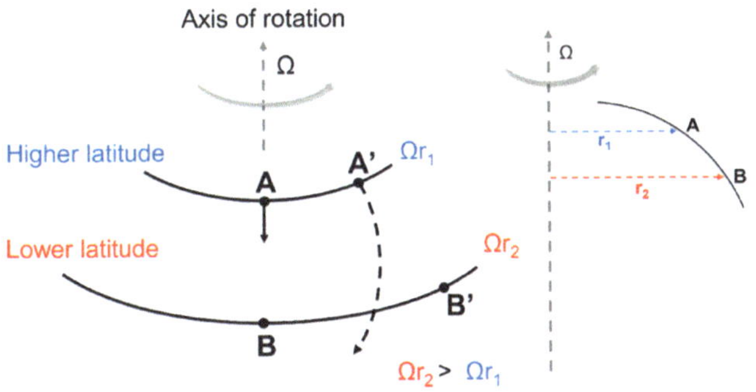

Figure 2.6 Schematic representation of the movement of an airmass from a point A at a higher latitude to a point B at a lower latitude under the influence of a pressure gradient as the earth is rotating with angular velocity Ω. The values r_1 and r_2 are the perpendicular distances from the axis of rotation to points A and B, respectively. This figure is assumed to be in the northern hemisphere.

that, by the time the air mass has moved to the lower latitude, it appears to be deflected to the right of B which is now at B'. If this were in the southern hemisphere, the deflection would appear to be to the left.

As the earth rotates, a balance is set up between the local pressure gradient forces, the Coriolis forces, and centrifugal forces resulting in the formation of air circulation systems around low- and high-pressure systems, as shown in Figure 2.7. The air flow (wind) at a height sufficiently far above the earth's surface to no longer feel its frictional drag effect will tend to flow parallel to the isobars (lines of equal pressure). This wind is the *gradient wind*. The magnitude of the centrifugal force is relatively small compared with the Coriolis and pressure gradient forces. As the curvature of the isobars reduces this force also reduces. In the limit of straight isobars, this force tends to zero and the resulting balance between the pressure gradient force and the Coriolis force is known as the *geostrophic wind*. This is commonly assumed to be the prevailing wind in the free atmosphere above the ABL.

The westerly (u_g) and southerly (v_g) components of the geostrophic wind can be calculated by knowing the local pressure gradient and the latitude which defines the Coriolis force

$$u_g = -\frac{1}{\rho f}\frac{\partial p}{\partial y} \tag{2.30}$$

$$v_g = +\frac{1}{\rho f}\frac{\partial p}{\partial x} \tag{2.31}$$

$$f = 2\Omega \sin\phi \tag{2.32}$$

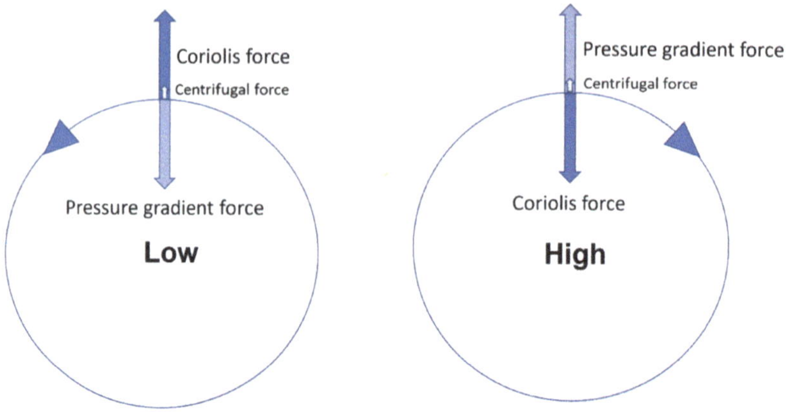

Figure 2.7 Circulation systems set up due to a balance between the Coriolis force, centrifugal force, and the pressure gradient force in the northern hemisphere. In the southern hemisphere the circulation is in the opposite sense.

where f is the Coriolis parameter and ϕ is the latitude. In these equations, the x coordinate is from west to east and the y coordinate is from south to north. Relating the geostrophic wind speed G and its angle α_0 to the surface layer wind speed remains a significant challenge. One such formulation which relates G to the friction velocity u_* is (Blackadar and Tennekes, 1968; Clarke and Hess, 1974; Yamada, 1976)

$$G = \frac{u_*}{\kappa}\sqrt{\left(\left(\ln\left(\frac{u_*}{fz_0}\right) - A\right)^2 + B^2\right)} \tag{2.33}$$

$$G = \sqrt{u_g^2 + v_g^2} \tag{2.34}$$

$$\sin\alpha_0 = \mp\frac{Bu_*}{\kappa G} \tag{2.35}$$

where A and B are empirically determined parameters, which for a neutral ABL are approximately 1.8 and 4.5, respectively (Hess and Garratt, 2002). Note that in Eq. (2.35) the angle α_0 between the geostrophic wind direction and the surface wind direction (sometimes called the *cross-isobar angle*) is negative in the northern hemisphere and positive in the southern hemisphere. For the non-neutral ABL, A and B are approximately functions of the surface stability parameter $\mu = \kappa u_* / (|f|L)$, where L is the Monin–Obukhov length. Several researchers have tried to determine empirical formulae for these parameters. One of the most well known is (Zilitinkevich, 1975)

For $\mu \lesssim -30$

$$A(\mu) = 10|\mu|^{-\frac{1}{2}} \tag{2.36}$$

$$B(\mu) = \frac{\ln|\mu|}{\kappa} \tag{2.37}$$

For $\mu \gtrsim 5$

$$A(\mu) = 4\mu^{\frac{1}{2}} \tag{2.38}$$

$$B(\mu) = \ln\mu - 4\mu^{\frac{1}{2}} \tag{2.39}$$

Note that the functions are discontinuous around $\mu = 0$ and the neutral values should be used in the region $-30 \lesssim \mu \lesssim 5$. There is significant scatter around these empirical fits to the data, and attempts have been made to provide more accurate formulae based on additional scaling parameters such as the stability above the ABL. At the top of the ABL (h_{ABL}), the potential temperature normally increases rapidly with height to give a stable layer known as an *inversion layer* with a depth of ~100 m, above which the potential temperature increases at a slower relatively constant rate (the free atmosphere stability). A schematic of the potential temperature variation with height is shown in Figure 2.8 comparing a convective (unstable), neutral, and stable ABL with progressively lower inversion heights. The height of the boundary layer is marked by the start of the upper inversion layer except for the stable case. In this case, the boundary layer height is not so easily identified but is normally defined as the height at which turbulence drops to negligible levels (less than 5% of surface levels). Note that a neutral boundary layer with a capping inversion layer is sometimes referred to as a *conventionally neutral boundary layer* (CNBL) as opposed to a *truly neutral boundary layer* (TNBL) with no capping inversion and neutral stability above the ABL, which does not occur in reality but is sometimes used in theoretical modelling.

The lapse rate in the free atmosphere is found to affect the values of A and B in addition to surface layer stability. Based on large eddy simulations (Zilitinkevich and Esau, 2005), for a neutral or stable ABL ($L > 0$), A and B are assumed to be functions of composite stability parameters m_A and m_B

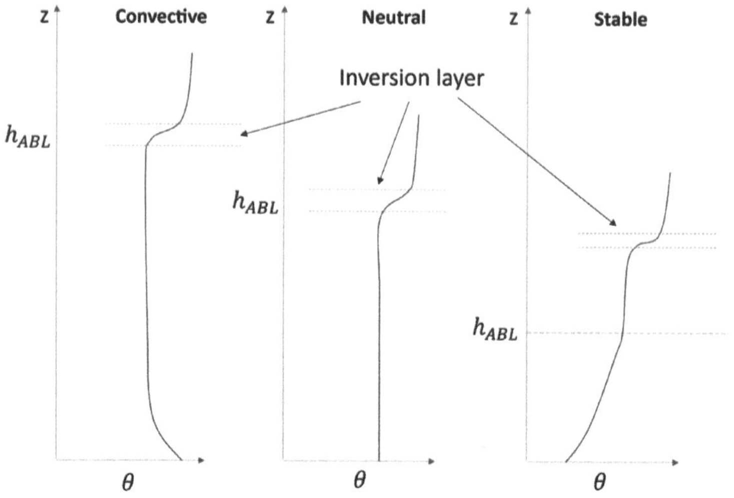

Figure 2.8 Variation in potential temperature with height for a convective (unstable), neutral, and stable boundary layer. Note that the definition of h_{ABL} is different for the stable ABL.

$$A(m_A) = -am_A + \ln(a_0 + m_A) \tag{2.40}$$

$$B(m_B) = b_0 + bm_B^2 \tag{2.41}$$

where

$$m_A = \left\{\left(\frac{h_{ABL}}{L}\right)^2 + \left(\frac{C_{NA}h_{ABL}}{L_N}\right)^2 + \left(\frac{C_{fA}h_{ABL}}{L_f}\right)^2\right\}^{\frac{1}{2}} \tag{2.42}$$

$$m_B = \left\{\left(\frac{h_{ABL}}{L}\right)^2 + \left(\frac{C_{NB}h_{ABL}}{L_N}\right)^2 + \left(\frac{C_{fB}h_{ABL}}{L_f}\right)^2\right\}^{\frac{1}{2}} \tag{2.43}$$

with constants: $a = 1.4$, $a_0 = 1.65$, $C_{NA} = 0.09$, $C_{fA} = 1$, $b = 10$, $b_0 = -2$, $C_{NB} = 0.15$, and $C_{fB} = 1$, and length scales: L is the surface layer Monin–Obukhov length defined in Section 2.2, L_f is the rotational length scale $(= u_* / |f|)$, and L_N is the stability scale in the free atmosphere $(= u_* / |N|)$ where N is the Brunt–Väisälä frequency

$$N = \left(\frac{g}{\overline{\theta}}\frac{\partial\overline{\theta}}{\partial z}\right)^{\frac{1}{2}} \tag{2.44}$$

When calculating N in the free atmosphere using Eq. (2.44), $\partial\overline{\theta} / \partial z$ is the mean lapse rate in the free atmosphere and $\overline{\theta}$ is the mean temperature within the layer used to calculate the free atmosphere lapse rate. A good approximation for h_{ABL} is given by (Zilitinkevich et al., 2007)

$$\left(\frac{u_*}{fh_{ABL}}\right)^2 = \begin{cases} C_R^{-2} + C_{CN}^{-2}\dfrac{N}{|f|} & |L| \geq 1000\,\mathrm{m} \\ C_R^{-2} + C_{NS}^{-2}\dfrac{u_*}{|fL|} & 200\,\mathrm{m} < L < 1000\,\mathrm{m} \end{cases} \tag{2.45}$$

where $C_R = 0.7 \pm 0.1$, $C_{CN} = 1.36 \pm 0.25$, and $C_{NS} = 0.51 \pm 0.06$.

Note that, as yet, there is no similar formulation based on composite stability parameters for the unstable surface layer, though A and B show somewhat less variation in a convective ABL: as L decreases, the value of A approaches zero and B tends to a value close to 5 based on observations (Zilitinkevich, 1975).

2.4 The Ekman Layer

As wind turbines become ever larger, blade tips at the maximum vertical extent for the largest turbines may extend beyond what is commonly thought of as the surface layer. This means that there is an interest in modelling the wind profile above the surface layer into the Ekman layer. One relatively simple approach for a neutral ABL is an empirically based polynomial extension to the neutral form of the logarithmic profile in Eq. (2.4) normalised to the ABL height h_{ABL} (Deaves and Harris, 1978; Harris, 1981)

$$u(z) = \frac{u_*}{\kappa}\left[\ln\left(\frac{z}{z_0}\right) + 5.75\left(\frac{z}{h_{ABL}}\right) - 1.88\left(\frac{z}{h_{ABL}}\right)^2 - 1.33\left(\frac{z}{h_{ABL}}\right)^3 + 0.25\left(\frac{z}{h_{ABL}}\right)^4\right] \tag{2.46}$$

Another approach complements the definition of composite stability parameters defined in Eqs. (2.42) and (2.43) with a composite length scale to give a wind speed profile valid for neutral and stable ABLs (Zilitinkevich and Esau, 2005)

$$u(z) = \frac{u_*}{\kappa}\left[\ln\left(\frac{z}{z_0}\right) + C_u\frac{z}{L_{CM}}\right] \tag{2.47}$$

where $C_u = 2.5$ and

$$L_{CM} = \left\{ \left(\frac{1}{L}\right)^2 + \left(\frac{C_{NM}}{L_N}\right)^2 + \left(\frac{C_{fM}}{L_f}\right)^2 \right\}^{\frac{1}{2}} \tag{2.48}$$

where $C_{NM} = 0.1$ and $C_{fM} = 1$.

In Emeis (2014), two approaches are described for non-neutral ABLs. The first assumes the logarithmic form of the surface layer profile described by Eq. (2.4) which is then matched to a second profile above, such that (Etling, 2002)

$$u(z) = G(-\sin\alpha_0 + \cos\alpha_0) \qquad z = z_p \tag{2.49}$$

$$u(z) = G\begin{bmatrix} 1 - 2\sqrt{2}e^{-\gamma_E(z-z_p)}\sin\alpha_0\cos\left(\gamma_E\left(z - z_p\right) + \pi/4 - \alpha_0\right) \\ +2e^{-2\gamma_E(z-z_p)}\sin^2\alpha_0 \end{bmatrix}^{\frac{1}{2}} \qquad z > z_p \tag{2.50}$$

where z_p is the height of the surface layer (also known as the *Prandtl layer*) and

$$\gamma_E = \sqrt{\frac{f}{2\kappa u_* z}} \tag{2.51}$$

In this case, the value of the cross-isobar angle α_0 is obtained by equating Eqs. (2.4) and (2.49) and their derivatives at $z = z_p$

$$\alpha_0 = \tan^{-1}\left(\frac{1}{1 + \frac{2\gamma_E z_p}{\phi_m\left(z_p/L\right)}\left(\ln\left(z_p/z_0\right) - \psi_m\left(z_p/L\right)\right)}\right) \tag{2.52}$$

where ϕ_m and ψ_m are defined in Eqs. (2.13)–(2.17). Note that α_0 and u_* need to be solved simultaneously by iteration.

The second approach in Emeis (2014), based on Gryning et al. (2007), modifies Eq. (2.4) so that it is applicable in both the surface and the Ekman layers. Two formulations based on this principle are presented; the first is detailed in Gryning et al. (2007)

$$u(z) = \frac{u_*}{\kappa}\left[\ln\left(\frac{z}{z_0}\right) + T\left(\frac{z}{L}\right) + \left(\frac{z}{L_M}\right) - \left(\frac{z}{h_{ABL}}\right)\left(\frac{z}{2L_M}\right)\right] \tag{2.53}$$

and the second is explained in (Peña et al., 2010a, 2010b)

$$u(z)=\frac{u_*}{\kappa}\left[\ln\left(\frac{z}{z_0}\right)+T\left(\frac{z}{L}\right)+\frac{1}{d}\left(\frac{\kappa z}{\eta}\right)^d-\left(\frac{1}{1+d}\right)\left(\frac{z}{h_{ABL}}\right)\left(\frac{\kappa z}{\eta}\right)^d-\left(\frac{z}{h_{ABL}}\right)\right] \tag{2.54}$$

where

$$\eta=\frac{\kappa h_{ABL}}{\left[d(1+d)\right]^{\frac{1}{d}}}\left[\frac{G\kappa}{u_*}+1-\ln\left(\frac{h_{ABL}}{z_0}\right)\right]^{-\frac{1}{d}} \tag{2.55}$$

G is the geostrophic wind speed given in Eq. (2.33), $d \sim 1$–1.25, $L_M=(u_*/f)(-2\ln(u_*/(fz_0))+55)^{-1}$, and $T(z/L)$ is a stability correction factor of the form

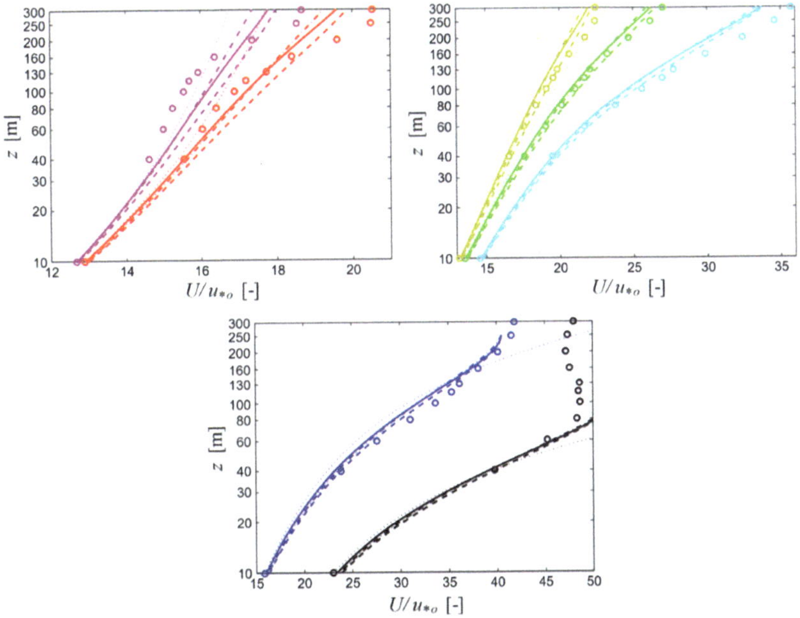

Figure 2.9 Comparison between observed wind speed data at Høvsøre in Denmark, the standard logarithmic profile, and two profiles modified to extend into the Ekman layer. Magenta: very unstable; red: stable; yellow: near unstable; green: neutral; cyan: near stable; blue: stable; and black: very stable. Dotted line: standard logarithmic profile in Eq. (2.4); dashed-dotted line: Eq. (2.53); dashed line: Eq. (2.54) with $d=1$; and solid line: Eq. (2.54) with $d=1.25$. *Source:* Peña et al. (2010a) / with permission of Springer Nature.

$$T(z/L) = -\psi_m(z/L) \qquad -2 \leq z/L \leq 0 \tag{2.56}$$

$$T(z/L) = -\psi_m(z/L)(1 - z/2h_{ABL}) \qquad 0 \leq z/L \leq 1 \tag{2.57}$$

Note that in Peña et al. (2010a), a simpler form of the stable boundary layer height is used based on (Rossby and Montgomery, 1935)

$$h_{ABL} = C\frac{u_*}{|f|} \tag{2.58}$$

where C is a constant with a value between 0.1 and 0.5. In Peña et al. (2010a), a value of 0.1 is suggested for very stable, 0.12 for stable, and 0.15 for neutral conditions. For unstable conditions, as mentioned in Section 2.3, there is no simple reliable formula that can be used and the boundary layer height may need to be inferred from measurements (e.g. using such as a ceilometer) or inferred from model data such as from a reanalysis (reanalysis datasets are described in Chapter 9).

A comparison between observed wind speed data at Høvsøre in Denmark is made with the standard logarithmic profile in Eq. (2.4) and the wind speed profiles determined using Eqs. (2.53) and (2.54), as shown in Figure 2.9 (Peña et al., 2010a). The comparisons are made as a function of surface layer stability. The two extended profiles are generally an improvement on the standard logarithmic profile, particularly under neutral or unstable conditions. The drawback of using these extended profiles is that additional parameters need to be determined, some of which may require additional observations (e.g. to determine the height of the ABL).

References

Blackadar, A.K., Tennekes, H., 1968. Asymptotic similarity in neutral barotropic planetary boundary layers. *J. Atmos. Sci.* 25, 1015–1020. https://doi.org/10.1175/1520-0469(1968)025<1015:asinbp>2.0.co;2.

Businger, J.A., Wyngaard, J.C., Izumi, Y., Bradley, E.F., 1971. Flux-profile relationships in the atmospheric surface layer. *J. Atmos. Sci.* 28, 181–189. https://doi.org/10.1175/1520-0469(1971)028<0181:FPRITA>2.0.CO;2.

Charnock, H., 1955. Wind stress on a water surface. *Q. J. R. Meteorol. Soc.* 81, 639–640. https://doi.org/10.1002/QJ.49708135027.

Clarke, R.H., Hess, G.D., 1974. Geostrophic departure and the functions A and B of Rossby-number similarity theory. *Boundary-Layer Meteorol.* 7, 267–287. https://doi.org/10.1007/BF00240832.

Davenport, A.G., 1960. Rationale for determining design wind velocities. *J. Struct. Div.* 86, 39–68.

Deaves, D.M., Harris, R.I., 1978. *A Mathematical Model of the Structure of Strong Winds*. London Construction Industry Research and Information Association.

Dyer, A.J., 1974. A review of flux-profile relationships. *Boundary-Layer Meteorol.* 7, 363–372. https://doi.org/10.1007/BF00240838.

Emeis, S., 2014. Current issues in wind energy meteorology. *Meteorol. Appl.* https://doi.org/10.1002/met.1472.

Etling, D., 2002. *Theoretische Meteorologie*, 3rd ed. Springer Berlin. https://doi.org/10.1007/978-3-662-10430-9_1.

Fairall, C.W., Bradley, E.F., Godfrey, J.S., Wick, G.A., Edson, J.B., Young, G.S., 1996. Cool-skin and warm-layer effects on sea surface temperature. *J. Geophys. Res. Ocean.* 101, 1295–1308. https://doi.org/10.1029/95JC03190.

Garratt, J.R., 1994. *The Atmospheric Boundary Layer*. Cambridge University Press.

Grachev, A.A., Fairall, C.W., 1997. Dependence of the Monin–Obukhov stability parameter on the bulk Richardson number over the ocean. *J. Appl. Meteorol.* 36, 406–414. https://doi.org/10.1175/1520-0450(1997)036<0406:DOTMOS>2.0.CO;2.

Gryning, S.E., Batchvarova, E., Brümmer, B., Jørgensen, H., Larsen, S., 2007. On the extension of the wind profile over homogeneous terrain beyond the surface boundary layer. *Boundary-Layer Meteorol.* 124, 251–268. https://doi.org/10.1007/s10546-007-9166-9.

Harris, R.I., 1981. The structure of strong winds, in: *Proc. CIRIA Conf. on Wind Engineering in the Eighties*. CIRIA.

Hess, G.D., Garratt, J.R., 2002. Evaluating models of the neutral, barotropic planetary boundary layer using integral measures: part I. *Boundary-Layer Meteorol.* 104, 333–358. https://doi.org/10.1023/A:1016521215844.

Högström, U., 1988. Non-dimensional wind and temperature profiles in the atmospheric surface layer: a re-evaluation. *Boundary-Layer Meteorol.* 42, 55–78. https://doi.org/10.1007/BF00119875.

Kaimal, J.C., Finnigan, J.J., 1994. *Atmospheric Boundary Layer Flows: Their Structure and Measurement*. Oxford University Press.

Monin, A.S., Obukhov, A.M., 1954. Basic laws of turbulent mixing in the surface layer of the atmosphere. *Contrib. Geophys. Inst. Acad. Sci. USSR.* 24, 163–187.

Peña, A., Floors, R., Sathe, A., Gryning, S.-E., Wagner, R., Courtney, M.S., Larsén, X.G., Hahmann, A.N., Hasager, C.B., 2016. Ten years of boundary-layer and wind-power meteorology at Høvsøre, Denmark. *Boundary-Layer Meteorol.* 158, 1–26. https://doi.org/10.1007/s10546-015-0079-8.

Peña, A., Gryning, S.E., Hasager, C.B., 2010a. Comparing mixing-length models of the diabatic wind profile over homogeneous terrain. *Theor. Appl. Climatol.* 100, 325–335. https://doi.org/10.1007/s00704-009-0196-8.

Peña, A., Gryning, S.E., Mann, J., Hasager, C.B., 2010b. Length scales of the neutral wind profile over homogeneous terrain. *J. Appl. Meteorol. Climatol.* 49, 792–806. https://doi.org/10.1175/2009JAMC2148.1.

Rossby, C.-G., Montgomery, R.B., 1935. The layer of frictional influence in wind and ocean currents. *Papers in Physical Oceanography and Meteorology*, 3, 1–101 https://doi.org/10.1575/1912/1157.

Sathe, A., Gryning, S.-E., Pena, A., 2011. Comparison of the atmospheric stability and wind profiles at two wind farm sites over a long marine fetch in the North Sea. *Wind Energy* 14, 767–780. https://doi.org/10.1002/we.456.

Wieringa, J., 1992. Updating the Davenport roughness classification. *J. Wind Eng. Ind. Aerodyn.* 41, 357–368. https://doi.org/10.1016/0167-6105(92)90434-C.

Yamada, T., 1976. On the similarity functions *A, B* and *C* of the planetary boundary layer. *J. Atmos. Sci.* 33, 781–793. https://doi.org/10.1175/1520-0469(1976)033<0781:OTSFAO>2.0.CO;2.

Zilitinkevich, S., Esau, I., Baklanov, A., 2007. Further comments on the equilibrium height of neutral and stable planetary boundary layers. *Q. J. R. Meteorol. Soc.* 133, 265–271. https://doi.org/10.1002/qj.27.

Zilitinkevich, S.S., 1975. Resistance laws and prediction equations for the depth of the planetary boundary layer. *J. Atmos. Sci.* 32, 741–752. https://doi.org/10.1175/1520-0469(1975)032<0741:RLAPEF>2.0.CO;2.

Zilitinkevich, S.S., Esau, I.N., 2005. Resistance and heat-transfer laws for stable and neutral planetary boundary layers: old theory advanced and re-evaluated. *Q. J. R. Meteorol. Soc.* 131, 1863–1892. https://doi.org/10.1256/qj.04.143.

3 Measurement

The most accurate way to estimate the wind speed at a site is to measure it directly. Although this will only give you information about a specific location at a specific point in time, wind speed measurements are, nevertheless, considered the gold standard when making a site assessment. A variety of instruments has been developed over the centuries for this purpose, and in this chapter we review how they have developed, how they are used, and the latest remote sensing technology, which is gaining popularity, particular where measurements are challenging, such as at great height or out at sea.

3.1 History of Wind Speed Measurement

The principal instrument still used today for the measurement of wind speed is the anemometer. The invention of this device is credited to the Italian Renaissance polymath Leon Battista Alberti in 1450 who developed a mechanical instrument consisting of a swinging plate which could move in a direction perpendicular to the wind and its angle of inclination would indicate the strength of the wind. In 1846, the Irish researcher John Thomas Romney Robinson at Armagh Observatory invented an anemometer consisting of four hemispherical cups which would spin at a rate proportional to the wind speed (see Figure 3.1). This was later developed as a three-cup instrument by a Canadian meteorologist John Patterson in 1926 and further refinements of the design were made by the American researchers M. J. Brevoort and U. T. Joyner (Brevoort and Joyner, 1934a, 1934b). This three-cup design is today the most frequently used form of the anemometer. This type of instrument only gives an indication of the magnitude of the wind, so an additional device such as a wind vane is required to indicate the

Handbook of Wind Resource Assessment, First Edition. Simon Watson.

Figure 3.1 An illustration of Robinson's four-cup anemometer. *Source:* Sean Linehan / Wikipedia Commons / Public Domain.

wind direction. Wind vanes have been in existence for centuries in the form of a plate designed in such a way as to point into the wind. It is then a relatively straightforward task to convert this movement into some form of automated measurement.

3.2 Types of Measurement

For the purposes of assessing the wind resource at a site, there are several parameters that are of importance. These are illustrated in Figure 3.2. In terms of determining energy density, the mean wind speed $\overline{u}$ over a given time period is of most relevance. However, over such a time period, the wind speed may deviate at any instant from the mean by a quantity u', such that an instantaneous wind speed $U = \overline{u} + u'$ will be recorded. In several such time periods, this instantaneous wind speed may reach some extreme value $\widehat{U}$. All of these quantities are of importance. The short-term fluctuations in the wind u' are of importance in assessing turbulence which has implications for potential wind turbine loading. The root-mean-squared

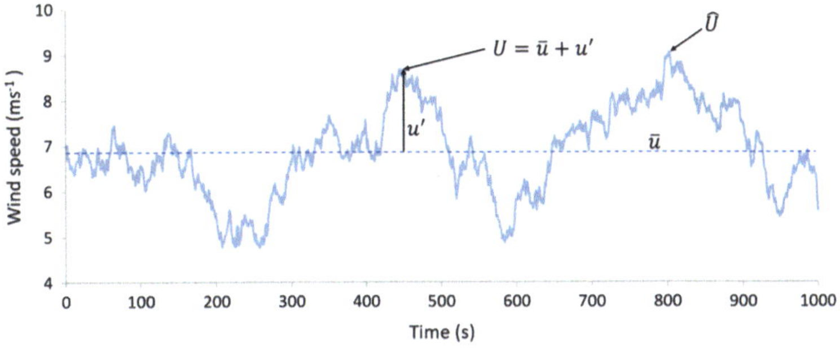

Figure 3.2 Wind speed measurements over a given period. The mean, $\bar{u}$, fluctuation from the mean, u', instantaneous value, U, and extreme value, $\hat{U}$, in the period are shown.

value of u' over a given time interval is frequently used to determine a quantity known as the *longitudinal turbulence intensity* I_u

$$I_u = \frac{\sqrt{\overline{u'^2}}}{\bar{u}} = \frac{\sigma_u}{\bar{u}} \tag{3.1}$$

where σ_u is the standard deviation of the streamwise or longitudinal turbulent fluctuations in each averaging period (noting that $\overline{u'} = 0$). As explained later in Section 4.2, this is frequently a period of 10 minutes. The typical sampling time within this 10-minute period is between 1–2 seconds.

The instantaneous wind speed U drives the potential output of a wind turbine at a given point in time. In terms of life-time design of a structure, it is important to know the extreme wind speed $\hat{U}$ that may be experienced over a given period. All of these measures may be recorded by a wind speed measuring instrument, but it is important to know how they are defined (e.g. sampling period, averaging period, return period) and also it is important to understand how a given wind speed measurement responds when making a measurement.

3.3 Terrestrial Measurements

3.3.1 Anemometers

In this section the main types of anemometer are introduced in terms of how they work, their accuracy and their response to changes in wind speed.

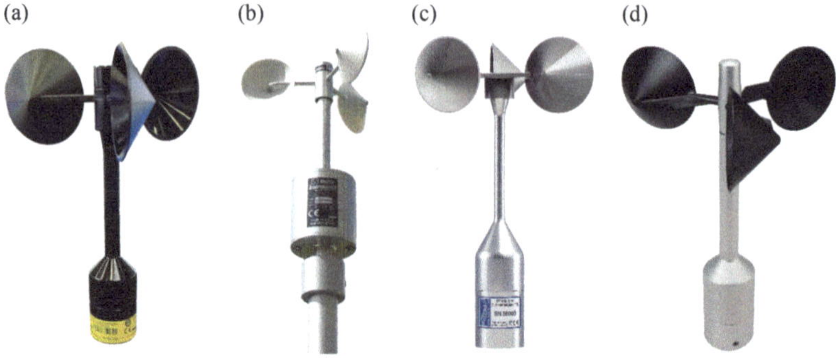

Figure 3.3 Popular cup anemometers for measuring wind speeds: (a) NRG S1 (*Source:* author's own photograph); (b) Vector A100 (*Source:* courtesy of D S Dicks, Vector Instruments); (c) WindSensor P2546D-OPR (*Source:* courtesy of WindSensor); (d) Thies FCA (*Source:* courtesy of Thomas Stadie, Adolf Thies GmbH & Co. KG).

3.3.1.1 Cup Anemometers

These types of wind measuring instruments were briefly introduced above and have become the instrument of choice for measuring wind speeds for a variety of meteorological applications including wind resource assessment. A selection of some commonly used brands for wind resource assessment is shown in Figure 3.3. The three-cup configuration with conical shape has become the most popular choice.

The conical cups create a differential drag force which causes the rotor to rotate at a speed proportional to the wind speed. The rotational movement is translated into a voltage or series of pulses by means of an electromechanical mechanism such as a reed switch or a slotted disc which interrupts an optical sensor.

Cup anemometers have become a popular choice due to their perceived linearity and accuracy. Their response has been the subject of much research over the last century to look at aspects such as an optimal cup and arm design and the aerodynamics of the cups (Brevoort and Joyner, 1934a, 1934b). One of the potential shortcomings of cup anemometers is their tendency to overspeed, as demonstrated by Wyngaard (1981). He showed that, by making certain ideal assumptions, the torque T produced on the rotor cups was of the form

$$T = \frac{\rho A r}{2}\left[C_d(\bar{u} + u' - \bar{s} - s')^2 - c_d(\bar{u} + u' + \bar{s} + s')^2\right] \tag{3.2}$$

where $\overline{s}$ is the mean rotor tangential speed at mean wind speed $\overline{u}$, s' is the deviation in tangential speed in response to a deviation in wind speed u', ρ is the local air density, r is the rotor radius of the cups (measured from the centre of the rotor to the middle of the cup), C_d is the drag coefficient of the rear cup face, c_d is the drag coefficient of the front cup face, and A is the cup area normal to the wind.

If steady state conditions are assumed such that $u' = s' = 0$ then the mean tangential cup speed is proportional to mean wind speed

$$\overline{s} = K\overline{u} \tag{3.3}$$

where K typically has a value of 0.3.

This leads to

$$T = \frac{2\rho r C_d A K \overline{u}^2}{(1+K^2)}\left[(1-K^2)\frac{u'}{\overline{u}} - (1-K^2)\frac{s'}{\overline{s}} + \left(\frac{u'}{\overline{u}}\right)^2 + K^2\left(\frac{s'}{\overline{s}}\right)^2 - (1+K^2)\left(\frac{s'}{\overline{s}}\right)\left(\frac{u'}{\overline{u}}\right)\right] \tag{3.4}$$

If the cups are initially in equilibrium so that $s' = 0$ and then accelerate in response to a change in wind speed, this implies

$$T = \frac{I}{r}\frac{ds'}{dt} = \frac{2\rho r C_d A K \overline{u}^2}{(1+K^2)}\left[(1-K^2)\frac{u'}{\overline{u}} + \left(\frac{u'}{\overline{u}}\right)^2\right] \tag{3.5}$$

where I is the polar moment of inertia of the cups. It is the second nonlinear term in the square brackets which shows that the magnitude of this acceleration is greater when $u' > 0$ which means that on average they will tend to measure a wind speed slightly higher than actual value (i.e. will tend to overspeed).

Wyngaard goes on to consider the rotor inertia in the limit where changes in wind and rotor tangential speed are small (i.e. $s' \ll \overline{s}$ and $u' \ll \overline{u}$)

$$\frac{I}{r}\frac{ds'}{dt} = \frac{2\rho r C_d A \overline{u}(1-K^2)}{(1+K^2)}\left[Ku' - s'\right] \tag{3.6}$$

which can be simplified to

$$\frac{L_d}{\overline{u}}\frac{ds'}{dt} + s' = Ku' \tag{3.7}$$

where $L_d = I/(\rho r^2 CA)$ and C is an effective drag coefficient. L_d is known as a *distance constant*, which is a function of the design of the cup anemometer, the air density, and the speed range. This leads to a time constant of the form $\tau = L_d / \bar{u}$ showing that the time constant is inversely proportional to the mean wind speed. For the purposes of wind resource measurements, the value of L_d should be less than 5 m in order that overspeeding does not introduce a significant error in the recording of site wind speeds. Most industry standard anemometers have distance constants of 2–3 m in practice. According to Kristensen (1993), the fractional overspeeding is the sum of four components

$$\frac{\delta\omega_c}{\Omega_c} = \delta_u + \delta_v + \delta_w + \delta_* \tag{3.8}$$

where $\delta\omega_c$ is the mean excess rotational speed of the cups due to wind speed fluctuations and Ω_c is the mean rotational speed of the cups for a steady state wind speed $\bar{u}$. The components δ_u, δ_v, and δ_w are biases due to turbulent wind speed fluctuations in the streamwise, transverse, and vertical directions, respectively. The component δ_* is the bias due to the covariance of fluctuations in the streamwise and vertical directions. The degree of overspeeding is dependent on the ratio of the distance constant to the peak length scale of the turbulence and is thus height and stability dependent. If the stability is known, an estimate of the fractional overspeeding can be made, assuming measurements are made in flat homogeneous terrain, using the expressions (Kristensen, 1993)

$$\delta_u = a\frac{\pi}{\sqrt{3}}\kappa^{\frac{4}{3}}\alpha_1 \frac{\phi_\epsilon^{\frac{2}{3}}}{(\ln(z/z_0) - \psi_m)^2}\left(\frac{L_d}{z}\right)^{\frac{2}{3}} \tag{3.9}$$

$$\delta_v = \frac{\kappa^2}{2}\frac{\phi_{u,v}{}^2}{(\ln(z/z_0) - \psi_m)^2} \tag{3.10}$$

$$\delta_w = 1.34\frac{\kappa^2}{2}\frac{\phi_w^2}{(\ln(z/z_0) - \psi_m)^2} \tag{3.11}$$

$$\delta_* = -0.018a\frac{\pi}{\sqrt{3}}\kappa^{\frac{2}{3}}\frac{\phi_m\phi_\epsilon^{\frac{1}{3}}}{(\ln(z/z_0) - \psi_m)^2}\left(\frac{L_d}{z}\right)^{\frac{2}{3}} \tag{3.12}$$

where κ is the von Kármán constant $a \approx 1$; α_1 is the Kolmogorov constant; (≈ 0.56), ϕ_m, and ψ_m are the stability dependent functions of z / L defined in Section 2.2; and ϕ_ϵ and ϕ_w are additional stability dependent functions which scale with z / L (Kaimal and Finnigan, 1994)

$$\phi_\epsilon(z / L) = \left(1 + 0.5\left|z / L\right|^{\frac{2}{3}}\right)^{\frac{3}{2}} \qquad -2 \le z / L \le 0 \tag{3.13}$$

$$\phi_\epsilon(z / L) = (1 + 5z / L)0 \qquad \le z / L \le 1 \tag{3.14}$$

$$\phi_w(z / L) = 1.25(1 + 3\left|z / L\right|)^{\frac{1}{3}} \qquad -2 \le z / L \le 0 \tag{3.15}$$

$$\phi_w(z / L) = 1.25(1 + 0.2z / L) \qquad 0 \le z / L \le 1 \tag{3.16}$$

The function $\phi_{u,v}$ scales as the ratio of z_i / L where z_i is the height of the convective boundary layer (typical values of which are ~1000 m, but in practice this may have to be measured separately using a ceilometer) (Panofsky and Dutton, 1984)

$$\phi_{u,v}(z_i / L) = (12 + 0.5\left|z_i / L\right|)^{\frac{1}{3}} \qquad -2 \le z / L \le 0 \tag{3.17}$$

Under stable conditions, there is no reliable formulation for ϕ_v, and the value for neutral conditions is a reasonable approximation (i.e. $\phi_{u,v} \approx 2.3$ when $z / L > 0$) (Garratt, 1994).

In Eq. (3.8), the covariance bias term δ_* is normally small compared with the other terms. The vertical bias term δ_w is usually less than or of the same magnitude as the streamwise bias δ_u. Under unstable conditions, directional fluctuations can be significant and the transverse bias δ_v can become the dominant term. Under stable conditions, all the bias terms are much reduced and the overall degree of overspeeding is small.

To further illustrate this, Figure 3.4(a) shows the contribution that the different bias components make to the total overspeeding as a function of stability. The dominance of the δ_v component during unstable conditions is clear. It can also be seen that the contribution from the covariance component is negligible for all stability conditions. Figure 3.4(b) shows the expected overspeeding as a function of measurement height and roughness length. The two roughness length cases contrast offshore $(z_0 = 0.0003\text{m})$ with onshore $(z_0 = 0.03\text{m})$. For the lower measurement height (10 m)

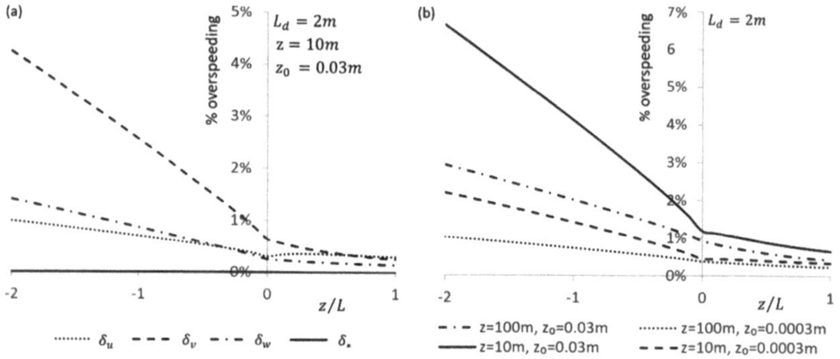

Figure 3.4 (a) Contribution to the fractional overspeeding of a cup anemometer from the different bias components. (b) Fractional overspeeding of a cup anemometer for different measurement heights and roughness lengths. N.B. $z_i = 1000\text{m}$.

onshore it can be seen that the overspeeding can be quite significant under very unstable conditions.

The expression for fractional overspeeding can be simplified using the peak turbulence length scale Λ and the three turbulence intensity components based on (Busch and Kristensen, 1976)

$$\frac{\delta\omega_c}{\Omega_c} = J_u I_u^{\ 2} + b I_v^{\ 2} + c I_w^{\ 2} \tag{3.18}$$

where $b = 0.5$, $c = 0.67$, $I_v = \sigma_v / \bar{u}$ (the transverse turbulence intensity where σ_v is the standard deviation of transverse velocity turbulence components, v'), $I_w = \sigma_w / \bar{u}$ (the vertical turbulence intensity where σ_w is the standard deviation of vertical velocity turbulent components, w'), and

$$J_u = \left(1 + \frac{\Lambda}{L_d}\right)^{-1} \tag{3.19}$$

In a *neutral atmosphere*, assuming $I_v \approx 0.69 I_u$ and $I_w \approx 0.53 I_u$ (Réthoré, 2009; Stull, 1988), Eq. (3.18) becomes

$$\frac{\delta\omega_c}{\Omega_c} \approx (J_u + 0.43) I_u^2 \tag{3.20}$$

An estimate of the Λ can be made using the expression (Kelly, 2018)

$$\Lambda \approx z \frac{I_u}{\alpha} \tag{3.21}$$

where α is the power law shear exponent in Eq. (2.2). This can be estimated if wind speed measurements are available at two heights.

Another factor which contributes to overspeeding is the response of the cup anemometer to steady wind speed components which are not perpendicular to the axis of rotation. This can result from measurements made in terrain which is sloping, resulting in upward or downward components of the wind speed. Ideally, a cup anemometer should only be sensitive to the horizontal component and thus have a cosine response to the wind speed as a function of the inclination angle ϕ_i, as shown in Figure 3.5(a). The actual response of a cup anemometer compared to a cosine response is shown in Figure 3.5(b). An overestimate compared with the cosine curve occurs for almost all inclination angles, thus adding to the overspeeding.

A further consideration in the measurement of fluctuations in the wind speed u' is the frequency response of the cup anemometer. The cups and

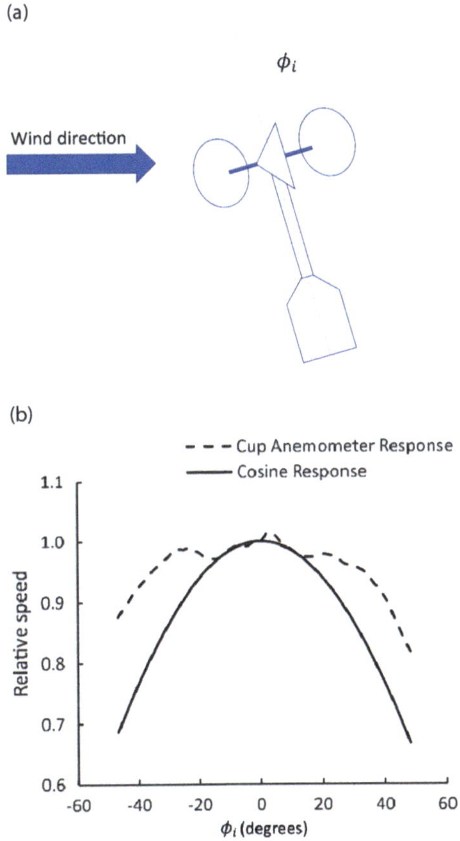

Figure 3.5 (a) Definition of cup inclination angle ϕ_i relative to wind direction where a positive angle implies the wind inclined from above. (b) Response of a typical cup anemometer to wind flow at an inclination angle ϕ_i, with an ideal cosine response shown for reference.

rotor have a particular inertia which limits how quickly a device can respond to changes in wind speed. In addition, at very high frequencies, the signal from the device is effectively dominated by white noise. Yahaya and Frangi (2004) compared the response of cup anemometers with a sonic anemometer as a function of wind frequency variation and determined the difference in turbulence intensity measure from the two devices. They state a theoretical cut-off frequency beyond which the amplitude of the measure of turbulence from a cup anemometer is diminished in comparison to a sonic anemometer, which is able to respond more faithfully to a much higher range of frequencies. This cut-off frequency f_{co} at a given wind speed U is given by

$$f_{co} = \frac{U}{2\pi L_d} \tag{3.22}$$

For a cup anemometer with a distance constant of 2 m, this would give, for example, at 10 ms^{-1} a cut-off frequency of 0.8 Hz. This would suggest a sampling interval for a cup anemometer to be no faster than around 1 Hz.

The sampling by the cup anemometer itself needs to be considered in the measurement of the turbulence intensity. Many cup anemometers produce pulses in response to the rotation of the cups. The number of pulses per revolution depends on the design of the device. If the number of pulses per revolution is relatively small, then this can lead to misleading measurements of the turbulence intensity I. An example of this is shown in Figure 3.6. In this case, wind speed measurements have been made on a mast with a cup anemometer. The figure shows a plot of the standard deviation of the wind speed measurements sampled every second and averaged over a 10-minute interval. It can be seen that up to 10 ms^{-1} there is some regular oscillating pattern in the data. This is due to the fact that the anemometer used produced a relatively low number of pulses per revolution (two pulses/revolution in this case). This means that at low levels of turbulence, the anemometer frequency varies very little and when sampling in one-second intervals this can lead to variation in the number of pulses recorded within each successive interval. This is particularly noticeable at low wind speeds when the rotational frequency is low and so anemometer and sampling frequencies are of the same order of magnitude. This in turn leads to cyclic variation in the standard deviation of wind speeds measured over a number of sampling periods. At wind speeds above around 10 ms^{-1}, the lowest levels of turbulence where this effect is seen are rarely observed and the frequency output of the anemometer is sufficiently high compared with the sampling interval, and so this artefact diminishes.

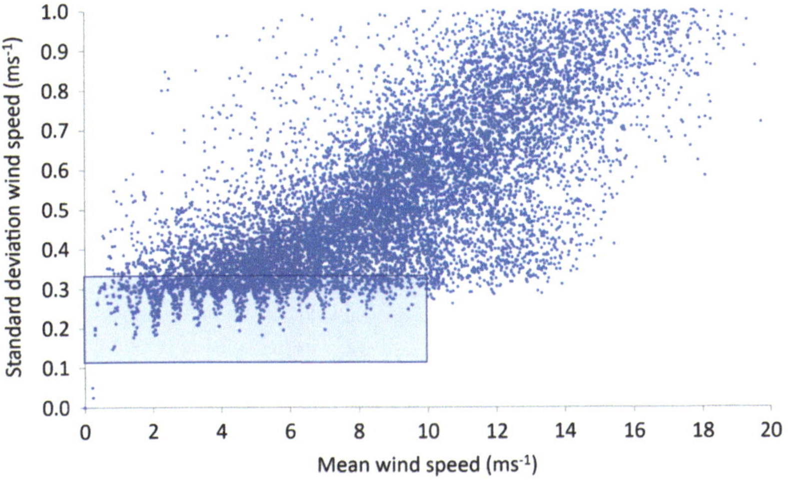

Figure 3.6 Standard deviation of wind speed measurements against mean wind speed for a 10-minute averaging period and a one-second sampling period. Artificial artefacts at low levels of turbulence due to the low number of pulses per revolution are highlighted by the box.

For anemometers that produce a much larger number of pulses per revolution, this is not such a problem.

Some caution is also required in the measurement of wind speed extremes (i.e. gusts). An example of this is shown in Figure 3.7 (Burt, 2022) where high wind speeds were recorded at an inland site during the passage of a very low pressure weather system followed later by squally showers. During this latter period, very high wind speeds measured by pulsed-type cup anemometers were observed at two heights (2 m and 10 m) that were not reflected in the measurements from identical instruments at other heights on the same mast (in the case of the 5 m anemometer) or on a mast a few metres away (0.56 m and 6.4 m). Careful analysis of the data ruled out such as a lightning strike, or instrument or wiring problems. Although it was not possible to confirm, the most likely explanation was the induction of high frequency pulses within the body of the anemometers resulting from possible corona discharges due to the high electrical fields measured in the vicinity of the area (lightning had been present in the area but not observed at the site itself). It was suggested that the relatively short-lived burst of high frequency pulses was registered by the data logger giving erroneously high wind speeds. Only careful analysis of this event ruled out the possibility of this being a true wind speed extreme, but the fact that it did occur during a period of high wind speeds meant that it may have been considered as

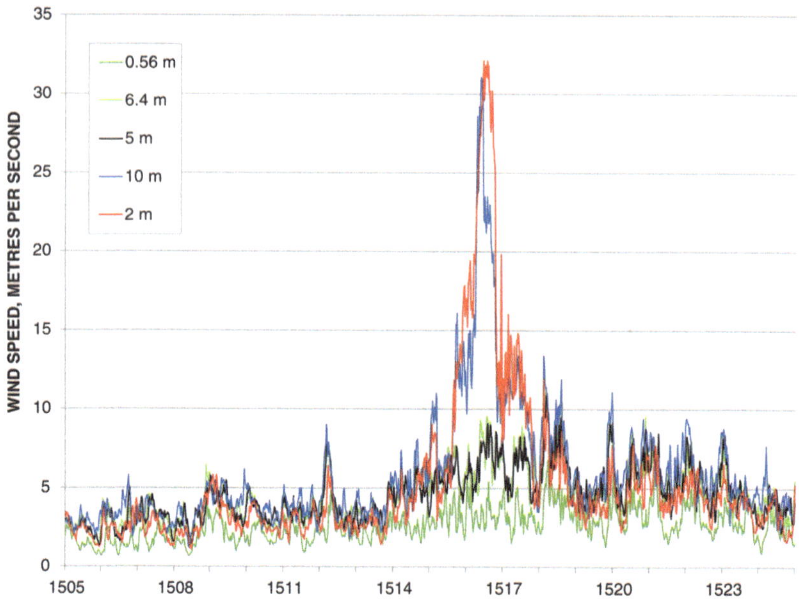

Figure 3.7 Pulsed-type cup anemometer measurements of one-second wind speeds at multiple heights indicating anomalously high wind speeds for selected heights (Burt, 2022). Time on the x-axis is in UTC. Measurements were made at the Reading Atmospheric Observatory on 28 March 2016. *Source:* Burt 2022 / Royal Meteorological Society / CC BY 4.0.

plausible had the measurements not been made at a well-maintained meteorological research site. Other high wind speed events have also been later questioned due to possible electrical interference (Aylott et al., 2020). Given the reliance on a relatively small number (by their very nature) of high wind speed measurements to parameterise the distribution of gusts at a site (this is covered in Section 4.8.2), particular attention is required in the quality assurance of such datasets.

Cup anemometers are primarily intended for measuring horizontal wind speed; however, they are sensitive to vertical wind speed, and this needs to be considered when measuring the wind resource. Wind speed measurements can also be adversely affected by environmental parameters such as icing of the cups. For this reason, anemometers are classified as to their suitability for use in different terrain and climate conditions (International Electrotechnical Commission, 2017). Five suitability classes are defined:

- A: flat terrain and moderate climate.
- B: complex terrain and moderate climate.

- C: flat terrain and cold climate.
- D: complex terrain and cold climate.
- S: special user-defined class.

Cup anemometers are also classified in terms of their accuracy by a k parameter. This parameter value can be used to provide an estimate of the operational uncertainty u_{ou} in the measured wind speed u

$$u_{ou} = k / \sqrt{3} \times (0.05 \text{ ms}^{-1} + 0.005u) \tag{3.23}$$

Cup anemometers are thus labelled using both the suitability and operational uncertainty classes (i.e. kA, kB, etc.).

3.3.1.2 Sonic Anemometers

If wind speed measurements at a frequency of higher than 1 Hz are required (e.g. to provide detailed turbulence measurements), a sonic anemometer can be used. This type of instrument can also provide simultaneous measurements of wind speed and direction and has the advantage of no moving parts. A 2D device can give a measurement of the horizontal components of the wind speed and a 3D device can also provide a measurement of the vertical wind speed component. This may be required in complex terrain where nonhorizontal flow may be significant. Example devices can be seen in Figure 3.8.

A sonic anemometer consists of two (for a 2D device) or three (for a 3D device) ultrasonic transmitter–receiver pairs. Ultrasonic pulses are sent

(a) (b)

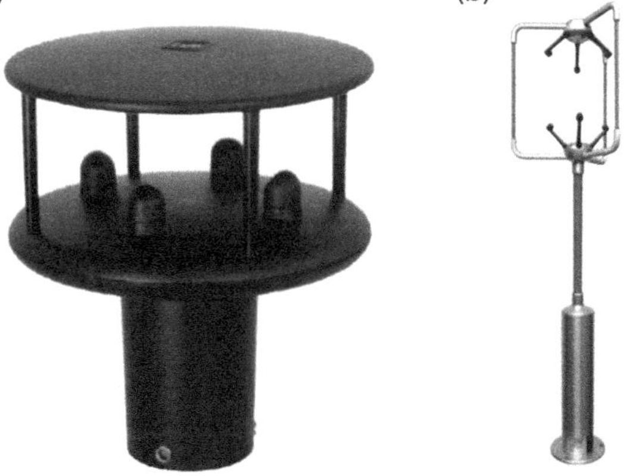

Figure 3.8 Examples of sonic anemometers: (a) Gill WindSonic 2D anemometer; (b) Gill WindMaster 3D anemometer. *Source:* courtesy of Gill Instruments, www.gillinstruments.com.

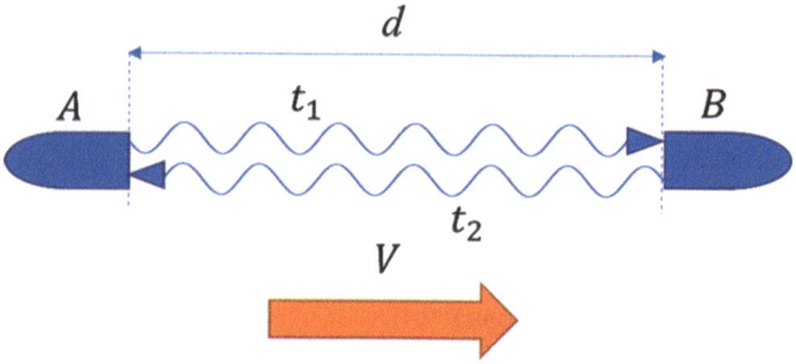

Figure 3.9 Schematic showing how ultrasonic pulses are transmitted and received by the transmitter–receiver pairs (*A* and *B*) for one axis of a sonic anemometer. Key: *d*: distance between transmitter/receiver; t_1: time for pulse to travel between transmitter *A* and receiver *B*; t_2: time for pulse to travel between transmitter *B* and receiver *A*; *V*: wind speed along the line of sight.

back and forth between each pair of sensors, as shown in Figure 3.9. The time taken for the pulse to travel between the sensors A and B depends on the wind speed and the ambient temperature. So, if the wind speed component in the line of sight between the sensors is V, the time $t_{1,2}$ taken to travel in each direction is given by

$$t_1 = \frac{d}{c_s + V} \quad t_2 = \frac{d}{c_s - V} \tag{3.24}$$

where c_s is the speed of sound and d is the distance between sensors A and B.

When rearranged, this results in

$$V = \frac{d}{2}\left(\frac{1}{t_1} - \frac{1}{t_2}\right) \quad c_s = \frac{d}{2}\left(\frac{1}{t_1} + \frac{1}{t_2}\right) \tag{3.25}$$

Knowing the speed of sound, it also possible to infer the ambient temperature through the formula

$$c_s = \sqrt{\gamma R T} \tag{3.26}$$

where γ is the ratio of specific heats (= 1.402 at standard temperature and pressure), R is the specific gas constant (usually the dry air value of $287.05\ \text{J kg}^{-1} \cdot \text{K}^{-1}$ is used for practical calculations), and T is the absolute temperature in K.

Sonic anemometers can sample the wind speed at frequencies greater than 30 Hz. Such instruments are significantly more expensive than cup anemometers (especially 3D devices) but may be beneficial where a robust device is required, such as in a hostile marine environment or where detailed turbulence measurements are required.

3.3.2 Wind Vanes

Where a cup anemometer is used for measuring wind speed, an additional instrument is required for measuring the wind direction. Devices for measuring wind direction have been in existence for many centuries in the form of weathervanes. Most modern instruments provide an electronic signal that may be recorded automatically and are in the form of a wind vane. Examples of these are shown in Figure 3.10. The measurement is usually made by one of two methods. The first uses a potentiometer that is attached to the vertical shaft of the device. The rotation of the vane changes the resistance of the potentiometer and a measurement of the voltage gives a measure of the wind direction. In this case, because the potentiometer resistance increases from no resistance to full resistance after close to one rotation, there is a small region between these extremes which cannot be measured. This is known as the *dead band*. This can be several degrees in size. The common approach is to align the wind vane so the dead band sector is in that direction from which the wind speed is least expected to blow. The second method uses a series of reed switches (usually eight or 16). A magnet is connected to the shaft of the wind vane and as the wind causes the

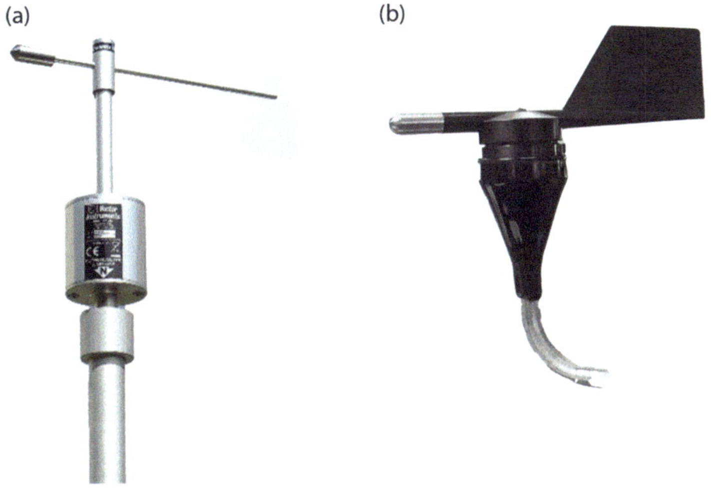

Figure 3.10 Popular wind vanes for measuring wind direction: (a) Vector W200P (*Source:* courtesy of D.S. Dicks, Vector Instruments), (b) NRG W200P.

shaft to rotate, the magnet aligns with one of the reed switches, thus completing a circuit. Each reed switch circuit normally has a different resistor so that the voltage measured across the resistor indicates the wind direction.

3.3.3 Pressure, Temperature, and Humidity

As the ultimate aim of any wind resource assessment is to calculate the estimated annual energy production of a wind farm, it is important to have the measurements of local pressure, temperature, and humidity. This is because a standard wind turbine power curve is determined at a reference air density with a specific standard value of pressure, temperature, and humidity. The power produced is linearly dependent on the air density, so the standard power curve may need to be corrected for any potential wind farm site. Following Dalton's law, the dependence of humid air density ρ_h on pressure, temperature, and humidity is given by

$$\rho_h = \frac{p_d}{R_d T} + \frac{p_v}{R_v T} \tag{3.27}$$

where p_d is the partial pressure of dry air, R_d is the specific gas constant of dry air, R_v is the specific gas constant for water vapour (461.52 J kg^{-1} · K^{-1}), T is the absolute temperature, and p_v is the pressure of water vapour given by

$$p_v = \phi_{RH} p_{sat} \tag{3.28}$$

where ϕ_{RH} is the relative humidity. The value of p_{sat} in Pa can be estimated using Tetens' formula

$$p_{sat} = 611 \times 10^{\frac{7.5(T-273.15)}{T-35.85}} \tag{3.29}$$

where T is in K.

The partial pressure of dry air, p_d, is given by

$$p_d = p - p_v \tag{3.30}$$

where p is the absolute pressure.

Thus, a measurement of pressure p, temperature T, and relative humidity ϕ_{RH} will yield an estimate of the air density.

Barometric pressure sensors are normally designed using a flexible diaphragm which is deformed by the pressure of the air. In the case of a resistive device, one side of the diaphragm is exposed to the air and the other has a strain gauge attached. As the diaphragm deforms under pressure the resistance of the strain gauge changes, which can be measured using an electronic circuit known as a *Wheatstone bridge*, and the atmospheric pressure inferred. In the case of a capacitive device, a fixed inflexible plate is mounted close to the diaphragm, and as the diaphragm deforms under pressure the distance between the diaphragm and plate changes, changing the capacitance of the pair, which can be measured electrically. Example devices are shown in Figure 3.11.

Most temperature sensing devices measure the change in resistance of a ceramic (a thermistor) or a metal (resistance temperature detector or RTD) as a function of temperature. An example of a thermistor device is shown in Figure 3.12(a). A type of RTD which is commonly used for air temperature measurement is a platinum resistance thermometer (PRT). Platinum has a well-known and stable linear resistance relationship over a wide range of temperatures making it ideal for this purpose. Typically, the change in resistance with temperature is measured by making the PRT one arm of a Wheatstone bridge and measuring the change in voltage.

In order to ensure the sensor measures the correct air temperature, it is normally enclosed in a ventilated sun shield to prevent direct radiation affecting the measurement. An example of such a device is shown in Figure 3.12(b).

Relative humidity is normally measured electrically by means of sensing resistance or capacitance changes with the level of humidity. Such a device is shown in Figure 3.12(c). They are often incorporated in the same housing as an air temperature sensor.

(a) (b)

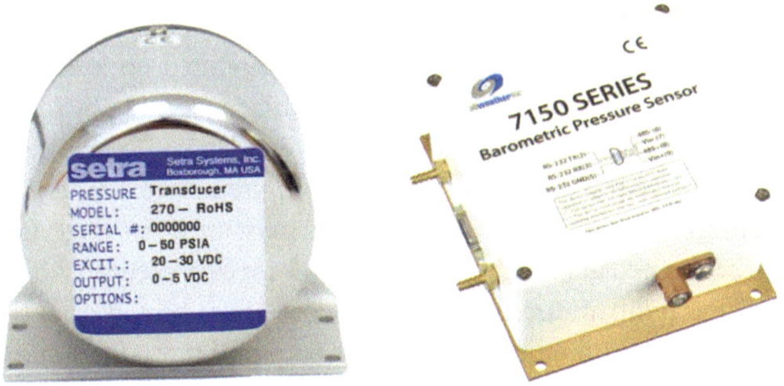

Figure 3.11 Example barometric pressure devices: (a) a Setra capacitive pressure sensor (*Source:* Setra Systems, Inc.); (b) an AWI resistive sensor (*Source:* All Weather, Inc.).

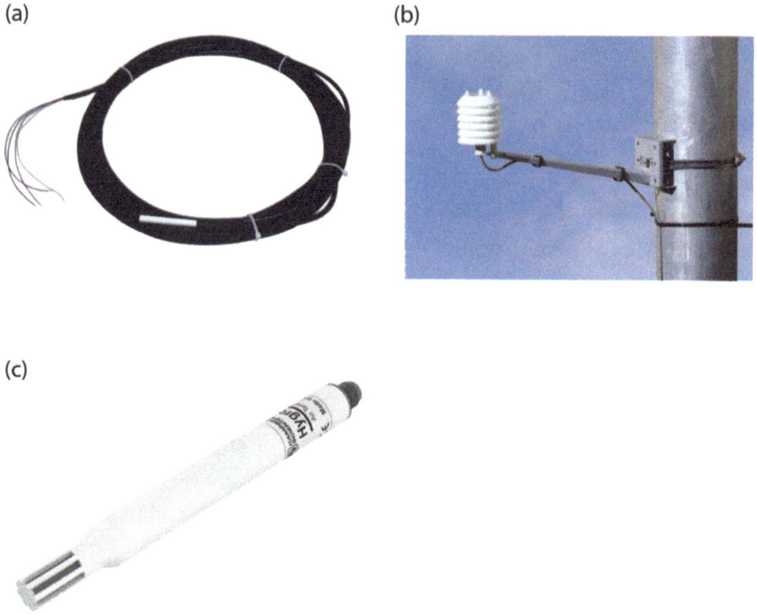

Figure 3.12 Air temperature and relative humidity measurement devices: (a) Campbell Scientific temperature probe (*Source:* Campbell Scientific, Inc.), (b) Vector Instruments T351 housing with internal temperature probe (*Source:* courtesy of D. S. Dicks, Vector Instruments, www.windspeed.co.uk, last accessed 5 August 2022), (c) Campbell Scientific combined air temperature and relative humidity sensor. (*Source:* reproduced with permission from Campbell Scientific, Inc.,www.campbellsci.com, last accessed 5 August 2022).

3.3.4 Instrument Mounting

In order to make measurements of meteorological parameters at multiple heights, the instruments described above are normally mounted on a mast. Depending on the height of the mast, it may be freestanding or secured by several guy wires. A typical mast is made of tubular steel in a lattice design and consists of multiple sections depending on the height required. An example of such a mast is shown in Figure 3.13. Instruments on the mast are connected via wires to a data logger mounted at the bottom of the tower for easy access. Wind speed and direction and temperature measuring devices are mounted on the mast itself. Temperature measurements may be carried out at more than one height to make atmospheric stability calculations. Pressure measurements are typically made at ground level within the logger enclosure. The logger is frequently accessed remotely through the GSM network and is normally powered using batteries charged up using a solar panel if no mains power is available.

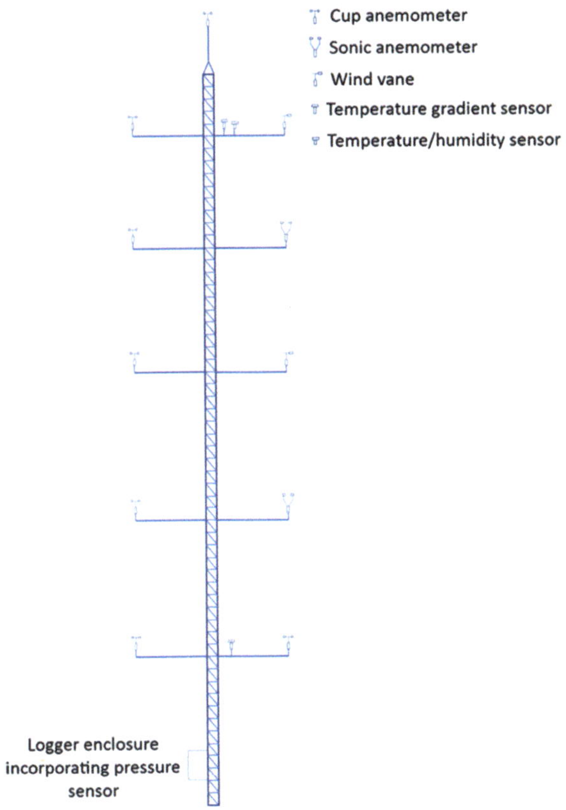

Figure 3.13 Typical mast mounting arrangement for measuring wind speed and wind direction at multiple heights as well as temperature gradient, relative humidity, and pressure.

In this example, wind measuring instruments such as cup and sonic anemometers are installed at several heights using booms. The reason for this becomes obvious when you consider the perturbation of the upstream wind by the mast itself. Detailed guidance on best practice for the measurements of wind speed and direction are available in International Electrotechnical Commission (2017). Figure 3.14 shows a computational fluid dynamics (CFD) simulation of the wind flow around a triangular section tubular lattice mast where the wind is flowing from left to right. The contours show how the wind speed varies in relation to the far upstream unperturbed wind speed. It can be seen the wind speed at around five mast diameters or 5D (assuming a circle encompassing the triangular section) is reduced by 1% relative to the 'true' wind speed. This is considered about the limit of

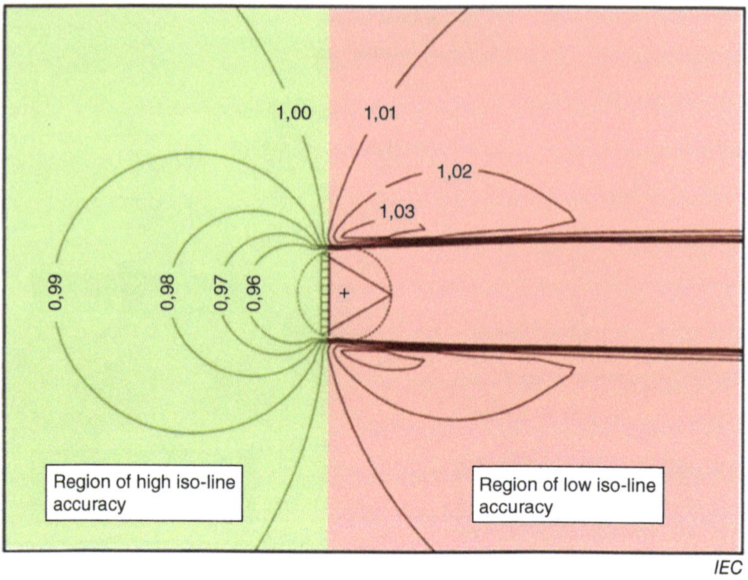

Figure 3.14 CFD simulation of the wind flow around a triangular section tubular lattice mast. Green represents the area of high confidence in the simulation (on the upwind side) and red represents low confidence (on the downwind side). The contours represent the speed-up of the wind relative to the far upwind, unperturbed wind speed. *Source:* taken with permission of the IEC from International Electrotechnical Commission (2017).

acceptable perturbation. The recommended distance of the anemometer from the mast (and thus the boom length) depends on a number of factors including the mast geometry, but typically it is recommended that the mounting distance is of the order of 5–6D. Needless to say, on the downwind side, the wake effect of the mast extends a long way downstream, as shown in Figure 3.14. For this reason, anemometers ae frequently mounted on pairs of booms either side of the tower (shown schematically in Figure 3.13). Depending on the wind direction, the measurement from the upwind device is used. In some cases, especially for triangular section towers, three booms may be mounted at 120° to each other to improve the quality of the measurements and further minimise the effect of what is termed *mast shading*.

The top of the mast and the booms themselves also have some effect on the wind flow, and it is recommended to mount anemometers on smaller tubes to provide clearance and minimise flow distortion effects. For example, it is recommended that the top anemometer be mounted using one or more tubes so that it is around 4 m above the top of the main lattice mast structure, and boom-mounted wind speed measuring devices should be mounted at least 20 boom diameters (20D) above the boom.

3.4 Remote Sensing

3.4.1 Surface-based Devices

With the advent of offshore wind farms, the requirement for lower cost measurements of wind speeds up to heights of more than 100 m has driven the development of surface-mounted remote sensing devices such as lidars and sodars. These provide an indirect measure of the wind flow through scattering of sound or light from features in a moving wind field. Such devices, known as *remote sensing devices* (RSDs), have been available for some years, but only in the last 10–15 years have they reached a level of maturity that has allowed them to be used in commercial wind farm development. Lidars, in particular, are becoming increasingly used offshore as the cost of tall masts becomes prohibitive. In an offshore environment, a lidar can be mounted on a fixed platform with a monopile foundation if the water is relatively shallow and the seabed geology is appropriate. However, in deeper waters, there is also the possibility to mount such a device on a floating platform. Surface remote sensing devices have also opened up the possibility to measure 2D and 3D wind fields and give much better insight into wind flow in areas of complex terrain and in wind turbine wakes. At the same time, the importance of atmospheric stability throughout the ABL in determining the resource is increasingly seen as important. Surface remote sensing devices also provide the opportunity to profile temperature and humidity profiles to high altitude.

3.4.1.1 Lidars

The name *lidar* has come about by analogy from radar, though strictly it is not an acronym in the same way as radar was when first developed in the 1940s. A radar is based on the principle of the scattering of radio waves from a moving (usually) metal object and the subsequent detection of backscattered radiation allowing the tracking of its movement. In the case of a lidar, laser light, typically in the infrared range (~1.5 μm) is projected normally, though not exclusively, upwards and the Doppler shifting of backscattered radiation is used to infer wind speed along the line of sight. Different scanning patterns can be employed, the most popular of which is the plan position indicator (PPI) scan. In this case, the laser beam is scanned at a fixed elevation angle ϕ around a range of azimuth angles θ, as shown in Figure 3.15(a). In the case of a velocity azimuth display (VAD) scanning pattern, a full 360° scan is used to retrieve the wind vector. This type of scanning pattern is convenient for measuring a wind profile. An alternative, simplified version of this employs a Doppler beam swinging (DBS) pattern, where the beam is projected alternately at just four azimuth angles at 90° to one another to retrieve the orthogonal wind speed vectors, though the accuracy is slightly less than the VAD scan where many more azimuthal measurements are available. A fifth beam may also be projected vertically to retrieve the vertical wind speed. A further, alternative scan pattern is the range height indicator (RHI) scan, which is shown in Figure 3.15(b). This is useful for measuring a 2D slice of wind speeds to assess wind flow over varying terrain, for example. The other popular scan pattern is known as an *arc sector scan*, as shown in Figure 3.15(c). This type of scan pattern has been used to assess the structure of a wind turbine wake, by siting a ground-based lidar close by pointing downstream and sampling the wake using a stack of arc sector scans at different elevation angles (Doubrawa et al., 2016).

3.4.1.1.1 Operating Principle

Doppler wind lidars use the principle that, when laser light is projected into the atmosphere, a small fraction of the light is back scattered from moving aerosol particles and returns to the lidar detector. The frequency of the backscattered laser light is Doppler-shifted due to the movement of the scattering aerosol particles, and this frequency shift $\delta\nu$ is related to the speed of the particles and thus the local wind speed

$$\delta\nu = \frac{2V_{LOS}}{\lambda_o} \tag{3.31}$$

where V_{LOS} is the velocity component in the line of sight of the laser beam and λ_o is the wavelength of the outgoing laser light.

This frequency shift is very small compared with the frequency of the outgoing laser light but can be determined accurately using a heterodyne or homodyne technique. In the case of a heterodyne, the outgoing laser light of frequency ν_o ($= c / \lambda_o$, where c is the speed of light) is emitted and the returning Doppler-shifted backscattered light of frequency ν_S is combined coherently with a local oscillator light source of frequency ν_{LO}, which is slightly different from ν_o. The local oscillator can be from a separate laser or split off the outgoing main beam using a beam splitter and then the frequency altered using a frequency shifter.

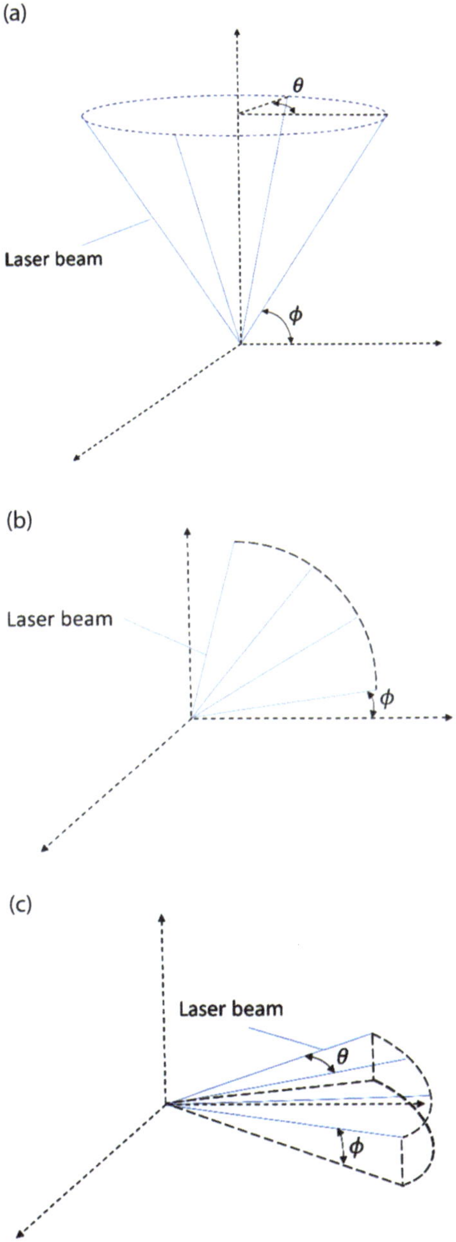

Figure 3.15 Popular lidar scanning patterns: (a) plan position indicator (PPI); (b) range height indicator (RHI); and (c) lidar arc.

The combined signal creates a 'beat' pattern at a much lower frequency to that of the laser light, and this beat frequency can be used to accurately determine the speed of the aerosol particles in the direction of the line of sight of the laser beam. The principle of this beat frequency can be shown simply if it is considered that the local oscillator is a harmonic oscillating field e_{LO} in time t of the form

$$e_{LO}(t) = E_{LO}\cos 2\pi\nu_{LO}t \quad (3.32)$$

where E_{LO} is the amplitude of the local oscillator. The Doppler-shifted backscattered beam can also be considered an oscillating field e_s of the form

$$e_s(t) = E_s\cos 2\pi\nu_s t \quad (3.33)$$

where E_s is the amplitude of the backscattered beam oscillation. The intensity i of the combined signal detected will then be of the form (Slinger and Harris, 2012)

$$i(t) \propto (E_{LO}\cos 2\pi\nu_{LO}t + E_s\cos 2\pi\nu t)^2 \quad (3.34)$$

This can be rewritten as

$$i(t) \propto (E_{LO}^2 + E_s^2) + 2E_{LO}E_s\cos 2\pi(\nu_s - \nu_{LO})t \quad (3.35)$$

The first term is constant and can be filtered out and as

$$\nu_s - \nu_{LO} = \delta\nu + \nu_o - \nu_{LO} \quad (3.36)$$

then the combined filtered signal i' becomes a function of the beat frequency

$$i'(t) \propto \cos 2\pi(\delta\nu + \nu_o - \nu_{LO})t \quad (3.37)$$

As ν_o and ν_{LO} are known, then the Doppler frequency shift $\delta\nu$ can be determined. This is usually done using a fast Fourier transform of many samples of the combined signals. Once $\delta\nu$ is determined, then the line-of-sight wind speed, V_{LOS}, can be calculated using Eq. (3.31). The offset frequency, $\nu_{off} = \nu_o - \nu_{LO}$, is chosen to be sufficient for the lidar to resolve the highest wind speeds of interest for the sampling rate used.

When using a homodyne technique, the outgoing beam and the local oscillator have the same frequency (i.e. after beam splitting, no frequency shifting is applied). Now the combined filtered signal becomes

$$i'(t) \propto \cos 2\pi(\delta\nu)t \quad (3.38)$$

This means that the cosine function is symmetric about the zero-frequency shift and that only the magnitude and not the sign of the frequency shift (and thus the line-of-sight wind speed) can be determined.

3.4.1.1.2 Continuous Wave and Pulsed Lidars

Figure 3.16(a) shows an example of a continuous wave (CW) lidar. As the name suggests, this type of lidar uses a continuous beam of laser light which is projected upwards at an angle of elevation ϕ of around 60°. Figure 3.17(a) shows a schematic illustration of a CW lidar performing a VAD scan at multiple heights. The beam is sequentially focussed at a series of heights scanning multiple azimuth points within ever larger circles, each circle taking around a second. The probe volume associated with each circle effectively becomes a ring which increases in size with height. From the combination of measurements along each circle focussed at several heights, a wind profile can be rapidly determined.

Figure 3.16(b) shows an example of a pulsed lidar and Figure 3.17(b) shows a schematic of this device performing a DBS scan. In this case, the laser sends out pulses of light at regular intervals. Normally, each pulse has a certain width which defines the spatial resolution of the wind velocity. The

(a) (b)

Figure 3.16 Example lidar systems: (a) ZX 300M CW lidar; (b) Vaisala WindCube v2.1 pulsed lidar. *Source:* Watson (author).

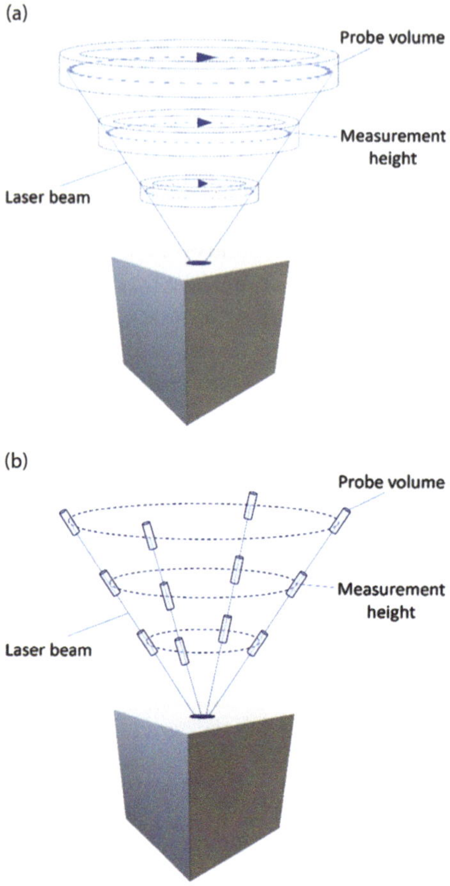

Figure 3.17 Laser probe volumes for (a) a continuous wave lidar executing a VAD scan and (b) a pulsed lidar executing a DBS scan.

device detects backscattered pulses at different times using a series of range gates such that radial wind speed can be inferred at several heights simultaneously. The 3D wind vector is determined by the device repeating the measurement at other angles of azimuth also using an elevation angle of around 60° where the beam will dwell for less than a second at each angle (Wächter et al., 2008), as shown in Figure 3.17(b).

Compared with CW lidars, pulsed lidars have an almost constant probe length with height, have a higher maximum range, and wind speed data can be gathered at multiple heights simultaneously. On the other hand, they are more complex in design, and have a higher minimum measurement height.

3.4.1.1.3 Wind Retrieval Patterns

In the case of a VAD scan, multiple measurements are made at a range of azimuth angles at a constant elevation for several heights. From the backscattered measurements, the line-of-sight wind speed V_{LOS} can be related to the wind speed components by fitting a cosine function of the form (Newman et al., 2016)

$$V_{LOS} = a + b\cos(\theta - \theta_{max}) \tag{3.39}$$

where the horizontal wind speed s is given by

$$s = \frac{b}{\cos\phi} \tag{3.40}$$

and the vertical wind speed w is given by

$$w = \frac{a}{\sin\phi} \tag{3.41}$$

The horizontal wind direction is given by θ_{max}.

In the case of a homodyne Doppler lidar, the cosine function is rectified

$$V_{LOS} = \left|a + b\cos(\theta - \theta_{max})\right| \tag{3.42}$$

The backscatter measurements can be visualised as a figure-of-eight plot, as shown in Figure 3.18. Here, the value of θ_{max} (+/−180 degrees) is determined by the primary axis of symmetry of the figure-of-eight which runs approximately NNE–SSW in Figure 3.18. The 180-degree ambiguity is normally solved by using a wind vane or similar, an example of which can be seen as the stalk on top of the lidar device in Figure 3.16(a). Note the degree of asymmetry in the figure-of-eight pattern denotes the amount of vertical flow (e.g. when the terrain is not flat), and the degree of scatter in the points around the figure-of-eight is related to the amount of turbulence.

Where a DBS scan is used, the wind speed components may be retrieved from four 90° beams thus

$$u = \frac{V_{LOS2} - V_{LOS4}}{2\cos\phi} \tag{3.43}$$

$$v = \frac{V_{LOS1} - V_{LOS3}}{2\cos\phi} \tag{3.44}$$

$$w = \frac{V_{LOS1} + V_{LOS2} + V_{LOS3} + V_{LOS4}}{4\sin\phi} \tag{3.45}$$

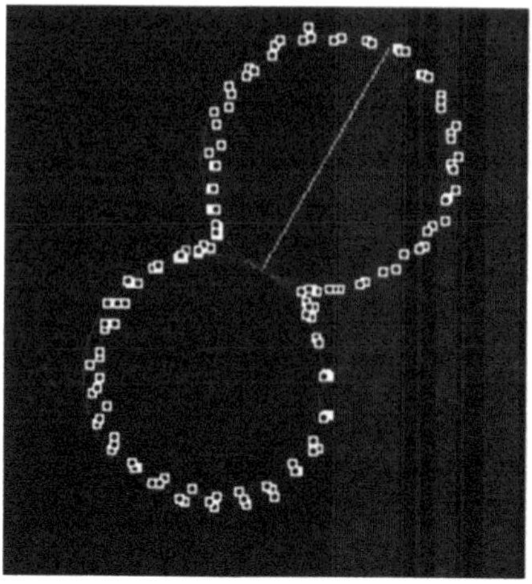

Figure 3.18 Figure-of-eight of processed beat frequencies as the lidar beam scans out a horizontal circle. *Source:* courtesy of ZX Lidars.

where V_{LOSn} is the line-of-sight wind speed for the beams $n = 1$–4 pointing along bearings 0°, 90°, 180°, and 270°, respectively. If a fifth vertical beam is used with line-of-sight wind speed V_{LOS5}, then, additionally, $w = V_{LOS5}$.

Both the VAD and DBS wind speed retrieval methods assume homogeneous flow with no variation in the wind speed vector within the circle bounding the four or more measurements. This has implications, as discussed further in Section 3.4.2.

3.4.1.1.4 Sodars and Radio Acoustic Sounding Systems

Sodars work on a similar principle to lidars, but instead of emitting infrared radiation they emit acoustic beeps (single frequency) or chirps (range of frequencies). Typically, a sodar device consists of several speakers which emit a pulse of sound at a frequency of around 4–5 kHz, which is alternately focused into one of three beams, as shown in Figure 3.19. This allows resolution of the wind vector into its three components. Small local variations in the refractive index of the air, caused by temperature fluctuations and driven by atmospheric turbulence, scatter the sound, some of which is backscattered to the device. Once again, the Doppler shift of the backscattered sound wave is used to measure the wind speed. Wind speed can be measured at multiple heights by means of measuring at fixed time delays between the emission and detection of the backscattered pulse. The wind speed measured corresponds to a volumetric average within slices through a cone, as seen in Figure 3.19. Figure 3.20 shows an example of two sodar devices. The device in Figure 3.20(a) uses three parabolic dishes to direct the sound chirps from three speakers, whereas the top image in Figure 3.20(b) shows a device which uses a phased array of multiple smaller

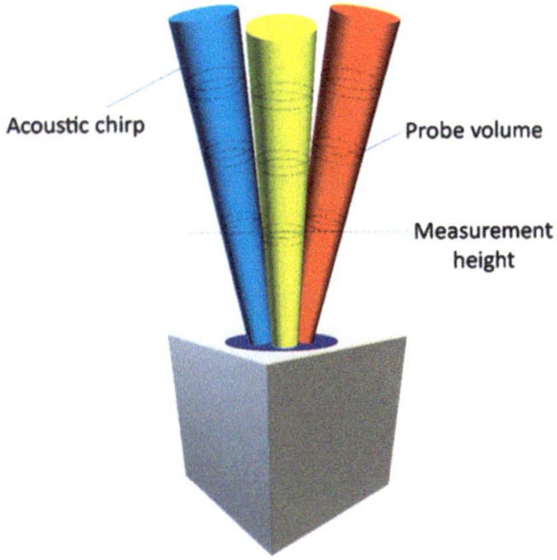

Figure 3.19 Schematic representation of the acoustic beams emitted by a sodar.

speakers where the beam can be steered electronically. In Figure 3.20(c), a radio acoustic sounding system (RASS) is also shown. This consists of several antennas which emit radio waves in the ultra-high frequency (UHF) range, typically 915 MHz in the US and 1290 MHz in Europe and Asia (Angevine, 2000). These are scattered off the fluctuations created by the sodar acoustic pulses, and the temperature can be determined from the frequency of the Doppler-shifted return signal, which depends on the wind speed along the line of sight of the acoustic beam (close to vertical) and the local speed of sound. In this way, collocated measurements of wind speed and temperature can be determined.

Some studies have indicated that wind speed measurements from sodars correlate less well with measurements from anemometers on meteorological masts compared with lidars. However, one reason for this is that, when making comparisons between measurements from a meteorological mast and a lidar or a sodar, the sodar needs to be further away from the mast than the lidar to prevent unwanted reflections of the acoustic signals. When the reduction in correlation due to spatial separation is taken into account, there is no real difference in the performance between lidars and sodars (Bradley et al., 2012).

3.4.1.1.4.1 Ceilometers

The height of the ABL, h_{ABL} as explained in Chapter 2, is an important parameter for modelling the wind resource, but is not straightforward to calculate. Ideally, a measurement should be made. This can be done using a device such as a ceilometer, as shown in Figure 3.21. Such devices use a vertically pointing pulsed laser and detect the backscattered light from aerosols and water droplets in clouds. The time of flight can be used to determine the height of the backscattered signal. They are frequently used to measure

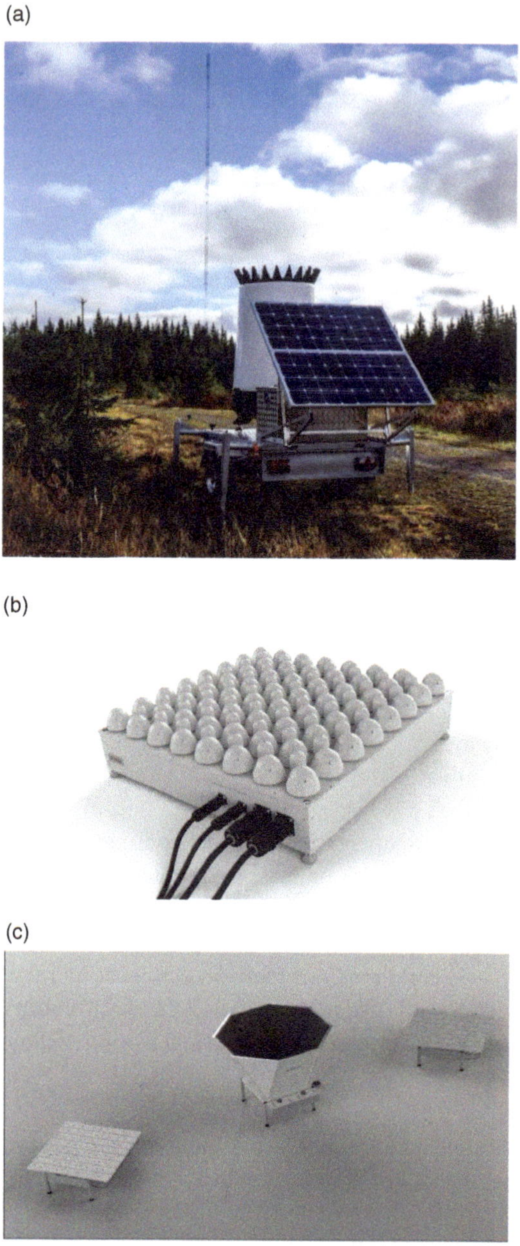

Figure 3.20 Example of sodar devices: (a) AQSystem AQ510 with photovoltaic charging power supply (*Source:* courtesy of Pelle Hurtig, AQSystem); (b) Scintec MFAS phased array (*Source:* Scintech AG); and (c) RASS extension (left and right) for simultaneous measurement of temperature profiles and an acoustic enclosure for the MFAS (centre) with sound-absorbing walls (*Source:* Scintech AG).

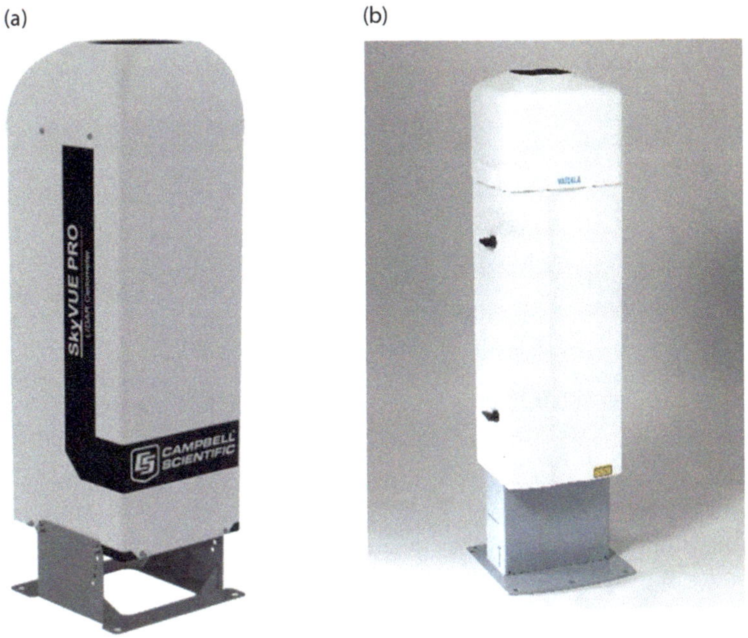

Figure 3.21 Ceilometers: (a) Campbell Scientific SkyVUEPRO (*Source*: Campbell Scientific, Inc., www.campbellsci.com, last accessed under 5 August 2022); (b) Vaisala CL31 (*Source*: Vaisala, www.vaisala.com, last accessed 5 August 2022).

cloud base heights. The boundary layer height is generally determined by a temperature inversion, which caps convection and is associated with a rapid change in aerosol concentration. A change in the backscatter associated with a change in aerosol concentration can then be used to determine the boundary layer height.

3.4.1.1.4.2 Microwave Radiometers

Microwave radiometers (MWRs) can be used to determine vertical profiles of temperature and humidity of the atmosphere by measuring the emission of microwave radiation from molecular oxygen and water vapour. The level of emission is dependent on the local concentration of these molecules, the pressure (and thus the height), and the air temperature. Water and oxygen have strong microwave radiation absorption peaks within the ranges 20–30 GHz and 50–60 GHz, respectively. MWRs profile the atmosphere by measuring the emission levels at several discrete frequencies within these bands (Ka and V bands) to determine the frequencies of peak absorption which can then be related to the air temperature and humidity at a given height. Typically, an MWR will carry out elevation scans at discrete angles

(a) (b)

Figure 3.22 Microwave radiometers (MWRs): (a) An RPG humidity and temperature profiler (HATPRO) (*Source:* courtesy of Danica Greve, RPG Radiometer physics GmbH); (b) a Radiometrics MP-3000A (*Source:* courtesy of Dave Patton, Radiometrics).

and some devices can also carry out full azimuth scans. The root-mean-squared measurement accuracy of such devices is typically 0.2–0.3 K, and they can profile the entire troposphere. Within the ABL, the vertical measurement resolution is typically 30–50 m. Example devices are shown in Figure 3.22. These devices are not routinely used for commercial wind resource assessment but have been deployed at research sites.

3.4.1.1.4.3 Radar Profilers

To date, radar profilers have been little used for wind resource assessment. However, they have been used for scanning wind fields in and around wind farms to better understand wake structure (Hirth et al., 2015, 2012). They work on a similar principle to lidar but use short pulses (tens of μs) of microwave radiation in the Ka band (35 GHz). They have the advantage over lidars in that they have a much longer range, but the equipment is bulkier and generally needs to be trailer mounted. In addition, they are normally only set up to provide single line-of-sight measurements and require multiple devices to fully resolve wind vectors. An example of a device which forms part of a dual Doppler radar system is shown in Figure 3.23.

3.4.2 Differences Between Spatial-averaged and Point Measurements

Single device lidars, sodars, and radar profilers effectively give a volume-averaged estimate of the line-of-sight wind speed in the vicinity of a given point. Wind profile measurements using RSDs are spatial averages of the wind speed, where the average is calculated based on several spatially separated points whose separation increases with height. This contrasts with a measurement made using a cup or sonic anemometer, where the

Figure 3.23 A radar profiling unit which forms part of a dual Doppler radar system used to provide 2D horizontal scanning profiles of wind speed. *Source:* Valldecabres et al. (2018), MDPI, CC BY 4.0.

measurement is based on a very small volume which can be considered, to all intents and purposes, a point. This is an important difference particularly when the wind field changes due to the underlying terrain. This can be illustrated by examining Figure 3.24 based on Stoelinga and LaWhite (2018). The figure shows a comparison between the measurement of wind flow over flat terrain and on a hill. For simplicity's sake, just two measurement beams are shown. An example is shown of the intended measurement of the wind speed at one height, X, inferred from the slices through the two beams at points A and B. In the case of flat terrain, the reconstruction of the average wind speed at X is accurate as the flow is uniform and horizontal. However, in the case of the hill, the flow is convex, and on both the upwind and downwind sides the line-of-sight components are less than the flat terrain case. The reconstructed wind speed is thus biased low. The opposite

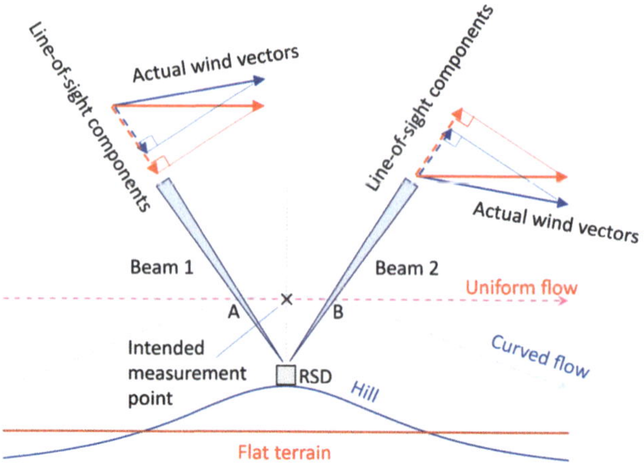

Figure 3.24 Illustration of a line-of-sight wind speed measurement by an RSD in flat and complex terrain. The point 'X' is the intended measurement point. *Source*: adapted from Stoelinga and LaWhite (2018).

case would be true if the RSD were placed at the bottom of a valley where the flow would be concave.

This shows that nonuniform or complex terrain can lead to a discrepancy in the measurement of wind speed by an RSD compared with that of a cup or sonic anemometer. One way around this would be to use multiple devices spaced out on the ground that produce beams which converge so the measured backscatter is from one point in space. However, this is generally impractical and prohibitively expensive. A less expensive alternative is to use a wind flow model to correct for the effect of the flow curvature (Bradley et al., 2015). For example, for curved flow over terrain, the RSD estimate of the streamwise wind speed $\hat{u}$ can be transformed to the horizontal wind speed u at a height z using

$$\hat{u} = u\left(1 - \frac{z}{R_c \cos\beta_0}\right) \tag{3.46}$$

where R_c is the radius of curvature of the streamlines and β_0 is the angle of the slope of the streamlines above the RSD. For a typical hill, this relationship would suggest errors of up to around 14%. The challenge is then to accurately estimate R_c and β_0. This can be done using a model of varying sophistication such as a linearised flow model or a more complete CFD model, which is covered in more detail in Section 5.2.4. Several studies have concluded that the use of a CFD model can reduce RSD wind speed errors in nonuniform terrain to within 1–2% (Bradley et al., 2015; Kim and Meissner, 2017).

3.4.3 Volume-averaged and Point Measurements of Turbulent Quantities

Section 3.4.2 suggests that there are important differences in the measurement of mean wind by point wind speed measuring devices and RSDs. The same is also true of turbulence measurements. The variance in the backscattered Doppler-shifted frequency for an RSD gives a measure of the turbulence, but again this is averaged over a large volume compared with a sonic or cup anemometer. This averaging effect is in both space and time, as it takes some time for the beams to move around for the backscattered measurements to be made. Because the beams are moving around to different locations in space, some of the variability seen in the wind speed will be due to spatial variation in addition to the temporal variation that would be seen at a point. Because the measurement is a line-of-sight measurement and because of the volume-averaging effect, multiple components of the turbulence are measured at the same time. This can be seen by considering the relationship between V_{LOS} and the wind flow within the probe volume for a lidar (Bonin et al., 2017)

$$V_{LOS} = u\cos\Delta\theta\cos\phi + v\sin\Delta\theta\cos\phi + w\sin\phi + \epsilon \tag{3.47}$$

where u, v, and w are the streamwise, crosswind, and vertical wind speeds, respectively; $\Delta\theta$ is the angle between u and the azimuth angle of the lidar beam; ϕ is the elevation angle; and ϵ is the uncorrelated random measurement error. If Eq. (3.47) is squared and the mean values of each component are removed, the variance of the line-of-sight wind speed and its dependence on the variance $(\overline{u'^2}, \overline{v'^2}\ \overline{w'^2})$ and covariance $(\overline{u'v'}, \overline{u'w'}, \overline{v'w'})$ of the turbulence components is obtained

$$\begin{aligned} \overline{V_{LOS}'^2} &= \overline{u'^2}\cos^2\Delta\theta\cos^2\phi + \overline{v'^2}\sin^2\Delta\theta\cos^2\phi + \overline{w'^2}\sin^2\phi \\ &+ 2\overline{u'v'}\sin\Delta\theta\cos\Delta\theta\cos^2\phi + 2\overline{u'w'}\cos\Delta\theta\cos\phi\sin\phi \\ &+ 2\overline{v'w'}\sin\Delta\theta\cos\phi\sin\phi + \overline{\epsilon}^2 \end{aligned} \tag{3.48}$$

Note that covariance terms involving ϵ disappear as the error term is uncorrelated.

The sensitivity to the different turbulence components contrasts with a cup anemometer which is sensitive predominantly to the streamwise turbulence component only, u'. Sonic anemometers (3D types) measure the variance and covariance of the three turbulence components by combining measurements from the three ultrasonic pulses.

There has been a significant amount of research on the use of RSDs for measuring turbulence, particularly within the meteorological community. There has also been some work recently to study the difference in turbulence measurements from RSDs and point measurements. A detailed study compared CW and pulsed lidar measurements of turbulence with that from mast-mounted sonic anemometers for a range of stability conditions at the Danish National Test Centre for Large Wind Turbines at Høvsøre, Denmark showing significant differences (A. Sathe et al., 2015, 2011a and b).

Figure 3.25 shows the results of that study for the streamwise turbulence. It can be seen the behaviour for the CW and pulsed lidar is quite different. The CW lidar underestimates the turbulence in general. The pulsed lidar also underestimates the turbulence except for unstable conditions, where there is the greatest level of cross-contamination from covariance terms in Eq. (3.48), so the measurement is no longer purely the streamwise term. The general underestimation of turbulence can be understood in that the volume averaging of the two devices is unable to capture the variance due to the smallest turbulence structures in the atmosphere, compared to the sonic anemometer. This also explains the increased underestimation of the turbulence with increasing stability as the turbulence length scales are larger under unstable conditions than stable conditions. The increased underestimation with height seen for the CW lidar results from the fact that the backscattering of laser light is never exactly from a point but from a circle of a certain diameter and along a certain proportion of the line of sight. This proportion is related to

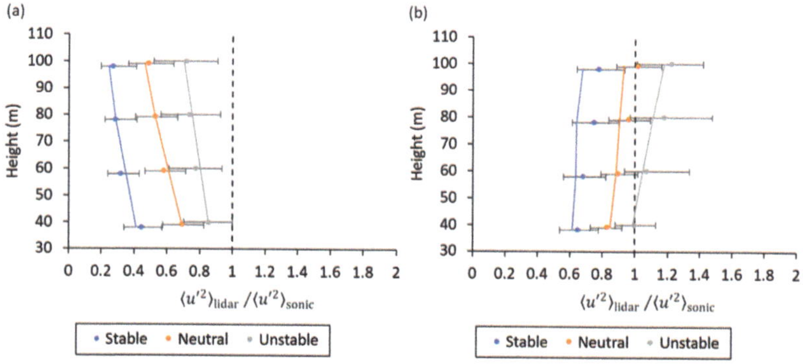

Figure 3.25 Streamwise turbulence as measured by (a) a CW lidar and (b) a pulsed lidar compared with measurements from sonic anemometers as a function of surface layer stability. Points are the measurements; bars represent the first and third quartiles and the solid lines represent a theoretical model estimate. Measurement heights for the three stability classes are slightly offset for clarity. *Source:* Sathe et al. (2015) / DTU Wind Energy.

the so-called Rayleigh length, which, in the case of the CW lidar being focused at specific heights, is proportional to the square of the distance from the source to the target. However, turbulence length scales increase linearly with height. This increasing mismatch means that the CW lidar tends to filter out more of the turbulent eddies at greater heights, reducing the magnitude of the total turbulence measured. By contrast, the pulsed lidar maintains a constant Rayleigh length with height. Now, as the turbulence length scales increase linearly with height, the pulsed lidar can resolve a greater proportion of the turbulence spectrum with height resulting in an increased estimate of turbulence at higher levels. In Figure 3.25, the solid lines represent a model of the expected lidar turbulence values based on scanning geometry and the backscattering characteristics of each lidar. In the case of the CW lidar, the fit is good implying that a relatively accurate estimate of the true turbulence intensity could be possible by applying a correction factor. The model is less good for the pulsed lidar but captures the trends correctly, and the values lie within the interquartile ranges.

Similar trends are seen for the other turbulence components, though the magnitude of the underestimation of vertical turbulence is significantly greater, due to the much smaller length scales of the vertical turbulence components. Studies to compare sodars have come to similar conclusions as for lidars (e.g. Thomas and Vogt, (1993)), showing a general underestimation of turbulence compared with sonic anemometers which increases with stability.

Floating lidars present a particular challenge for the measurement of turbulence intensity. Although the motion of the floating platform has a relatively modest effect on the measurement of mean wind speed, the motion effectively adds a significant component to the turbulence, and thus the turbulence intensity will be significantly overestimated if there is no motion compensation. Fortunately, applying an algorithm to correct for the motion is relatively straightforward assuming that the floating lidar has a motion reference unit installed to measure the motion in the six degrees of freedom. Figure 3.26 shows (a) the mean wind speed and (b) the turbulence intensity measured at several heights by a continuous wave lidar mounted on a buoy, comparing uncompensated values against those which have been compensated and measurements from a nearby coastal fixed lidar of a similar type (Kelberlau et al., 2020) when measuring offshore winds. The motion of the buoy in the case of the mean wind speed makes only a 1.5% difference compared with the fixed lidar, but without motion compensation, the turbulence intensity is overestimated by around 50%. It can also be seen that the motion compensation performs well in correcting the values in both cases.

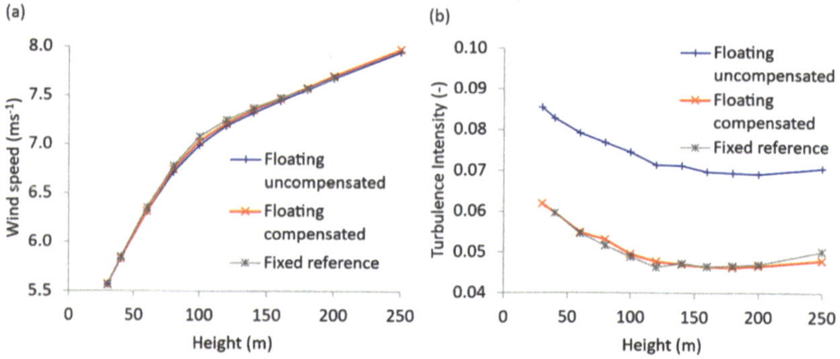

Figure 3.26 Continuous wave lidar measurements of (a) mean wind speed (*Source*: reproduced from (Kelberlau et al., 2020) / MDPI / Public Domain CC BY 4.0) and (b) turbulence intensity, comparing floating lidar measurements with and without motion compensation with those from a nearby coastal fixed lidar of a similar type (*Source*: adapted from Kelberlau et al. (2020).

3.4.4 Radiosondes

The measurement of upper air parameters is important to provide boundary conditions for weather forecasting models. This type of information is less important for wind resource assessment purposes, but increasingly it is becoming clear that measurements well beyond the surface layer (>200 m) are important to understand mesoscale phenomena that can affect wind farm output.

Since the early twentieth century, unmanned helium-filled weather balloons with meteorological instruments have been used to make upper air measurements. Such devices are known as *radiosondes*. Frequently, these are tracked using radio direction finders or more recently global positioning system (GPS) satellites. These tracked devices are also called *radar wind sondes* or *rawinsondes*. Other similar devices can be dropped from a plane (dropsondes) or launched using a rocker motor (rocketsondes). The tracking of these devices as they travel up through the atmosphere and drift with the wind can provide an estimate of the wind profile (speed and direction) up to around 40 km for balloons and 145 km for rockets. An example of a radiosonde is shown in Figure 3.27.

The drawback of using data from these devices is that there are a limited number of sites where they are launched, they are generally only launched on land, ascents are only made a few times a day (up to four usually), and the number of launch sites is reducing over time as other remote sensing techniques are becoming more widely used (e.g. satellite measurements). Furthermore, the instrumentation is only rarely recovered, so their sustainability is open to question.

Figure 3.27 A radiosonde about to be launched. *Source:* National Oceanic and Atmospheric Administration (NOAA), https://esrl.noaa.gov/, accessed 5 August 2022.

3.4.5 Satellite Measurements

In recent years, there has been an increased interest in the use of satellite data to infer wind speeds. Most satellite measurements are not direct measurements of wind speed but rather provide a proxy measurement and normally only over sea areas. Measurements are made by sending out a beam of radio waves to the sea surface and measuring the radiation which is scattered back to the satellite. As the roughness of the sea increases, the amount of backscattered radiation also increases. As the roughness of the sea surface increases as the wind speed increases, this means there is a relationship between the fraction of radiation which is backscattered to the satellite and the surface wind speed. This is the principle of a *scatterometer*. Normally, a satellite has several antennae to send and receive signals and uses polarised radiation to provide resolution of both wind speed and direction. The relationship between the wind vector and the backscattered radiation is empirically determined and relatively simple algorithms have been developed for this purpose. One such well-used algorithm or transfer or geophysical model function (GMF) is CMOD2 (Stoffelen and Anderson, 1997a)

$$\sigma^0 = B_0(1 + B_1 \cos\phi + B_2 \cos 2\phi). \tag{3.49}$$

This function relates the normalised radar cross-section σ^0 to the 10 m wind speed V and wind direction ϕ, assuming a neutral surface layer, where $B_0 = 10^{\alpha} V^{\gamma}$, B_1, and B_2 are linear functions of V and parabolic functions of the angle of incidence of the radar beam θ with respect to the sea surface. The exponents α and γ are also parabolic functions of θ. The various parameters of the GMF are tuned using training data to provide the best estimate of wind speed and direction. Various improvements have been made to this algorithm, including the CMOD4 GMF (Stoffelen and Anderson, 1997b).

Several satellites have been launched over the years with onboard scatterometer devices operating in the C band (5.4 GHz) or Ku band (13.4 GHz), including ESCAT, ASCAT, ERS-1/2, QuickSCAT, and more recently the International Space Station (ISS). The spatial resolution of the wind speed measurements is relatively low, being of the order of tens of kilometres (Frison et al., 2016) and the accuracy of σ^0 is around 0.1 dB.

The spatial resolution of radar backscattered measurements can be significantly improved using a synthetic aperture radar, or SAR. This can give a spatial resolution potentially down to 1 m, but because it scans a relatively narrow swath of the sea, temporal resolution is significantly less than for a scatterometer device. SAR devices generally have only one viewing angle (in contrast to scatterometers) and thus require another independent source of information to infer wind direction, such as a numerical weather prediction (NWP) model. The accuracy of scatterometer and SAR devices when compared with other sources of information, including in-situ measurements and NWP data, is typically of the order of 1–1.5 ms^{-1} and a typical bias of 0.5 ms^{-1}. In addition, errors are wind speed dependent with larger errors at higher wind speeds (Yang et al., 2011). Errors in wind direction are typically around 20° (Dagestad et al., 2013).

An alternative measure of wind speed (and wave height) can be made using a satellite *altimeter*, which relies on a slightly different physical process to scatterometers or SARs. This instrument sends a beam of microwaves vertically down to the sea surface and the strength of the specular reflected power gives a measure of the wind speed. There is little to choose in terms of accuracy compared with backscatter measurements, and the typical accuracy of altimeter measurements is around 1 ms^{-1}.

Direct measurements of atmospheric wind speeds from a satellite have recently become available. In 2018, the Aeolus satellite was launched with an onboard lidar. For the first time, this provided the capability to measure wind speeds globally and vertically throughout the atmosphere giving a vertical spatial resolution of between 250 m and 2 km, depending on height and horizontal resolution of 3.6 km. This is done using the Atmospheric

Laser Doppler Instrument (ALADIN) which uses the backscatter (Rayleigh scattering from molecules and Mie scattering from aerosols and cloudy air) from a pulsed ultraviolet (355 nm) lidar along an 87 km beam. In principle, wind speed measurements are capable of being made from the ground to 30 km above the surface. The purpose of the Aeolus satellite was to demonstrate the ability of a space lidar to provide accurate global wind speed profiles which could be used in weather forecasting models.

Though validation with ground-based measurements has been limited to date, a comparison between ALADIN, NWP model data, and a ground-based prototype of the same instrument has been made (Lux et al., 2020). This showed that the satellite measurements had an estimated bias of 1.7 ms^{-1} and an accuracy of 2.5 ms^{-1}. Figure 3.28 shows a comparison between the satellite and airborne measurements. Although there is a clear correlation, the scatter is quite high, particularly for measurements close to the ground. Although the accuracy is not high for wind resource assessment purposes, the space-based instrument was a first demonstration and later missions will undoubtedly increase measurement accuracy and resolution.

The use of satellite data for wind resource assessment is something which should be treated with caution. It is mostly only available over sea areas. Because the measurements rely on data collected in swaths over the earth's surface, this limits either spatial or temporal resolution (or both). Measurements close to the coastline can be unreliable as the GMFs are no longer accurate in these regions. The GMFs have been empirically determined to produce 10 m wind speeds under the assumption of neutral atmospheric stability. Conditions offshore are rarely neutral and an accurate

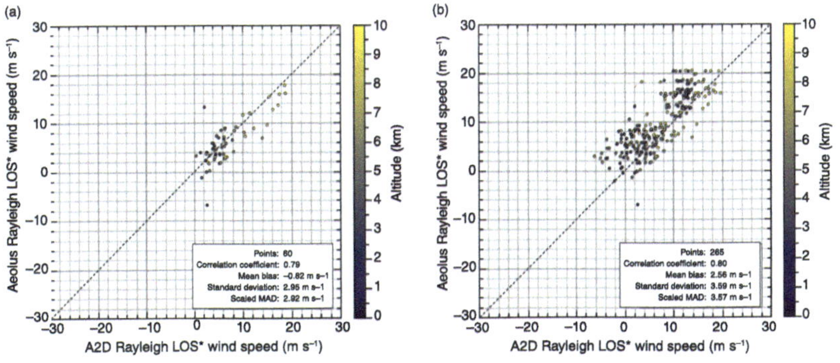

Figure 3.28 Comparison between the Aeolus ALADIN line-of-sight (LOS*) Doppler lidar wind speed measurements and a similar airborne device (A2D) corrected to the same line of sight for (a) one swath under-flight on 22 November 2018 and (b) for the entire three-month measurement campaign. *Source:* Lux 2020 / Copernicus Publications / CC BY 4.0.

estimate of wind speed is frequently required at heights significantly higher than 10 m. This involves a significant degree of extrapolation and the potential for increased error. Errors of the order of 1–2 ms^{-1} are generally not satisfactory for an accurate wind resource assessment. Satellite lidar measurements have the potential to provide more comprehensive direct measurements of wind speed, but early indications are that their accuracy is not better than a scatterometer or SAR measurements.

On the one hand, satellite measurements can give a very wide spatial indication of wind speed variability over the sea, something which would be difficult to achieve with terrestrially based instruments. They can be used to visualise phenomena – such as gravity waves, storm fronts, squalls, etc. – which may be of relevance when assessing the variability in the wind resource at a site. Examples are shown in Figure 3.29. On the other hand, their accuracy and resolution in time and space limit their use. At best they can provide a broad indication of wind conditions offshore but can only be as supplement to more accurate measurements or models. Satellite lidar measurements may provide better data, but this is yet to be proven as such measurements are at an early stage.

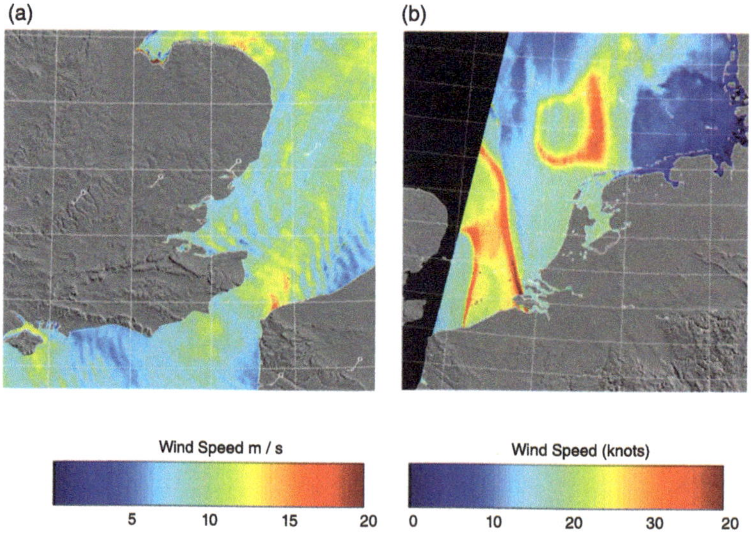

Figure 3.29 Processed SAR images showing 10 m wind speeds for (a) gravity waves in the English Channel and (b) gust fronts associated with thunderstorms in the North Sea. Wind fields from synthetic aperture radar are provided by the Technical University of Denmark, Department of Wind Energy (DTU Wind Energy).

3.5 Meteorological Measurement Stations

3.5.1 Background

Historically, wind speed and direction have been measured, along with a range of other meteorological variables, at purpose-built meteorological stations or on mobile sites such as ships and more recently aircraft. The siting of these stations is determined by the purpose for which the measurements were made, which has not typically been for the purposes of determining wind resource in advance of constructing a wind farm. The main reasons for recording such data are listed here.

3.5.1.1 Aviation

An airport requires detailed meteorological data, particularly wind speed, to ensure the safe and efficient take-off and landing of aircraft. For example, depending on the wind direction, a different runway may be used to minimise the speed of an aircraft relative to the ground. The quality of data from such sites is generally good, given that airports are normally in flat open areas with few obstacles and trees in the immediate vicinity. Aircraft themselves also make measurements of meteorological quantities, including wind speed, using pitot tubes. Due to the number of aircraft flying around the world, this provides a rich network of upper air measurements in areas where fixed sites cannot be located.

3.5.1.2 Astronomy

The ability to make astronomical observations and the quality of such observations is affected by atmospheric conditions. For this reason, major observatories, both old and new, frequently have associated meteorological stations. The quality of data can vary depending on the location of the observatory. Recordings from older sites in urban areas may be affected by changes in the local built environment.

3.5.1.3 Shipping

Knowledge of present and expected weather conditions is very important to shipping for reasons of safety and, during the era of sailing ships, to chart the most efficient course depending on wind direction. For this reason, meteorological stations are frequently to be found at coastal locations, particularly ports. Offshore, measurements are made on lighthouses, lightships, buoys, and offshore platforms (e.g. oil rigs). Ships also measure meteorological data onboard. The quality of offshore observations, such as wind speed, may be quite variable. For example, anemometers mounted on lighthouses and offshore platforms may be affected by structural shading.

Ship-borne measurements are especially problematic, again, due to shading from objects on the ship but also by the movement of the vessel.

3.5.1.4 Scientific Research

Atmospheric measurements may be required for the purposes of research. This may include a number of different areas, including air quality, agricultural meteorology, hydrology, or climatology. For obvious reasons, the quality of such observations is normally high, though site exposure may not be ideal for the purposes of wind resource assessment (e.g. if a site is located close to an area used for experimental crop growing).

3.5.2 Data Quality

Although meteorological stations may have measurement records over a period of several decades, the use of such data for long-term correction should be approached with caution as the quality and continuity of these data may be questionable. Some of the factors which can affect data quality are listed here.

3.5.2.1 Exposure

Over time, the area surrounding a meteorological mast can change due to the growth or removal of trees and the construction or demolition of buildings. This affects the wind speed measured at a mast particularly where the wind comes from a direction where such changes have occurred.

3.5.2.2 Continuity

Instruments and masts may be changed a number of times at a site over time. Effective measurement heights can change, which can cause significant bias in wind speed values (Watson et al., 2015). Masts may be re-sited sufficiently far from the original location to cause a change in exposure or, indeed, the whole site may be relocated some kilometres from its original site. Instrument types may change over time with different operating characteristics (e.g. a cup anemometer may be changed for a device with a different rotor inertia). These factors can cause gradual or sudden changes in mean wind speed trends or wind direction distributions that, if unaccounted for, can bias long-term corrections.

3.5.2.3 Data Recording

Modern meteorological stations use digital equipment to record measurements, which leaves little room for error. However, instruments can fail, which may result in erroneous values if not properly quality controlled. In the past, measurements will have been made manually, which left room for human error and for some latitude in terms of conventions used by a meteorologist to record data. For example, measurements were (and indeed still are) rounded to the nearest measurement unit (e.g. knot or metre per second). However, whether the value is rounded up or rounded down may not necessarily be consistent from site to site.

There are quality assurance methods which can be used to assess the accuracy of measurements (e.g. Abbott, 1986), and data from meteorological stations, even if run by well-respected government agencies, should not be taken at face value.

3.5.3 Geographical Spread

A map showing the locations of present-day worldwide operating meteorological stations is presented in Figure 3.30. It can be seen that the distribution of stations is far from uniform. The density of meteorological stations reflects the local population density and the wealth of the country in question. For this reason, there is a high density of stations in Europe and the United States, whereas there are large areas such as much of northern Africa, parts of China, eastern Russia, Greenland, central Australia, northern Canada, Mexico, and large areas of South America with relatively few stations. In these areas, should the long-term wind resource need to be assessed, this represents a challenge in terms of correcting local mast measurements for long-term variation. In these areas and for many sites offshore, alternative sources of information to adjust

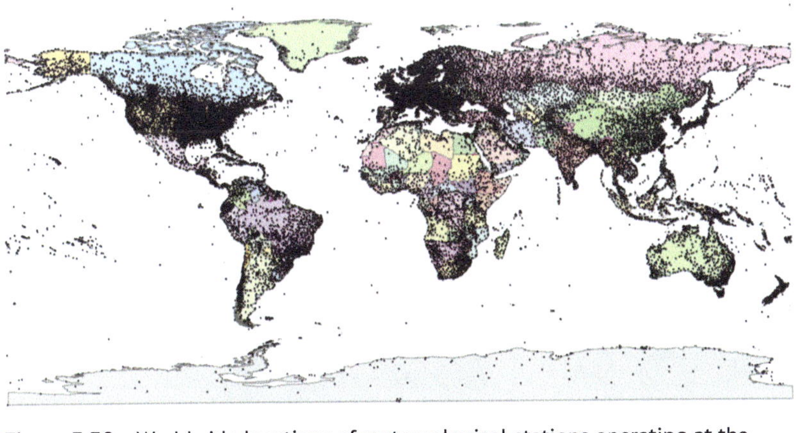

Figure 3.30 Worldwide locations of meteorological stations operating at the end of 2021. *Source:* map of the countries produced using an ESRI GIS file based on data from Garmin International Inc. and the US Central Intelligence Agency. Meteorological station locations provided by Weather Graphics.

for climatological variation must be sought. In recent years, this role has been filled by reanalysis datasets, which are discussed in more detail in Chapter 9.

3.6 The Effect of Climatic Conditions on Wind Speed Measurements

Although the instruments described in this chapter can be used in a variety of climatic conditions, there are some limitations that should be considered, which are summarised in this section.

3.6.1 Cold Climates

In very cold climates, the impact of icing is an obvious impediment to the measurement of wind speed and special measures are required (Bredesen et al., 2017). Figure 3.31 shows an example of where both the mounting structure and the instrumentation are affected by the build-up of ice. For mechanical instruments, such as cup anemometers, progressive icing slows the cups and eventually causes them to stop. In the case of a sonic anemometer, a build-up of ice between the transmitter/receiver pairs locally affects the wind flow, blocking the flow under extreme icing conditions. Additionally, icing on the mast structure can cause significant flow distortion and erroneous measurements of wind speed even if wind speed measuring instruments are working correctly. Where ice accumulation is expected, such as in high latitudes, heating the instruments and mounting structure may, therefore, be required.

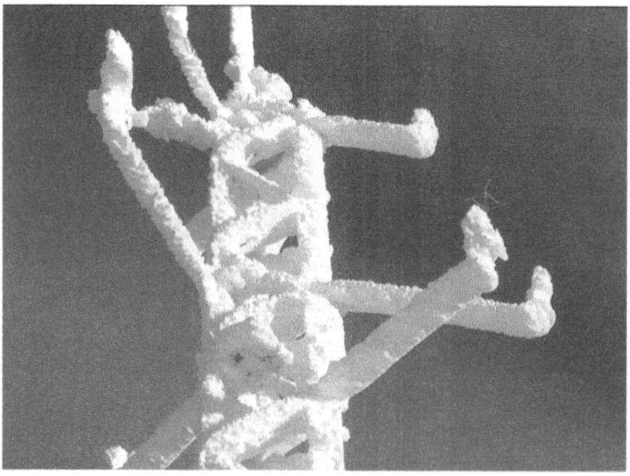

Figure 3.31 Icing affecting the structure and instruments of a meteorological mast at Mount Needle, Murdochville, Canada © Nergica 2022. *Source:* with permission from Nergica, Quebec, CA / Bredesen et al. (2017).

This will reduce periods when measurements may be ice affected but can never completely eliminate them. The downside to this is that the additional equipment required (e.g. heating elements) affects the accuracy of the instruments through added inertia and changes to the shape of the cups. For sites where icing is only occasionally a problem, unheated equipment can be used, and the data filtered accordingly for when icing is expected. Contrary to what might be expected, using measurements of relative humidity and temperature is not a reliable indicator of icing as it tends to overestimate periods of when icing might be expected. A more reliable indicator has been shown to be based on temperature and the variation of the standard deviation of the wind direction or the difference in wind speed from a heated sonic anemometer and an unheated cup anemometer (Wadham-Gagnon et al., 2015). Alternatively, a dedicated ice accumulation sensor can be used based, for example, on the load detected on a rotating cylinder or ultrasonic measurements.

Although lidars and sodars have no moving parts that can be affected by icing, snow accumulation on the instruments can affect operation (Bredesen et al., 2017). The availability of lidar measurements, in particular, has been shown to be affected by the relatively low concentration of aerosol particles in the cleaner air present in colder climates and, for this reason, a sodar may be preferred in such locations.

3.6.2 Effect of Precipitation, Low Clouds, and Fog

The availability of lidars is affected by the presence of precipitation, low clouds, and fog, all of which attenuate the laser beam and hamper the detection of Doppler-shifted, backscattered light. The effect of these conditions is also height dependent, as shown in Figure 3.32. Precipitation is seen to significantly affect availability at lower levels, particularly when precipitation

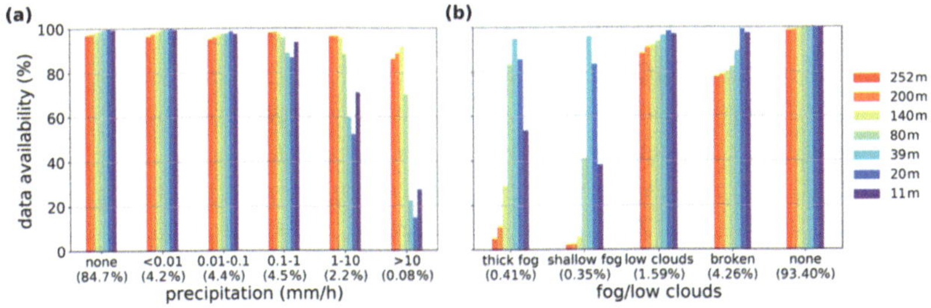

Figure 3.32 Data availability of a continuous wave lidar as a function of (a) precipitation and (b) fog/low clouds at different heights based on measurements at Cabauw, Netherlands (Knoop et al., 2021). The percentages in brackets represent the amount of the time a particular condition was observed. *Source*: Knoop et al. 2021 / Copernicus Publications / CC BY 4.0.

exceeds 1 mm · h^{-1}. On the other hand, fog and low clouds affect availability more at higher levels, particularly in the case of foggy conditions.

References

Abbott, P.F., 1986. World Climate Data Programme Guidelines on the Quality Control of Surface Climatological Data. *WCP-85, WMO/TD-No. 111.*

Angevine, W.M., 2000. Radio acoustic sounding system (RASS) applications and limitations. *Int. Geosci. Remote Sens. Symp.* 3, 1180–1182. https://doi.org/10.1109/IGARSS.2000.858060.

Aylott, L., Burt, S., Saunders, M., 2020. Northern Ireland's longstanding record wind gust is almost certainly incorrect. *Weather* 75, 8–13. https://doi.org/10.1002/WEA.3613.

Bonin, T.A., Choukulkar, A., Brewer, W.A., Sandberg, S.P., Weickmann, A.M., Pichugina, Y.L., Banta, R.M., Oncley, S.P., Wolfe, D.E., 2017. Evaluation of turbulence measurement techniques from a single Doppler lidar. *Atmos. Meas. Tech.* 10, 3021–3039. https://doi.org/10.5194/amt-10-3021-2017.

Bradley, S., Strehz, A., Emeis, S., 2015. Remote sensing winds in complex terrain: a review. *Meteorol. Zeitschrift* 24, 547–555. https://doi.org/10.1127/metz/2015/0640.

Bradley, S., Perrott, Y., Behrens, P., Oldroyd, A., Bradley, S., Perrott, Y., Behrens, P., Oldroyd, A., 2012. Corrections for wind-speed errors from sodar and lidar in complex terrain. *Boundary-Layer Meteorol.* 1431 143, 37–48. https://doi.org/10.1007/S10546-012-9702-0.

Bredesen, R.E., Cattin, R., Clausen, N.-E., David, N., Jordaens, P.J., Khadiri-Yazami, Z., Klintström, R., Krenn, A., Lehtomäki, V., Ronsten, G., Wadham-Gagnon, M., Wickman, H., 2017. Wind energy projects in cold climates. *IEA Wind TCP Recommended Practice 13.* https://euagenda.eu/upload/publications/untitled-102011-ea.pdf (accessed 4 October 2022).

Brevoort, M.J., Joyner, U.T., 1934a. Aerodynamic characteristics of anemometer cups. *NACA Technical Note No. 489.*

Brevoort, M.J., Joyner, U.T., 1934b. Experimental investigation of the Robinson-type cup anemometer. *NACA Report No. 513.*

Burt, S., 2022. The gust that never was: a meteorological instrumentation mystery. *Weather* 77, 123–126. https://doi.org/10.1002/WEA.3830.

Busch, N.E., Kristensen, L., 1976. Cup anemometer overspeeding. *J. Appl. Meteorol.* 15, 1328–1332. https://doi.org/10.1175/1520-0450(1976)015<1328:cao>2.0.co;2.

Dagestad, K.-F., Horstmann, J., Mouche, A., Perrie, W., Shen, H., Zhang, B., Li, X., Monaldo, F., Pichel, W., Lehner, S., Badger, M., Hasager, C.B., Furevik,

B., Foster, R.C., Falchetti, S., Caruso, M.J., Vachon, P., 2013. Wind retrieval from synthetic aperture radar: an overview. *Proceedings of SEASAR 2012, 4th SAR Oceanography Workshop (SEASAR 2012).* Tromsø, Norway, 18-22 June 2013. ESA SP709. https://backend.orbit.dtu.dk/ws/portalfiles/portal/59267621/SeaSAR2012_whitepaper_wind.pdf (accessed 4 October 2022).

Doubrawa, P., Barthelmie, R.J., Wang, H., Pryor, S.C., Churchfield, M.J., Hasager, C.B., Peña, A., Li, X., Thenkabail, P.S., 2016. Wind turbine wake characterization from temporally disjunct 3-D measurements. *Remote Sens.* 8 939 8, 939. https://doi.org/10.3390/RS8110939.

Frison, P.-L., Jarlan, L., Mougin, E., 2016. Using satellite scatterometers to monitor continental surfaces. *L. Surf. Remote Sens. Cont. Hydrol.* 79–113. https://doi.org/10.1016/B978-1-78548-104-8.50003-6.

Garratt, J.R., 1994. *The Atmospheric Boundary Layer.* Cambridge University Press.

Hirth, B.D., Schroeder, J.L., Gunter, W.S., Guynes, J.G., 2012. Measuring a utility-scale turbine wake using the TTUKa mobile research radars. *J. Atmos. Ocean. Technol.* 29, 765–771. https://doi.org/10.1175/JTECH-D-12-00039.1.

Hirth, B.D., Schroeder, J.L., Gunter, W.S., Guynes, J.G., 2015. Coupling Doppler radar-derived wind maps with operational turbine data to document wind farm complex flows. *Wind Energy* 18, 529–540. https://doi.org/10.1002/we.1701.

International Electrotechnical Commission, 2017. *IEC 61400-12-1:2017:* wind energy generation systems: power performance measurements of electricity producing wind turbines. International Electrotechnical Commission, Geneva.

Kaimal, J.C., Finnigan, J.J., 1994. *Atmospheric Boundary Layer Flows: Their Structure and Measurement.* Oxford University Press.

Kelberlau, F., Neshaug, V., Lønseth, L., Bracchi, T., Mann, J., 2020. Taking the motion out of floating lidar: turbulence intensity estimates with a continuous-wave wind lidar. *Remote Sens.* 12 898 12, 898. https://doi.org/10.3390/RS12050898.

Kelly, M., 2018. From standard wind measurements to spectral characterization: turbulence length scale and distribution. *Wind Energy Sci.* 3, 533–543. https://doi.org/10.5194/WES-3-533-2018.

Kim, H.-G., Meissner, C., 2017. Correction of LiDAR measurement error in complex terrain by CFD: case study of the Yangyang pumped storage plant. *Wind Eng.* 41, 226–234. https://doi.org/10.1177/0309524X17709725.

Knoop, S., Bosveld, F.C., De Haij, M.J., Apituley, A., 2021. A 2-year intercomparison of continuous-wave focusing wind lidar and tall mast wind measurements at Cabauw. *Atmos. Meas. Tech.* 14, 2219–2235. https://doi.org/10.5194/AMT-14-2219-2021.

Kristensen, L., 1993. The cup anemometer and other exciting instruments. *Risoe National Laboratory Report No. Risoe-R-615(EN).* https://www.windsensor.com/application/files/7714/2694/4640/The_Cup_Anemometer_and_Other_Exciting_Instruments.pdf (accessed 4 October 2022).

Lux, O., Lemmerz, C., Weiler, F., Marksteiner, U., Witschas, B., Rahm, S., Geiß, A., Reitebuch, O., 2020. Intercomparison of wind observations from the European Space Agency's Aeolus satellite mission and the ALADIN airborne demonstrator. *Atmos. Meas. Tech.* 13, 2075–2097. https://doi.org/10.5194/amt-13-2075-2020.

Newman, J.F., Klein, P.M., Wharton, S., Sathe, A., Bonin, T.A., Chilson, P.B., Muschinski, A., 2016. Evaluation of three lidar scanning strategies for turbulence measurements. *Atmos. Meas. Tech.* 9, 1993–2013. https://doi.org/10.5194/AMT-9-1993-2016.

Panofsky, H.A., Dutton, J.A., 1984. *Atmospheric Turbulence: Models and Methods for Engineering Applications.* Wiley Blackwell.

Réthoré, P.-E., 2009. Wind turbine wake in atmospheric turbulence. PhD thesis. Risø National Laboratory for Sustainable Energy. Risø-PhD No. 53(EN). Aalborg University.

Sathe, A., Banta, R., Pauscher, L., Vogstad, K., Schlipf, D., Wylie, S., 2015. Estimating turbulence statistics and parameters from ground- and nacelle-based lidar measurements: IEA Wind Expert Report. DTU Wind Energy. https://orbit.dtu.dk/files/116998413/Estimating_Turbulence_Statistics.pdf (accessed 4 October 2022).

Sathe, A., Mann, J., Gottschall, J., Courtney, M.S., 2011a. Can wind lidars measure turbulence? *J. Atmos. Ocean. Technol.* 28, 853–868. https://doi.org/10.1175/JTECH-D-10-05004.1.

Sathe, A., Gryning, S.-E., Pena, A., 2011b. Comparison of the atmospheric stability and wind profiles at two wind farm sites over a long marine fetch in the North Sea. *Wind Energy* 14, 767–780. https://doi.org/10.1002/we.456.

Slinger, C., Harris, M., 2012. Introduction to continuous-wave Doppler lidar. https://breeze.colorado.edu/ftp/RSWE/Chris_Slinger.pdf (accessed 4 October 2022).

Stoelinga, M., LaWhite, N., 2018. Validation of Triton wind profiler measurements in complex terrain, using WindSim CFD-based flow curvature correction. https://www.vaisala.com/sites/default/files/documents/TritonComplexTerrainValidation_0.pdf (accessed 4 October 2022).

Stoffelen, A., Anderson, D., 1997a. Scatterometer data interpretation: measurement space and inversion. *J. Atmos. Ocean. Technol.* 14, 1298–1313. https://doi.org/10.1175/1520-0426(1997)014<1298:SDIMSA>2.0.CO;2.

Stoffelen, A., Anderson, D., 1997b. Scatterometer data interpretation: estimation and validation of the transfer function CMOD4. *J. Geophys. Res. Ocean.* 102, 5767–5780. https://doi.org/10.1029/96JC02860.

Stull, R., 1988. *An Introduction to Boundary Layer Meteorology*. Springer, Dordrecht, Netherlands. https://doi.org/10.1007/978-94-009-3027-8.

Thomas, P., Vogt, S., 1993. Intercomparison of turbulence data measured by SODAR and sonic anemometers. *Boundary-Layer Meteorol.* 62, 353–359. https://doi.org/10.1007/BF00705564.

Valldecabres, L., Nygaard, N.G., Vera-Tudela, L., von Bremen, L., Kühn, M., 2018. On the use of dual-Doppler radar measurements for very short-term wind power forecasts. *Remote Sens.* 10 1701. https://doi.org/10.3390/RS10111701.

Wächter, M., Rettenmeier, A., Kühn, M., Peinke, J., 2008. Wind velocity measurements using a pulsed LIDAR system: first results. *IOP Conf. Ser. Earth Environ. Sci.* 1, 012066. https://doi.org/10.1088/1755-1315/1/1/012066.

Wadham-Gagnon, M., Swytink-Binnema, N., Bolduc, D., Tete, K., Arbez, C., 2015. Ice detection methods and measurement of atmospheric icing. in: *Proceedings of the International Workshop on Atmospheric Icing of Structures* 2015. pp. 151–156.

Watson, S.J., Kritharas, P., Hodgson, G.J., 2015. Wind speed variability across the UK between 1957 and 2011. *Wind Energy* 18, 21–42.

Wyngaard, J.C., 1981. Cup, propeller, vane, and sonic anemometers in turbulence research. *Annu. Rev. Fluid Mech.* 13, 399–423. https://doi.org/10.1146/annurev.fl.13.010181.002151.

Yahaya, S., Frangi, J.P., 2004. Cup anemometer response to the wind turbulence-measurement of the horizontal wind variance. *Ann. Geophys.* 22, 3363–3374.

Yang, X., Li, X., Pichel, W.G., Li, Z., 2011. Comparison of ocean surface winds from ENVISAT ASAR, MetOp ASCAT scatterometer, buoy measurements, and NOGAPS model. *IEEE Trans. Geosci. Remote Sens.* 49, 4743–4750. https://doi.org/10.1109/TGRS.2011.2159802.

4

Wind Speed Variability and Distributions

When considering the wind resource at a site, there is a very large range of temporal conditions that can be considered. If a wind farm site is to operate for a period of at least 20 years then the degree of interannual variation needs to be considered; indeed, in any long-term assessment, an understanding of decadal variation may be appropriate. For the purposes of planning operational maintenance or in medium-term energy trading, it is important to consider seasonal variation in wind conditions at a site. Short-term energy trading requires an understanding of the daily and hourly variation in the wind. In the selection process of a wind turbine for a site, it is necessary to understand the very short-term turbulent conditions on a timescale of seconds and indeed a wind turbine designer may require an understanding of subsecond variation to ensure that a machine will survive a particular level of fatigue loading during its lifetime. There are a very few sites for which detailed measurements have been made over a long enough period to allow a comprehensive analysis of variation across all of these scales. However, there has been a significant amount of research to understand variability over selected timescales. This chapter considers the characteristics of this variability at difference timescales and the distributions that are commonly used to parameterise wind speed variation on the long and short term.

4.1 Power Spectrum

One way to understand the degree of variation in wind speed at different temporal scales is to calculate a power spectrum based on wind speed measurements. Perhaps the most-well-known and possibly the first example of this was the Van der Hoven spectrum (Van der Hoven, 1957), as shown in Figure 4.1. The spectrum determined from measurements at

Handbook of Wind Resource Assessment, First Edition. Simon Watson.

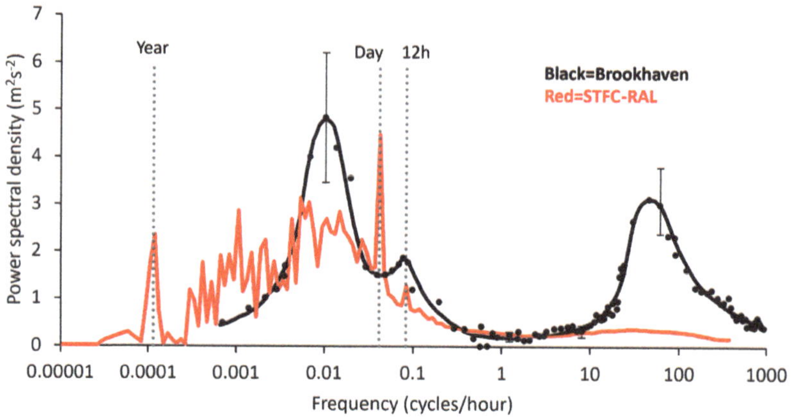

Figure 4.1 Wind speed power spectra: Black: the Van der Hoven spectrum based on wind speed measurements made at 91 m, 108 m, and 125 m at the Brookhaven National Laboratory, USA using data from (Van der Hoven, 1957); Red: based on wind speed data collected at 19 m at the wind test site at STFC-RAL, UK.

the Brookhaven National Laboratory is shown as black dots, with a continuous black trend line added. This was produced by stitching together a series of measurements taken on a meteorological mast at the site over different timescales and at different heights close to 100 m to determine wind speed variability between 0.001 and 1000 cycles per hour. This spectrum was shown to have particular features, namely: (i) a broad synoptic peak at around four days representing the passage of weather systems, (ii) a smaller peak at around 12 hours suggesting a twice daily maximum, and (iii) a peak at around one minute representing turbulent processes. Synoptic variation is associated with large-scale phenomena (e.g. pressure systems with a scale of several hundred kilometres), whereas turbulent variation is associated with local small-scale variation on a typical scale of less than 100 km. Another feature confirmed by measurements at two other sites was that of a spectral gap between 1 and 10 cycles per hour. The magnitude and structure of the turbulent peak was deliberately exaggerated since it was based on wind speed measurements made during a day when a hurricane passed through the site. This peak is around $3\,m^2 \cdot s^{-2}$, whereas during 'normal' weather situations it would be expected to be around $1\,m^2 \cdot s^{-2}$, suggesting that a ratio of around 5 to 1 for synoptic versus turbulent peak would be more normal. The existence of a two-day peak is postulated (Soberanis and Merida, 2015), though this has to be questioned given the short periods of data collection upon which this is based (up to six days).

The red line in Figure 4.1 shows a similar spectrum derived from wind speed measurements made at the STFC Rutherford Appleton Laboratory in the UK at 19 m height (Watson, 2014). Some similar features are seen in this spectrum, though the turbulent peak is of a much smaller magnitude compared to the synoptic peak. Indeed, its magnitude would seem to be consistent with a ratio of 5 to 1. There is a well-defined 24-hour peak not seen in the Van der Hoven spectrum, possibly due to the lower measurement height, where diurnal heating effects may have more influence. Harmonics of the 24-hour peak can also be seen. The synoptic peak is noisier than that of the Van der Hoven spectrum and, due to the fact that a longer data series was collected, an annual peak can be seen, though of a lower amplitude to the diurnal peak. There is evidence of a spectral gap, though perhaps not as defined as that seen in the Van der Hoven spectrum. In Section 4.2, we examine the evidence for the general existence of a spectral gap.

4.2 The Spectral Gap

The existence of a spectral gap suggested by Van der Hoven is often used as a justification for averaging wind speed measurements at intervals between ten minutes and one hour with the intention of preserving synoptic level fluctuation whilst only providing a measure of the magnitude of the turbulence at a site. This is generally sufficient for the purposes of activities such as wind resource assessment. However, this split between synoptic variation and turbulence may not be so clear. Analysis of long-term data from coastal sites (Heggem et al., 1998) suggests the absence of a gap around one hour and the presence of a dip in the spectrum around 10 minutes is only seen at heights of less than 40 m. It was also observed that the magnitude of any dip in the power spectrum at around 10 minutes was very dependent on the surface layer stability, as shown in Figure 4.2. Under stable conditions, the dip was visible, but as the surface layer became increasingly unstable, the dip disappeared and a monotonic reduction in power was seen with an increase in frequency. This would suggest that convective activity can contribute variance within what is traditionally thought of as the spectral gap and indeed spill over into the so-called turbulent range of frequencies. This was also seen in an analysis of data from the coast of central Norway (Gjerstad et al., 1995). An analysis of data from the Horns Rev offshore wind farm came to very similar conclusions (Vincent et al., 2011) noting that wind speed variability showed a strong dependence on atmospheric stability underlining the non-stationary nature of wind speeds.

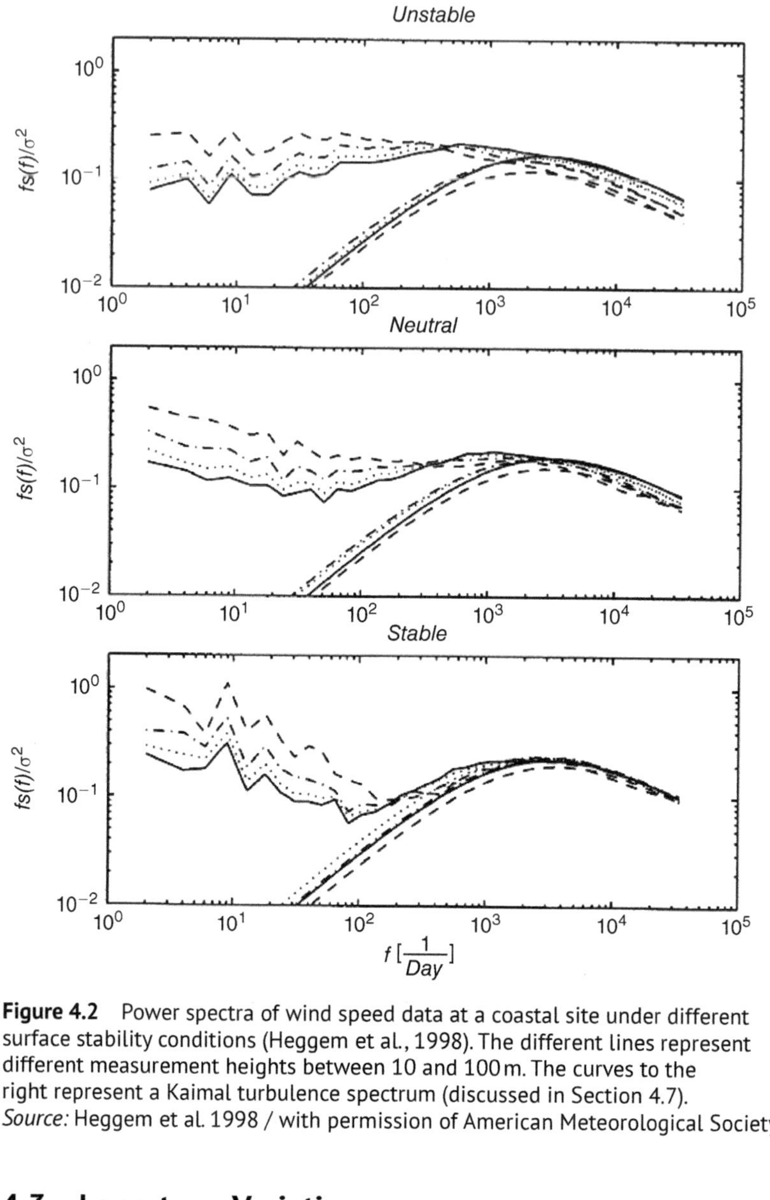

Figure 4.2 Power spectra of wind speed data at a coastal site under different surface stability conditions (Heggem et al., 1998). The different lines represent different measurement heights between 10 and 100 m. The curves to the right represent a Kaimal turbulence spectrum (discussed in Section 4.7). *Source:* Heggem et al. 1998 / with permission of American Meteorological Society.

4.3 Long-term Variation

A typical wind farm will be built to operate for a period of around 20 years. In addition, the data used to infer the expected long-term wind speed at a potential wind farm site may stretch back 20–30 years. It can be seen that an understanding of the long-term wind speed variability at the decadal level is important. However, it is not always straightforward to reconstruct a homogeneous time series of measured wind speeds which can be used to study long-term variation. Wind speed measuring instruments have a finite lifetime and will thus be periodically replaced. They may also be moved if a site is developed over time. Meteorological stations may be subject to change in exposure over time, such as increased urbanisation or changes to surrounding woodland. All these factors will affect the wind speed measured and give rise to variation which is not related to the climate and thus not truly representative of the long-term variation in wind speed of the local area. Reconstruction of long-term wind speeds must take this into account by careful selection of sites through analysis of logbooks and records or using other proxy variables which are not so prone to these changes. An alternative to long-term surface wind speed measurements is the output from a numerical weather prediction model. In addition to producing forecasts of future weather, these models may be used to reconstruct a range of meteorological variables on a regular three-dimensional grid at fixed time intervals, based on a large variety of input observations, including upper air and satellite observations. This is known as *reanalysis* and a number of such global (and a few regional) reanalysis datasets have been produced in recent times. These datasets have the advantage of removing spatial inhomogeneity and in many cases do not use surface station wind speed measurements as an input. Their disadvantage is the lack of spatial resolution, but they are useful to provide a long-term reference when trying to infer the temporal variability of a potential wind farm site. A more detailed description of reanalysis datasets is given in Chapter 9.

To try to quantify long-term variability of wind speeds over a particular region, a wind index may be constructed. This is done by calculating the mean wind speed over a region averaged over a particular time interval (e.g. a year) normalised by the long-term mean over that area, typically over a period of several decades. The area average can be constructed using wind speed observations or values from a global reanalysis dataset. An example of a long-term wind index is shown in Figure 4.3, where an index has been calculated for the UK based on data from seven meteorological stations over a period of 55 years (Watson et al., 2015). Swings in annual wind speed of >10% can be observed with a standard deviation of the annual wind speeds over the long-term period of 4%. There are prolonged periods of increases and decreases in annual wind speed, most notably the increase during the period of 1958 to 1967. Such an increase highlights the importance of averaging wind speeds over a long enough period to smooth out climatological variation when attempting to infer the long-term wind resource. In addition, such indices as these give a valuable insight into the expected year-to-year variation in mean wind speeds at a potential wind farm site. Wind indices are discussed further in Section 8.4.

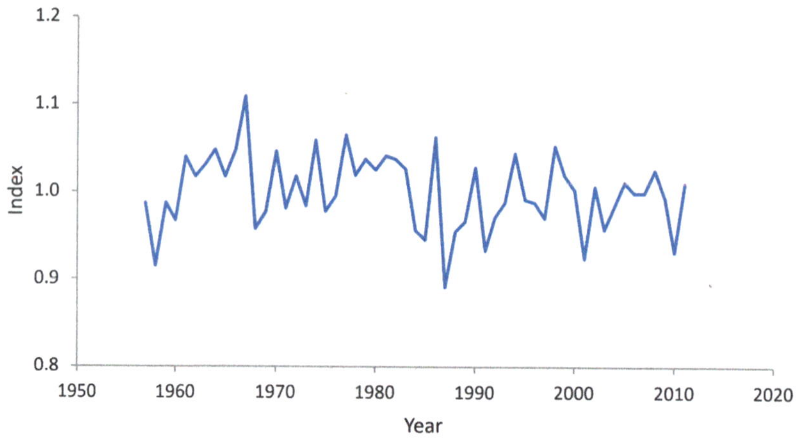

Figure 4.3 A UK wind index constructed using data from long-term wind speed measurements at seven meteorological stations.

4.4 Synoptic Scale Weibull Wind Speed Distributions

As mentioned in Section 3.2, it is common practice to measure wind speed every 1–2 seconds and average these wind speed measurements over a period of ten minutes when assessing a site for wind energy purposes. Once these ten-minute averages have been collected over a certain period (e.g. a year), it is possible to determine a wind speed distribution of the values. Such a distribution is shown in Figure 4.4. For many sites in the world, the wind speed distribution can be parameterised by a theoretical distribution which is a function of two variables. This is the *Weibull distribution*, so-called because it was first described in detail by the Swedish mathematician Ernst Hjalmar Waloddi Weibull (Weibull, 1951). The two-parameter Weibull probability distribution is of the form

$$p_W(u;C,k) = \left(\frac{k}{C}\right)\left(\frac{u}{C}\right)^{k-1} \exp\left(-\left(\frac{u}{C}\right)^{k}\right) \tag{4.1}$$

where u is the parameter of interest (in this case wind speed), C is the scale parameter, and k is the shape parameter. In Figure 4.4, a Weibull distribution, shown in red, is fitted to an observed distribution of ten-minute wind speed values averaged in bins of $1\,\mathrm{ms}^{-1}$ and has the parameter values $C = 9.2\ \mathrm{ms}^{-1}$ and $k = 2.19$. Note that the fit is made above $0\,\mathrm{ms}^{-1}$ as the fraction of the time that there is no wind (calms) is usually nonzero and also very low wind speed values may not cause an anemometer to rotate and thus be recorded as $0\,\mathrm{ms}^{-1}$.

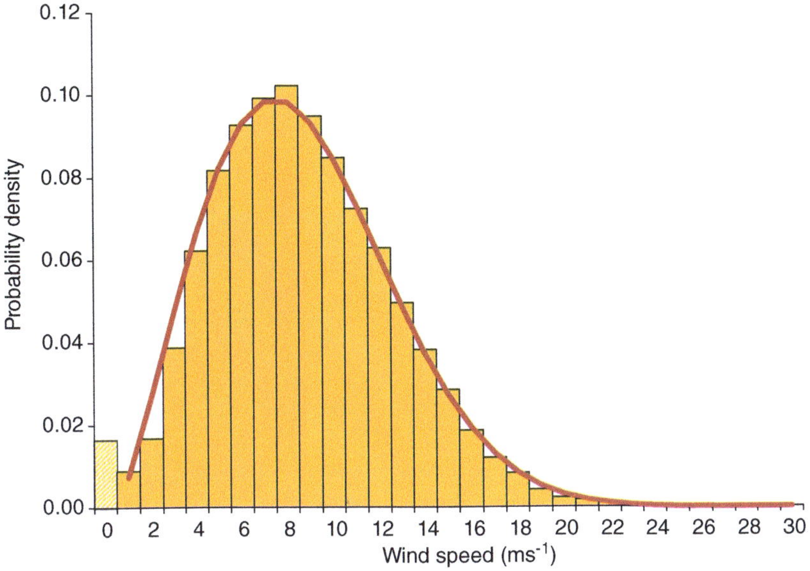

Figure 4.4 Example wind speed distribution with Weibull fit in red using a maximum likelihood estimation where $C = 9.2\,\mathrm{ms}^{-1}$ and $k = 2.19$. Note that the bin marked *0* represents calms (wind speed too low to be measured) and all other bins span wind speed intervals *u* to *u* + 1 where the bin label is *u*. The fit is plotted at the midpoint of each bin.

C and k are quite closely related to the mean $\bar{u}$ and the variance σ^2, respectively, of the wind speed values

$$\bar{u} = C\Gamma(1 + 1/k) \tag{4.2}$$

$$\sigma^2 = C^2\left[\Gamma(1 + 2/k) - \Gamma^2(1 + 1/k)\right] \tag{4.3}$$

where Γ is the gamma function. To a good approximation $(1 \leq k \leq 10)$

$$k \approx \left(\frac{\sigma}{\bar{u}}\right)^{-1.086} \tag{4.4}$$

For many wind speed distributions, k is close to 2. When $k = 2$, then Eq. (4.1) reduces to a *Rayleigh distribution*. In this case, the relationship between the mean wind speed and the scale parameter is simplified

$$C = \frac{2\bar{u}}{\sqrt{\pi}} \tag{4.5}$$

A useful form of the Weibull distribution is the cumulative Weibull distribution, where the probability F_W of *exceeding* a wind speed value v is given by

$$F_W(v;C,k) = \exp\left(-\left(\frac{v}{C}\right)^k\right) \tag{4.6}$$

This can be used to estimate the probability of being between two wind speeds, v_1 and v_2

$$p_W(v_1 < v < v_2;C,k) = F_W(v_1;C,k) - F_W(v_2;C,k)$$
$$= \exp\left(-\left(\frac{v_1}{C}\right)^k\right) - \exp\left(-\left(\frac{v_2}{C}\right)^k\right) \tag{4.7}$$

This is useful, as a typical wind turbine power curve is normally calculated based on averaging the power output of the turbine in 1 ms^{-1} bins across the operational wind speed range of the machine. Knowing the C and k values for a site and the power curve of a wind turbine thus allows a rapid calculation of the average power output of a turbine and its theoretical annual energy production (AEP).

Although the Weibull distribution is found to be a good fit for many sites, discrepancies have been found. For example, some sites are better fitted using a bimodal distribution consisting of two Weibull distributions. An example of this is shown in Figure 4.5, where there is a relatively large number of low wind speeds (Carta and Ramírez, 2007). As well as a normal Weibull distribution (W), a fit is made to the measurements using a two-component mixture Weibull distribution (WW) which is a function of two sets of scale (C_1, C_2) and shape (k_1, k_2) parameters and a weighting parameter w

$$p_{WW}(u;C_1,k_1,C_2,k_2) = wp_W(u;C_1,k_1) + (1-w)p_W(u;C_2,k_2) \tag{4.8}$$

The associated cumulative distribution is of the form

$$F_{WW}(v;C_1,k_1,C_2,k_2) = wF_W(v;C_1,k_1) + (1-w)F_W(v;C_2,k_2) \tag{4.9}$$

A fit to a third distribution known as a singly *truncated from below normal Weibull* (TNW) distribution is also shown which is a function of a set of Weibull parameters (C_0 k_0), a set of normal distribution parameters (a, b), and a weighting parameter w

$$p_{TNW}(u;a,b,C_0,k_0) = wq_N(u;a,b) + (1-w)p_W(u;C_0,k_0) \tag{4.10}$$

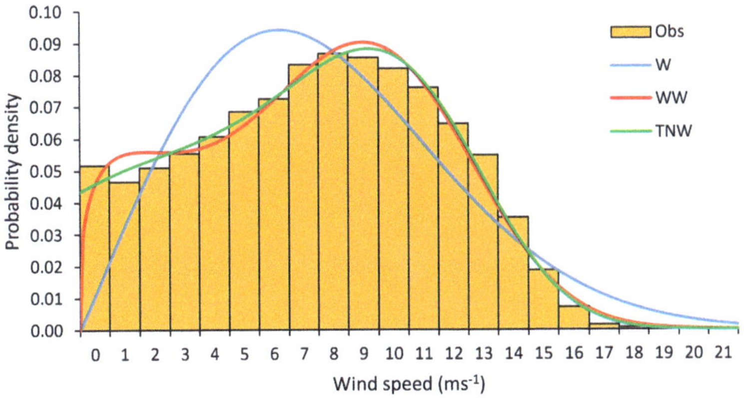

Figure 4.5 A comparison between a measured wind speed distribution and three theoretical distributions fitted to the data: Obs = binned observations; W = a normal Weibull distribution (C = 9.26 ms^{-1}, k = 2.05); WW = a two-component mixture Weibull distribution (C_1 = 10.90 ms^{-1}, k_1 = 3.46, C_2 = 4.56 ms^{-1}, k_2 = 1.31, w = 0.669); TNW = a singly truncated from below normal Weibull distribution (a = 2.73 ms^{-1}, b = 4.76 ms^{-1}, C_0 = 11.32 ms^{-1}, k_0 = 3.82, w = 0.437). Note that the x-axis bin labels span intervals such that '0' is 0–1, '1' is 1–2, etc. *Source:* Carta and Ramírez (2007) / with permission of Elsevier.

where q_N is a normal distribution

$$q_N(u;a,b) = \frac{1}{\sqrt{2\pi}bq_0(a,b)}\exp\left[-\frac{(u-a)^2}{2b^2}\right] \tag{4.11}$$

and

$$q_0(a,b) = \frac{1}{\sqrt{2\pi}b}\int_0^\infty \exp\left[-\frac{(u-a)^2}{2b^2}\right]du \tag{4.12}$$

where q_N has the cumulative form (probability of *exceeding* v)

$$G_N(v;a,b) = 1 - \frac{1}{\sqrt{2\pi}bq_0(a,b)}\int_0^v \exp\left[-\frac{(v-a)^2}{2b^2}\right]dv \tag{4.13}$$

so that the cumulative form of the TNW distribution is

$$F_{TNW}(v;a,b,C_0,k_0) = wG_N(v;a,b) + (1-w)F_W(v;C_0,k_0) \tag{4.14}$$

The WW distribution provides a better fit to the data than the W distribution, and the TNW distribution provides a slightly better fit again, though only marginally so in this case. The advantage of the TNW distribution is that it provides a better fit for sites where there is a large fraction of calms.

4.5 Weibull Parameter Variation with Height

As the Weibull scale parameter C is closely related to the mean wind speed, it is generally extrapolated with height following the logarithmic profile

$$C(z) = C(z_{ref}) \frac{\left[\ln(z/z_0) - \psi_m(z/L)\right]}{\left[\ln(z_{ref}/z_0) - \psi_m(z_{ref}/L)\right]} \tag{4.15}$$

where z_{ref} is the reference height at which C is initially determined.

The vertical dependence of the shape parameter k is rather more involved to determine as it does not typically increase monotonically with height even in flat homogeneous terrain. Normally, it is seen to initially increase with height until it reaches a maximum value at some reversal height, z_{rev} before decreasing aloft. A commonly used formula to predict k with height is (Wieringa, 1989)

$$k(z) = k(z_s) + c_k(z - z_s)e^{-((z-z_s)/(z_{rev}-z_s))} \tag{4.16}$$

The height z_s is a near surface height usually taken to be around 10 m, c_k is an empirical constant (≈ 0.022), and z_{rev} has a value of around 50–60 m near the coast and around 80–90 m inland. More recent research suggests a modified version of this formulation requiring only knowledge of the reversal height (Kelly et al., 2014)

$$k(z) = k(z_{ref}) + \frac{1 + (z/z_{rev})e^{-(z/z_{rev})}}{1 + (z_{ref}/z_{rev})e^{-(z_{ref}/z_{rev})}} \tag{4.17}$$

One such formula for z_{rev} is

$$z_{rev} = \alpha_{rev} z_0 (Ro_0)^{0.9} \tag{4.18}$$

where $\alpha_{rev} \approx 0.003$ and Ro_0 is the surface Rossby number ($= G/fz_0$, where G is the geostrophic wind speed described in Section 2.3). Figure 4.6 shows a comparison between a prediction of k using Eq. (4.17) and measurements made on the mast at Høvsøre in Denmark. It can be seen that there is reasonable agreement between the prediction and the data.

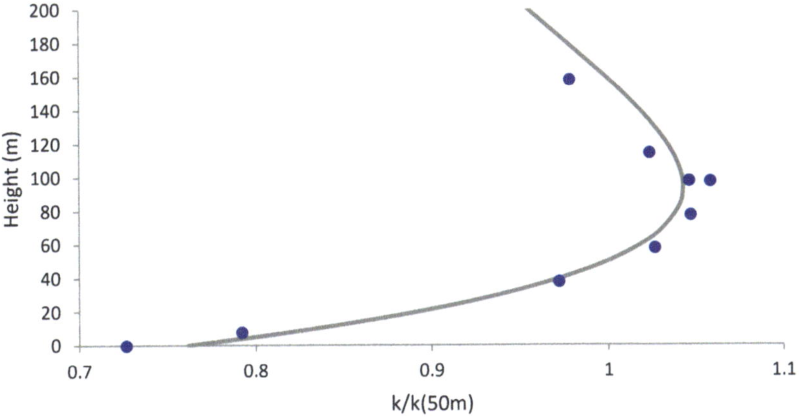

Figure 4.6 Vertical profile of *k* at the meteorological mast at Høvsøre in Denmark normalised to the observation at 50 m. Blue circles: observed values; grey solid line: prediction based on Eq. (4.17). *Source:* Kelly et al. (2014) / Springer Nature / CC BY 4.0.

4.6 Synoptic Scale Wind Direction Distributions

As wind direction is driven by a variety of parameters, including weather systems and local orography, parameterising its distribution is more complicated than for wind speed. One potential parameterisation uses a weighting of N von Mises distributions (Carta et al., 2008)

$$p_\theta(\theta) = \sum_{j=1}^{N} w_j M_j(\theta) \tag{4.19}$$

where $p_\theta(\theta)$ is the probability density function of the wind direction θ, $M_j(\theta)$ is the value of the jth von Mises distribution for direction θ, and w_j is the weighting value for the jth von Mises distribution where

$$\sum_{j=1}^{N} w_j = 1 \tag{4.20}$$

Each von Mises distribution is defined as

$$M_j(\theta;\kappa_j,\mu_j) = \frac{1}{2\pi I_0(\kappa_j)} \exp\left[\kappa_j \cos(\theta - \mu_j)\right] \tag{4.21}$$

where $0 \le \theta \le 2\pi$, $\kappa_j \ge 0$, and $0 \le \mu_j \le 2\pi$. The latter two parameters represent the concentration (width) and mean direction of each distribution, respectively. $I_0(\kappa_j)$ is the modified Bessel function given by

$$I_0(\kappa_j) = \frac{1}{2\pi}\int_0^{2\pi} \exp\left[\kappa_j \cos\theta\right] d\theta = \sum_{k=0}^{\infty} \frac{1}{(k!)^2}\left(\frac{\kappa_j}{2}\right)^{2k} \tag{4.22}$$

The cumulative form of Eq. (4.19) is given by

$$F_\theta(\theta; \kappa_j, \mu_j) = \sum_{j=1}^{N} \frac{w_j}{2\pi I_0(\kappa_j)} \int_0^{\theta} \exp\left[\kappa_j \cos(\theta - \mu_j)\right] d\theta \tag{4.23}$$

In Carta et al. (2008), a maximum of $N = 6$ was shown to be sufficient to parameterise the wind direction distributions at two sites on Gran Canaria. The three parameters w_j, κ_j, and μ_j can be fitted using a nonlinear least squares method. Alternatively, or as a starting point for least squares optimisation, the three parameters can be estimated analytically from the observed directional data values θ_i (Carta et al., 2008; Heckenbergerova et al., 2015). First of all, the observed directional values should be subdivided into N sectors, which can be of equal size for simplicity – though this is not a requirement – (i.e. values should be binned by sector: $0^0 - \theta_1'$, $\theta_1' - \theta_2'$, $\theta_2' - \theta_3'$, …, $\theta_{N-1}' - \theta_N'$, where θ_j' is the upper boundary of the jth sector). The mean wind direction in each sector is then calculated using the mean directional sine and cosine values for each sector

$$\mu_j = \begin{cases} \arctan\left(\dfrac{\overline{s}_j}{\overline{c}_j}\right); & \overline{s}_j \ge 0,\ \overline{c}_j \ge 0 \\ \dfrac{\pi}{2}; & \overline{s}_j \ge 0,\ \overline{c}_j = 0 \\ \pi + \arctan\left(\dfrac{\overline{s}_j}{\overline{c}_j}\right); & \overline{c}_j < 0 \\ \pi; & \overline{s}_j = 0,\ \overline{c}_j = -1 \\ 2\pi + \arctan\left(\dfrac{\overline{s}_j}{\overline{c}_j}\right); & \overline{s}_j < 0,\ \overline{c}_j > 0 \\ \dfrac{3\pi}{2}; & \overline{s}_j < 0,\ \overline{c}_j = 0 \end{cases} \tag{4.24}$$

The mean sine and cosine values are given, respectively, by

$$\bar{s}_j = \frac{\sum_{i=1}^{n_j} \sin\theta_i}{n_j} \tag{4.25}$$

$$\bar{c}_j = \frac{\sum_{i=1}^{n_j} \cos\theta_i}{n_j} \tag{4.26}$$

where n_j is the number of directional values falling within sector j.

The κ_j coefficients can be determined by solving

$$\frac{I_1(\kappa_j)}{I_0(\kappa_j)} = \left(\bar{s}_j^2 + \bar{c}_j^2\right)^{\frac{1}{2}} \tag{4.27}$$

where $I_1(\kappa_j)$ is the modified Bessel function of the first kind and order one

$$I_1(\kappa_j) = \frac{1}{2\pi}\int_0^{2\pi} \cos\theta \exp\left[\kappa_j \cos\theta\right] d\theta = \sum_{k=0}^{\infty} \frac{1}{\Gamma(k+2)\Gamma(k+1)}\left(\frac{\kappa_j}{2}\right)^{2k+1} \tag{4.28}$$

To a reasonable approximation, κ_j can be calculated as (Carta et al., 2008)

$$\kappa_j = \left(\begin{array}{l} 23.29041409 - 16.8617370(\bar{s}_j^2 + \bar{c}_j^2)^{\frac{1}{4}} \\ -17.4749884\exp\left[-(\bar{s}_j^2 + \bar{c}_j^2)\right] \end{array}\right)^{-1} \tag{4.29}$$

The weights can be approximated as

$$w_j = \frac{n_j}{\sum_{j=1}^{N} n_j} \tag{4.30}$$

Figure 4.7 shows an example fit to an observed wind direction distribution at an offshore site in the North Sea using the generalised reduced gradient nonlinear method. Three examples are shown (i.e. $N = 2$, 3, and 6). In this case, $N = 3$ provides an acceptable fit, but $N = 6$ fits the observed distribution well. The fitted values of w_j, κ_j, and μ_j are given in Table 4.1.

As an alternative to plotting a graph as in Figure 4.7, it is common to summarise wind direction statistics using a *wind rose*. Normally, ten-minute

Table 4.1 Optimised parameters used to fit the wind direction distributions in Figure 4.7.

$N = 2$	$j = 1$	$j = 2$
w_j	0.622	0.378
κ_j	0.29	2.68
$\mu_j(^0)$	36.2	220.6

$N = 3$	$j = 1$	$j = 2$	$j = 3$
w_j	0.653	0.030	0.317
κ_j	0.095	14.35	3.04
$\mu_j(^0)$	0.0	59.9	219.7

$N = 6$	$j = 1$	$j = 2$	$j = 3$	$j = 4$	$j = 5$	$j = 6$
w_j	0.105	0.040	0.042	0.206	0.249	0.358
κ_j	2.95	38.58	29.42	1.56	4.08	0.69
$\mu_j(^0)$	0.0	49.3	77.1	154.8	223.3	268.7

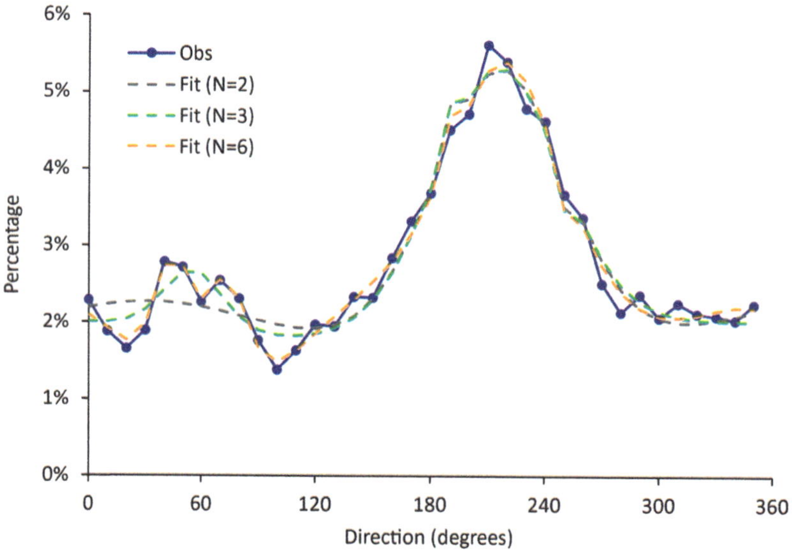

Figure 4.7 Example fits of weightings of von Mises distributions ($N = 2, 3, 6$) to an observed wind direction distribution (Obs).

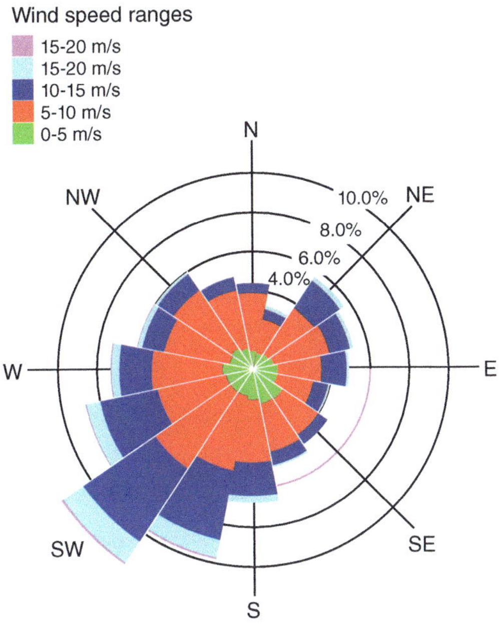

Figure 4.8 Example wind rose.

averaged wind direction measurements are binned by direction intervals of 10, 30, or 45 degrees. The percentage of measurements is then plotted by direction in the form of a polar plot. The distribution by direction is sometimes subdivided by wind speed range (e.g. intervals of 5 ms^{-1}). An example of this is shown in Figure 4.8. From such a plot the most frequent wind direction is immediately apparent. Phenomena such as channelling, which may be observed in a valley, or winds that have a strong diurnal character where the wind comes from one direction by day and another by night are also easy to determine.

4.7 Turbulent Scale Distributions

Since the advent of equipment and computing power capable of making such measurements, there has been a great deal of work devoted to assessing the structure of short-term variation in wind speed, normally defined as *turbulence*. From a wind resource assessment point of view, the precise spectral structure of the turbulence is less important than its magnitude, though wind turbine designers have an interest in the frequency components that may drive machine fatigue. One point of debate is at what frequency of variation does synoptic scale variation end and thus at what frequency does what is thought of as turbulence begin.

One of the best-known turbulence spectra was proposed by von Kármán (von Kármán, 1948), which, in normalised form for the three orthogonal wind speed components, u, v, w, is

$$\frac{nS_u(n)}{\sigma_u^2} = \frac{4n_u}{(1+70.8n_u^2)^{\frac{5}{6}}} \tag{4.31}$$

$$\frac{nS_v(n)}{\sigma_v^2} = \frac{4n_v(1+755.2n_v^2)}{(1+283.2n_v^2)^{\frac{11}{6}}} \tag{4.32}$$

$$\frac{nS_w(n)}{\sigma_w^2} = \frac{4n_w(1+755.2n_w^2)}{(1+283.2n_w^2)^{\frac{11}{6}}} \tag{4.33}$$

where n is the frequency, $S_i(n)$ is the spectral density of longitudinal velocity fluctuations of the ith spectral component (where $i = u, v, w$) at frequency n, σ_i^2 is the variance of turbulent fluctuations of the ith spectral component, $n_i = n\,{}_i^x L / U$, ${}_i^x L$ is the integral length scale of the ith component for separation in the x-direction, and U is the longitudinal wind speed.

Another well-known model for the turbulence spectrum is that proposed by Kaimal (Kaimal et al., 1972), which is based on earlier work by Kolmogorov (Kolmogorov, 1941) and uses Taylor's hypothesis (Taylor, 1938) to convert wave number to frequency, giving in its most general form

$$\frac{nS_u(n)}{u_*^2} = \frac{\alpha_1}{(2\pi\kappa)^{\frac{2}{3}}}\left(\frac{\kappa z \epsilon}{u_*^3}\right)^{\frac{2}{3}}\left(\frac{nz}{U}\right)^{\frac{-2}{3}} \tag{4.34}$$

where α_1 is the Kolmogorov constant (≈ 0.56), κ is the von Kármán constant, z is the height above the ground, u_* is the friction velocity, and ε is the turbulence dissipation rate.

By defining the dimensionless dissipation rate as

$$\phi_\epsilon = \frac{\kappa z \epsilon}{u_*^3} \tag{4.35}$$

and the nondimensional frequency

$$f = \frac{nz}{U} \tag{4.36}$$

Table 4.2 Kaimal spectrum constants determined by empirical fits to data for neutral stability.

i	*a*	*b*
u	105	33
v	17.0	9.5
w	2.0	5.3

then

$$\frac{nS_u(n)}{u_*^2} = \frac{\alpha_1}{(2\pi\kappa)^{\frac{2}{3}}}\phi_\epsilon^{2/3} f^{\frac{-2}{3}} \tag{4.37}$$

Note that ϕ_ϵ can be estimated for nonneutral conditions using the empirical stability functions defined in Eqs (4.13) and (4.14).

Kaimal et al. (1972) produced a number of empirical fits to observations giving, under neutral atmospheric conditions, spectra of the form

$$\frac{nS_i(n)}{u_*^2} = \frac{af}{(1+bf)^{\frac{5}{3}}} \tag{4.38}$$

where a and b are empirical constants for the different spectral wind speed components: $i = u, v, w$. From empirical fits (Kaimal et al., 1972), the values of a and b have been determined for neutral stability to a good approximation, as detailed in Table 4.2.

Where the simulation of spatial turbulence fields in time is required, more sophisticated turbulence models may be used such as that of Veers (1988) or Mann (1994), but these are not normally needed for resource assessment purposes.

4.8 Extreme Value Statistics

As described in Section 4.4, the most common distribution for parameterising wind speeds is the Weibull distribution. However, this distribution is less suitable for describing extreme wind speeds. When designing or selecting a wind turbine for a specific site, it is important to have a good estimate of the maximum expected wind speed during the lifetime of the machine. This requires the application of an extreme value distribution.

4.8.1 Modelling and Predicting Extremes

Generalised extreme value (GEV) distributions are normally used to model extreme values of wind speed. The most common of these used is the type-I extreme value distribution, or Gumbel distribution, which has the form

$$p_G(u;\alpha,\beta) = \frac{1}{\alpha}\exp\left(-\left(\frac{u-\beta}{\alpha}\right)\right)\exp\left(-\exp\left(-\left(\frac{u-\beta}{\alpha}\right)\right)\right) \tag{4.39}$$

where β is the mode of the distribution and α is the dispersion. The probability of *not* exceeding an extreme value v is given by the cumulative form of the Gumbel distribution (Palutikof et al., 1999)

$$F_G(v;\alpha,\beta) = \exp\left(-\exp\left(-\left(\frac{v-\beta}{\alpha}\right)\right)\right) \tag{4.40}$$

If a series of N extreme wind speed values is available for a site, an estimate can be made of F_G, by rank ordering the values (Gringorten, 1963)

$$F_G(v;\alpha,\beta) \approx \frac{m-0.44}{N+0.12} \tag{4.41}$$

where m is the rank order in ascending value.

A common metric to estimate is the average return period of a particular wind speed value. The quantile wind speed u_T with return period T (in years) is given by (Palutikof et al., 1999)

$$F_G(u_T) = 1 - \frac{1}{T} \tag{4.42}$$

For example, if F_G represents the distribution of annual extreme wind speeds then Eq. (4.42) would give the one in T year extreme wind speed. The corresponding quantile wind speed is then (Palutikof et al., 1999)

$$u_T = \beta - \alpha \ln\left(-\ln\left(1-\frac{1}{T}\right)\right) \tag{4.43}$$

The corresponding risk r of exceeding a one-in-T-year extreme wind speed in Y years is

$$r = 1 - \left(1 - \frac{1}{T}\right)^Y \tag{4.44}$$

Assuming $T >> 1$ year this gives

$$r \approx \frac{Y}{T} \tag{4.45}$$

The challenge in estimating F_G is having enough extreme values to be representative of the distribution. It has been suggested that at least 20 years of extreme wind speed values are required for a reliable distribution and that no less than 10 years of values should be used (Cook, 1985). This is often impractical for a site to be developed for wind power where measurements may only be made for a period of 12–18 months. Alternative approaches include:

- Selecting the k largest extreme values per year (assuming several years' worth of data are available).
- Selecting extreme values associated with individual storms (which requires a method to ensure separation between storm events). This is known as the *method of independent storms.*
- Selecting a threshold wind speed above which a wind speed event is assumed to be an extreme. This is known as a *peak-over-threshold approach.*

All these methods presuppose that data sampled are representative of the true extreme value distribution for the site of interest. Needless to say, a short period of 1–2 years may not provide a sufficient sample of extreme events from which to estimate a reliable extreme value distribution. An approximate approach under such circumstances is to relate the parameters of the Gumbel distribution to the Weibull distribution at the site, suitably adjusted for the long term (ESDU, 1987, 1988). Assuming measurements are available at a site for two years, Weibull parameters can be fitted to the data for each year separately. Assuming the availability of long-term data from a nearby meteorological station, or data from a nearby grid point from a reanalysis dataset, comparable Weibull parameters can be determined to adjust the short-term values.

Table 4.3 illustrates the values of the Weibull parameters for the two different years and the long term (>10 year) average. Based on the average of the two years, the long-term adjusted site values, k_{sl} and C_{sl}, are estimated to be, respectively

$$k_{sl} = \frac{\frac{1}{2}(k_{s1} + k_{s2})}{\frac{1}{2}(k_{m1} + k_{m2})} k_{ml} \tag{4.46}$$

Table 4.3 Weibull parameters for site at which extreme wind speeds are to be estimated and at a nearby meteorological station.

	Site		Meteorological station	
	k	C (m · s^{-1})	k	C (m · s^{-1})
Year 1	k_{s1}	C_{s1}	k_{m1}	C_{m1}
Year 2	k_{s2}	C_{s2}	k_{m2}	C_{m2}
Long-term average	—	—	k_{ml}	C_{ml}

Source: adapted from ESDU (1988).

$$C_{sl} = \frac{\frac{1}{2}(C_{s1} + C_{s2})}{\frac{1}{2}(C_{m1} + C_{m2})} C_{ml} \tag{4.47}$$

An estimate of the mode β can be made

$$\beta = C_{sl}^{\ 2} \left[n_0 + \left(\frac{k_{sl}}{2} - 1 \right) \frac{2.5}{k_{sl}} \ln n_0 \right]^{2/k_{sl}} \tag{4.48}$$

where

$$n_0 = \ln(2.33\nu_0) \tag{4.49}$$

The value ν_0 is the up-crossing rate, which in practice is hard to determine but, based on an analysis of many meteorological sites in the United Kingdom, a value of 1500 cycles yr^{-1} for wind speed extremes over all direction sectors was found to provide a reasonable fit. A reasonable estimate of the dispersion value α is given by

$$\alpha = \frac{\beta}{5} \tag{4.50}$$

It should be noted that this method will not give as accurate an estimate as the situation where the Gumbel parameters can be estimated from 20 years of data but will give adequate estimates where only a short-term dataset is available.

Using 40 years (1979–2018) of hourly ERA5 reanalysis wind speed data, an effective 'world wind atlas' of the hourly mean wind speed at 100 m with a 50-year return period (U_{50}) was determined based on annual maximum wind speeds. Figure 4.9 shows the atlas where the values were

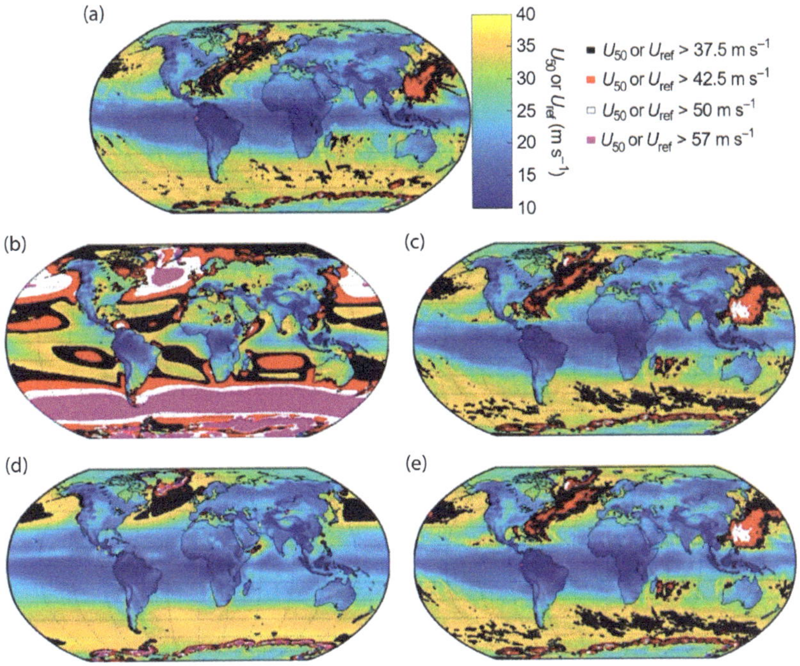

Figure 4.9 Extreme wind speeds at 100 m based on hourly ERA5 wind speed data between 1979 and 2018: (a) U_{50} (Gumbel-MLE); (b) IEC61400-1 U_{ref} (five times mean annual wind speed); (c) U_{50} (Gumbel-MoM); (d) U_{50} (Gumbel-Weibull), (e) U_{50} (Gumbel-graphical). N.B. colour scales truncated at 40 ms^{-1}. *Source:* Pryor and Barthelmie, 2021) / with permission of Springer Nature.

calculated using four methods to determine the Gumbel parameters (maximum likelihood estimation, MLE; method of moments, MoM; Gumbel–Weibull linking functions and graphical) and compared with the IEC61400-1 standard reference wind speed (U_{ref}) (IEC 61400-1: 2019, 2019), which is based on five times the mean annual wind speed. Although the atlas will not allow an assessment of detailed site-specific extreme wind speeds, it nonetheless highlights general areas where these values are high and when they are likely to exceed a value recommended in the wind turbine design standard.

4.8.2 Gusts

Section 4.8.1 covered the modelling of extreme wind speeds; however, extreme wind speed distributions can be based on values sampled over different time periods from seconds to minutes or even hours. To assess the structural safety of a wind turbine, it may be necessary to have an estimate of a maximum expected wind speed on a scale of a few seconds. If only ten-minute or one-hour averaged data are available then a method is required to estimate the corresponding second-scale wind speed or *gust*. One of the simplest ways to do this is to use a t-second *gust factor G*

$$G(t) = \frac{\hat{u}}{\bar{u}} \tag{4.51}$$

where $\bar{u}$ is the mean wind speed averaged over a period T and $\hat{u}$ is the maximum wind speed averaged over a shorter period $t < T$. Based on the turbulence intensity I defined in Eq. (4.1), the gust factor can be estimated for $T/t \geq 7$ as (Wieringa, 1973)

$$G(t) = 1 + I(1.42 + 0.3013\ln[(T/t) - 4]) \tag{4.52}$$

Based on observations during typhoons, an alternative, though similar expression for G has been proposed (Cao et al., 2009)

$$G(t) = 1 + k_1 I^{k_2} \ln[T/t] \tag{4.53}$$

Suggested values of k_1 and k_2 are given in Table 4.4.

As an alternative to using the turbulence intensity to determine the gust factor, the measurement height and surface roughness length may be used instead. Assuming neutral conditions, it has been observed that (Lumley and Panofsky, 1964)

$$\frac{\sigma_u}{u_*} \approx 2.5 \tag{4.54}$$

Combining this with the expression for the logarithmic wind profile in Eq. (4.4) (assuming $\psi_m = 0$) gives

Table 4.4 Suggested values of the constants k_1 and k_2 in Eq. (4.53) based on observations.

k_1	k_2	**References**
0.5	1.15	(Cao et al., 2009)
0.5	1.0	(Ishizaki, 1983)
0.62	1.27	(Choi, 1983)

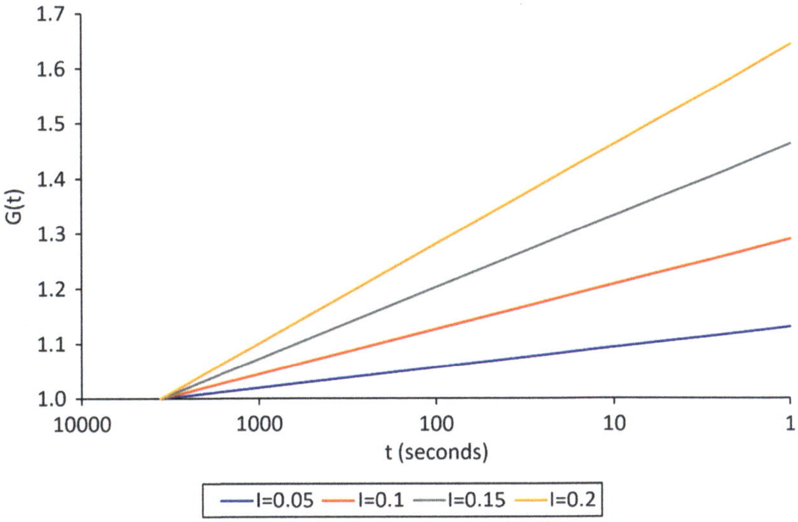

Figure 4.10 Gust factor based on hourly data (T = 3600 s) for four levels of turbulence intensity (I = 0.05, 0.1, 0.15, and 0.2) using Eq. (4.53) with k_1 = 0.5 and k_2 = 1.15.

$$I = \frac{\sigma_u}{\bar{u}} \approx \frac{1}{\ln(z/z_0)} \tag{4.55}$$

This can then be substituted into Eq. (4.52) or (4.53).

An example of the gust factor using Eq. (4.53) for different levels of turbulence intensity is shown in Figure 4.10.

References

Cao, S., Tamura, Y., Kikuchi, N., Saito, M., Nakayama, I., Matsuzaki, Y., 2009. Wind characteristics of a strong typhoon. *J. Wind Eng. Ind. Aerodyn.* 97, 11–21. https://doi.org/10.1016/j.jweia.2008.10.002.

Carta, J.A., Bueno, C., Ramírez, P., 2008. Statistical modelling of directional wind speeds using mixtures of von Mises distributions: case study. *Energy Convers. Manag.* 49, 897–907. https://doi.org/10.1016/j.enconman.2007.10.017.

Carta, J.A., Ramírez, P., 2007. Use of finite mixture distribution models in the analysis of wind energy in the Canarian Archipelago. *Energy Convers. Manag.* 48, 281–291. https://doi.org/10.1016/j.enconman.2006.04.004.

Choi, E.C.C., 1983. *Wind Loading in Hong Kong: Commentary on the Code of Practice on Wind Effects Hong Kong*. Hong Kong Institution of Engineers.

Cook, N., 1985. *The Designer's Guide to Wind Loading of Building Structures.* Butterworths, Boston.

ESDU, 1987. *ESDU 87034: world-wide extreme wind speeds: part 1: origins and methods of analysis.* http://www.esdu.com/cgi-bin/ps.pl?t=doc&p=esdu_87034b-r1 (accessed 4 November 2022)

ESDU, 1988. *ESDU 88037: world-wide extreme wind speeds: part 2: examples using various methods of analysis.*http://www.esdu.com/cgi-bin/ps.pl?t=doc&p=esdu_88037-r1 (accessed 4 November 2022).

Gjerstad, J., Aasen, S.E., Andersson, H.I., Brevik, I., Lovseth, J., 1995. An analysis of low-frequency maritime atmospheric-turbulence. *J. Atmos. Sci.* 52, 2663–2669. https://doi.org/10.1175/1520-0469(1995)052<2663:AAOLFM>2.0.CO;2.

Gringorten, I.I., 1963. A plotting rule for extreme probability paper. *J. Geophys. Res.* 68, 813–814. https://doi.org/10.1029/jz068i003p00813.

Heckenbergerova, J., Musilek, P., Kromer, P., 2015. Optimization of wind direction distribution parameters using particle swarm optimization. *Adv. Intell. Syst. Comput.* 334, 15–26. https://doi.org/10.1007/978-3-319-13572-4_2.

Heggem, T., Lende, R., Lovseth, J., 1998. Analysis of long time series of coastal wind. *J. Atmos. Sci.* 55, 2907–2917. https://doi.org/10.1175/1520-0469(1998)055<2907:AOLTSO>2.0.CO;2.

IEC 61400-1: 2019, 2019. *Wind energy generation systems part 1: design requirements.* International Electrotechnical Commission.

Ishizaki, H., 1983. Wind profiles, turbulence intensities and gust factors for design in typhoon-prone regions. *J. Wind Eng. Ind. Aerodyn.* 13, 55–66. https://doi.org/10.1016/0167-6105(83)90128-9.

Kaimal, J.C., Izumi, Y., Wyngaard, J.C., Cote, R., 1972. Spectral characteristics of surface-layer turbulence. *Q. J. R. Meteorol. Soc.* 98, 563–589. https://doi.org/10.1002/qj.49709841707.

Kelly, M., Troen, I., Jørgensen, H.E., 2014. Weibull-k revisited: 'tall' profiles and height variation of wind statistics. *Boundary-Layer Meteorol.* 152, 107–124. https://doi.org/10.1007/s10546-014-9915-5.

Kolmogorov, A.N., 1941. The local structure of turbulence in incompressible viscous fluid for very large Reynolds numbers. *Proc. USSR Acad. Sci.* 30, 299–303.

Lumley, J.L., Panofsky, H.A., 1964. *The Structure of Atmospheric Turbulence, Monographs and Texts in Physics and Astronomy Vol. XII.* Wiley.

Mann, J., 1994. The spatial structure of neutral atmospheric surface-layer turbulence. *J. Fluid Mech.* 273, 141–168. https://doi.org/10.1017/S0022112094001886.

Palutikof, J.P., Brabson, B.B., Lister, D.H., Adcock, S.T., 1999. A review of methods to calculate extreme wind speeds. *Meteorol. Appl.* 6, 119–132. https://doi.org/10.1017/S1350482799001103

Pryor, S.C., Barthelmie, R.J., 2021. A global assessment of extreme wind speeds for wind energy applications. *Nat. Energy* 1–9. https://doi.org/10.1038/s41560-020-00773-7.

Soberanis, M.A.E., Merida, W., 2015. Regarding the influence of the Van der Hoven spectrum on wind energy applications in the meteorological mesoscale and microscale. *Renew. Energy* 81, 286–292. https://doi.org/10.1016/j.renene.2015.03.048.

Taylor, G.I., 1938. The spectrum of turbulence. *Proc. R. Soc. London Ser. A: Mathematical Phys. Eng. Sci.* 164, 476–490.

Van der Hoven, I., 1957. Power spectrum of horizontal wind speed in the frequency range from 0.0007 to 900 cycles per hour. *J. Meteorol.* 14, 160–164.

Veers, P.S., 1988. Three-dimensional wind simulation. *Sandia report SAND-88-0152.*

Vincent, C.L., Pinson, P., Giebel, G., 2011. Wind fluctuations over the North Sea. *Int. J. Climatol.* 31, 1584–1595. https://doi.org/10.1002/joc.2175.

von Kármán, T., 1948. Progress in the statistical theory of turbulence. *Proc. Natl. Acad. Sci.* 34, 530–539. https://doi.org/10.1073/pnas.34.11.530.

Watson, S., 2014. Quantifying the variability of wind energy. *Wiley Interdiscip. Rev. Environ.* 3, 330–342. https://doi.org/10.1002/wene.95.

Watson, S.J., Kritharas, P., Hodgson, G.J., 2015. Wind speed variability across the UK between 1957 and 2011. *Wind Energy* 18, 21–42.

Weibull, W., 1951. A statistical distribution function of wide applicability. *J. Appl. Mech.* 18, 293–297.

Wieringa, J., 1973. Gust factors over open water and built-up country. *Boundary-Layer Meteorol.* 3, 424–441. https://doi.org/10.1007/BF01034986.

Wieringa, J., 1989. Shapes of annual frequency distributions of wind speed observed on high meteorological masts. *Boundary-Layer Meteorol.* 47, 85–110. https://doi.org/10.1007/BF00122324.

5

Numerical Modelling

One of the principal ways in which the spatial variation of wind speed is predicted in wind resource assessment is to use the fundamental equations of fluid flow which are discretised on a three-dimensional (3D) grid. These equations are then solved either analytically or, more frequently, iteratively to produce a solution at each cell within the grid. Model predictions are generally calibrated using a known value at one or more locations and used to extrapolate the wind field over a wide area for a range of heights. This chapter reviews the governing equations and considers the use of numerical models of different complexity based on these equations.

5.1 The Governing Equations

Wind can be treated as any other fluid and as such is governed by several basic equations (Versteeg and Malalasekera, 2007). A full derivation of these equations is beyond the scope of this book, but the main salient features and their application in wind resource assessment are explained.

The first of the basic equations relates to continuity or conservation of mass

$$\frac{\partial \rho}{\partial t} + \nabla.(\rho \boldsymbol{u}) = 0 \tag{5.1}$$

where ρ is the fluid (air in the case of the wind) density, t is time, and $\boldsymbol{u}$ is the vector of the velocity components u, v, and w, defined in the streamwise (x), cross-wise (y), and vertical (z) directions, respectively. If, as is commonly the case when modelling flow, the flow into and out of an element of fluid is considered then, in words, Eq. (5.1) can be written as

Handbook of Wind Resource Assessment, First Edition. Simon Watson.

The rate of increase of mass *in* the fluid element	+	The net rate of flow of mass *out of* the fluid element	= 0	

The second group of equations relate to the conservation of any scalar property ϕ of the fluid

$$\frac{\partial(\rho\phi)}{\partial t} + \nabla.(\rho\phi\boldsymbol{u}) = \nabla.(\Gamma\nabla\phi) + S_\phi \tag{5.2}$$

where Γ is the diffusion coefficient for the scalar quantity and S_ϕ is a source term that can combine several factors (e.g. a pressure gradient, gravity, or the Coriolis force) when considering atmospheric flows.

In words, this equation can be considered as

The rate of increase of ϕ of the fluid element	+	The net rate of flow of ϕ *out of* the fluid element	=	The rate of increase of ϕ due to diffusion	+	Rate of increase of ϕ due to sources

The final group of equations relates to the thermodynamic equations of state which, for an ideal gas, are Eq. (2.23) and

$$e_i = C_v T \tag{5.3}$$

where e_i is the internal energy and C_v is the specific heat capacity at constant volume. When applying Eq. (5.2) to model wind flow, the scalar quantities of most relevance are the velocity components u, v, and w, the internal energy e_i (or T) and the specific humidity q.

The group of equations defined above are known as the *Navier–Stokes equations*, named after the two scientists who derived them around the same time. Solving the equations precisely would require resolving all levels of turbulence down to the smallest levels before viscous dissipation. This is only feasible for very simple flow problems and is the approach known as *direct numerical simulation* (DNS). For anything but these simple problems, this is not feasible, and an alternative approach is required. One such approach (and the one most used) is to separate out the relevant variables into their mean terms and fluctuating components, known as *Reynolds' decomposition*

$$u = \overline{u} + u' \tag{5.4}$$

$$v = \overline{v} + v' \tag{5.5}$$

$$w = \overline{w} + w' \tag{5.6}$$

$$\rho = \overline{\rho} + \rho' \tag{5.7}$$

$$p = \overline{P} + p' \tag{5.8}$$

$$T = \overline{T} + T' \tag{5.9}$$

$$q = \overline{q} + q' \tag{5.10}$$

where the fluctuation (′) is significantly smaller than the mean (¯) term. If this is done then Eq. (5.2) can be expanded thus

$$\frac{\partial(\rho\overline{u})}{\partial t} + \mathbf{\nabla}.(\rho\overline{u}\,\mathbf{u}) = -\frac{\partial p}{\partial x} + \mathbf{\nabla}.(\mu\,\nabla\overline{u}) + \left[-\frac{\partial\left(\rho\overline{u'^2}\right)}{\partial x} - \frac{\partial\left(\rho\overline{u'v'}\right)}{\partial y} - \frac{\partial\left(\rho\overline{u'w'}\right)}{\partial z}\right] + S_{Mx} \tag{5.11}$$

$$\frac{\partial(\rho\overline{v})}{\partial t} + \mathbf{\nabla}.(\rho\overline{v}\mathbf{u}) = -\frac{\partial p}{\partial y} + \mathbf{\nabla}.(\mu\,\nabla\overline{v}) + \left[-\frac{\partial\left(\rho\overline{u'v'}\right)}{\partial x} - \frac{\partial\left(\rho\overline{v'^2}\right)}{\partial y} - \frac{\partial\left(\rho\overline{v'w'}\right)}{\partial z}\right] + S_{My} \tag{5.12}$$

$$\frac{\partial\,(\rho\overline{w})}{\partial t} + \mathbf{\nabla}.\,(\rho\overline{w}\,\mathbf{u}) = -\frac{\partial p}{\partial z} + \mathbf{\nabla}.\,(\mu\,\nabla\overline{w}) + \left[-\frac{\partial\left(\rho\overline{u'w'}\right)}{\partial x} - \frac{\partial\left(\rho\overline{v'w'}\right)}{\partial y} - \frac{\partial\left(\rho\overline{w'^2}\right)}{\partial z}\right] + S_{Mz} \tag{5.13}$$

$$C_v\frac{\partial(\rho\overline{T})}{\partial t} + C_v\mathbf{\nabla}.(\rho\overline{T}\mathbf{u}) = \mathbf{\nabla}.(k_T\nabla\overline{T}) + C_v\left[-\frac{\partial\left(\rho\overline{u'T'}\right)}{\partial x} - \frac{\partial\left(\rho\overline{v'T'}\right)}{\partial y} - \frac{\partial\left(\rho\overline{w'T'}\right)}{\partial z}\right] + S_{e_i} \tag{5.14}$$

$$\frac{\partial(\rho\overline{q})}{\partial t} + \mathbf{\nabla}.(\rho\overline{q}\,\mathbf{u}) = \mathbf{\nabla}.(k_q\nabla\overline{q}) + \left[-\frac{\partial\left(\rho\overline{u'q'}\right)}{\partial x} - \frac{\partial\left(\rho\overline{v'q'}\right)}{\partial y} - \frac{\partial\left(\rho\overline{w'q'}\right)}{\partial z}\right] + S_q \tag{5.15}$$

where S_{Mx}, S_{My}, and S_{Mz} are source terms for momentum in the x, y, z directions, respectively; S_{e_i} is the source term for internal energy; S_q is the source term for specific humidity; k_T is thermal conductivity; and $k_q = \rho\nu_q$ where ν_q is the molecular diffusivity for water vapour in air. These equations, along with the continuity equation, are known as *Reynolds-averaged Navier–Stokes (RANS) equations*. To incorporate the effect of the Coriolis force, the following source terms are added to Eqs. (5.11) and (5.12), respectively

$$S_{Mxcor} = +\rho f \overline{u} \tag{5.16}$$

$$S_{Mycor} = -\rho f \overline{v} \tag{5.17}$$

It can be seen that the Reynolds' decomposition has led to a number of variance and covariance terms involving the turbulent fluctuations. These terms are the *turbulent stress transport* terms which present some challenges to the solution of these equations, as is discussed in the following sections. Before considering the application of these equations to wind flow modelling, there are a number of general simplifications to be aware of that are made when applying these equations in the atmosphere boundary layer (Businger, 1984):

- The viscosity and thermal conductivity of air is assumed to be constant.
- The flow is assumed to be incompressible.
- All the fluctuating components are small compared to their mean values in Eqs. (5.4)–(5.10).
- The Boussinesq approximation is applied, namely: (i) relative pressure fluctuations are much smaller than relative density and temperature fluctuations and (ii) density fluctuations can be neglected unless the density term in the relevant equation is multiplied by g, the acceleration due to gravity (i.e. density changes resulting from changes in temperature are only important in terms of how they affect buoyancy).

In addition, the divergence of the mean viscous stresses $\nabla \cdot (\mu \nabla \overline{u})$, etc., for turbulent air flow are negligible compared to the divergence of the Reynolds turbulent stresses and so these terms are normally neglected.

In the rest of the chapter we review the different types of models which are used in the wind resource assessment process in terms of how they apply the governing equations taking into account the simplifications described in this section.

5.2 Microscale Models

The siting of wind turbines is done primarily at the microscale level. Typically, this means over distances of kilometres considering the local terrain conditions which have a significant influence on the wind conditions.

A microscale model is frequently used to determine, within the proposed boundary of a wind farm, where best to place the wind turbines in order to optimise their energy yield within a number of physical and other constraints. The sorts of models to do this vary significantly in their level of complexity. In the wind farm layout process, a relatively rapid model is required to simulate the output of a range of turbine layouts in a reasonable timeframe. There may be some compromise in terms of accuracy, but once an optimised layout is found, a more sophisticated, but computationally intensive, model may be used to fine-tune the layout to get a final accurate estimate of the wind farm yield. Wind farm layout optimisation incorporates several submodels, including a model of wind turbine wake losses. These submodels are beyond the scope of this book, but this section reviews the microscale models which are commonly used in the wind resource assessment process. The purpose of these microscale models is to predict the local wind conditions which can then be used as the input to other submodels such as those which account for turbine wake losses. All of the microscale models described here are based on the governing equations described in Section 5.1 with varying degrees of simplification.

5.2.1 Mass Consistent Models

Mass consistent models are the simplest form of physical flow model where only the continuity equation is solved. These types of model are generally used as diagnostic models to interpolate spatially dispersed measurements of wind speed into a regular mesh of cells which form a domain above an area of terrain (Ratto et al., 1994). Assuming mean wind vector measurements are available from n sites, $\boldsymbol{u}_1$, $\boldsymbol{u}_2$,..., $\boldsymbol{u}_n$, the first step is to interpolate, using some procedure ℶ, the measurements onto a regular mesh such that:

$$(\boldsymbol{u}_1,\ \boldsymbol{u}_2,\ldots,\ \boldsymbol{u}_n)\overset{\beth}{\rightarrow}\boldsymbol{u}^0(x,y,z) \tag{5.18}$$

where $\boldsymbol{u}^0$ is the first estimate of the interpolated wind speed field which most likely will not satisfy the requirement of continuity. A further optimisation procedure ℵ is then used to adjust the interpolated wind field the minimum amount required to ensure continuity

$$\boldsymbol{u}^0(x,y,z)\overset{\aleph}{\rightarrow}\boldsymbol{u}^a(x,y,z) \tag{5.19}$$

where $\boldsymbol{u}^a$ is the adjusted matrix of wind vectors such that continuity is satisfied as the wind flows from one cell to another. This process is shown schematically in Figure 5.1.

Assuming the incompressible assumption, Eq. (5.1) becomes

$$\nabla.(\rho\boldsymbol{u})=0 \tag{5.20}$$

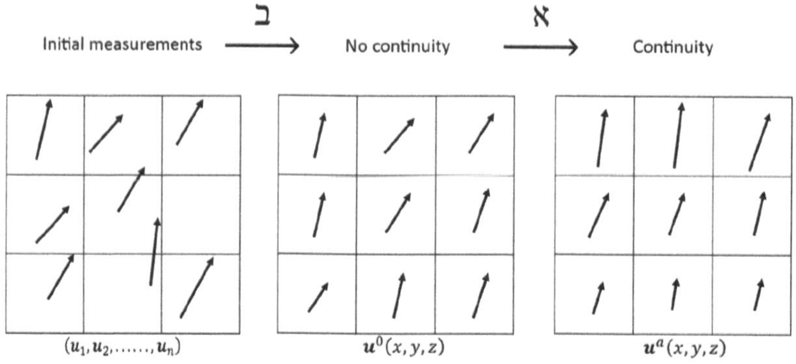

Figure 5.1 Steps followed by a diagnostic mass consistent flow model.

Mass consistent models minimise the variance of the difference E between the adjusted and initial wind fields, which satisfies continuity, as in Eq. (5.20)

$$E(\boldsymbol{u}) = \iiint_{V} \left[\alpha_1^2 (u - u^0)^2 + \alpha_2^2 (v - v^0)^2 + \alpha_3^2 (w - w^0)^2\right] dxdydz \quad (5.21)$$

where u, v and w are the components of the adjusted wind field $\boldsymbol{u^a}$; u^0, v^0, and w^0 are the components of the initial interpolated wind field $\boldsymbol{u^0}$; and the integral is over all cells in the volume V which make up the 3D domain above the region where the wind flow is being modelled. The variables α_1, α_2, and α_3 are values which weight the minimisation function in the three dimensions. The converged solution is sensitive to the relative values of these variables. Normally, α_1 and α_2 are set to the same value. The relative values of $\alpha_1 (= \alpha_2)$ and α_3 are adjusted to reflect the prevailing stability conditions using the parameter

$$\alpha^2 = \frac{\alpha_1^2}{\alpha_3^2} \quad (5.22)$$

Quite different values have been used by different researchers, for example in Guo and Palutikof (1990), a value of $\alpha = 1.0$ was used for neutral conditions, $\alpha = 0.1$ for very stable conditions, and $\alpha = 5$ for very unstable conditions. Several researchers have suggested that $\alpha = 0.01$ should be used to correctly balance horizontal and vertical adjustments based on the measured magnitude of w / u (Davis et al., 1984; Kitada et al., 1983; Mathur and Peters, 1990; Sherman, 1978). In general, it is accepted that α should be significantly smaller than one under stable conditions and significantly larger than one under unstable conditions.

The 3D mesh of cells in the volume above the terrain of interest needs to take account of the underlying orography. A simple way of doing this is to block off cells close to the ground where more than a certain fraction of the volume is occupied by the terrain. This is known as the *stair step terrain* approach. A better way is to use conformal or *terrain following coordinates* (TFC), also known as a *sigma* (σ) coordinate grid, which adapts to the shape of the terrain. The transformed coordinates take the form

$$\xi = x \tag{5.23}$$

$$\eta = y \tag{5.24}$$

$$\sigma = \frac{z - h(x,y)}{H - h(x,y)} \tag{5.25}$$

where $h(x,y)$ is the terrain height relative to some datum and H is the top of the height of the domain. This value is frequently chosen as the inversion height where vertical motions are suppressed acting as a lid on the flow. An example cross-section through such a terrain following grid is shown in Figure 5.2 for equally spaced σ coordinates, though they need not be equally spaced. For example, more detail may be required closer to the ground, particularly if the terrain is complex with large changes in orography. The ξ and η coordinates are normally equally spaced, but these two can be scaled if more detail is required at one point in the domain.

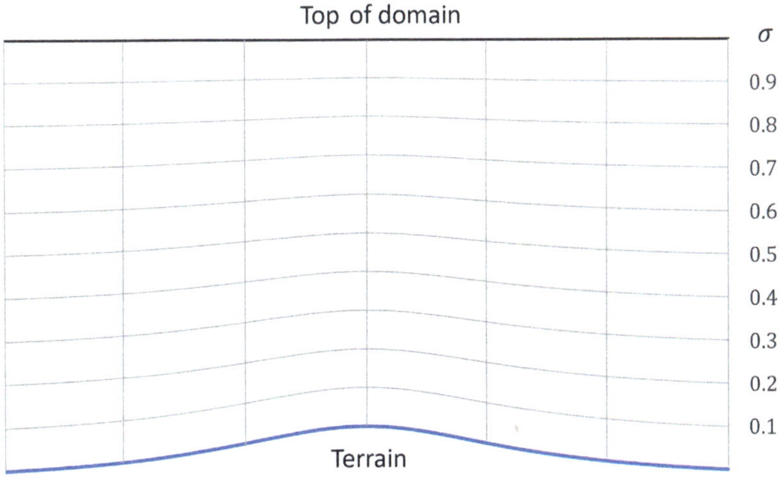

Figure 5.2 An example of terrain following sigma (σ) coordinates.

Models are normally initialised assuming a suitable wind profile for the point(s) where observations are available. An example of a well-known mass consistent model is NOABL (Numerical Objective Analysis of Boundary Layer) (Phillips, 1979; Traci et al., 1978). This was used to produce an early wind atlas for the UK based on measurements at several meteorological sites (Burch and Ravenscroft, 1992). The resulting wind atlas is shown in Figure 5.3. The figure shows that the effect of terrain features is captured, with the upland areas showing the highest wind speeds.

Such models can produce a rapid assessment of the wind resource over a wide area, but because they only solve continuity, their accuracy is limited. For resource assessment, they give a reasonable indication of the wind resource in a region, especially its spatial variability, but more detailed modelling is required to give an accurate estimate suitable for a wind farm energy yield analysis.

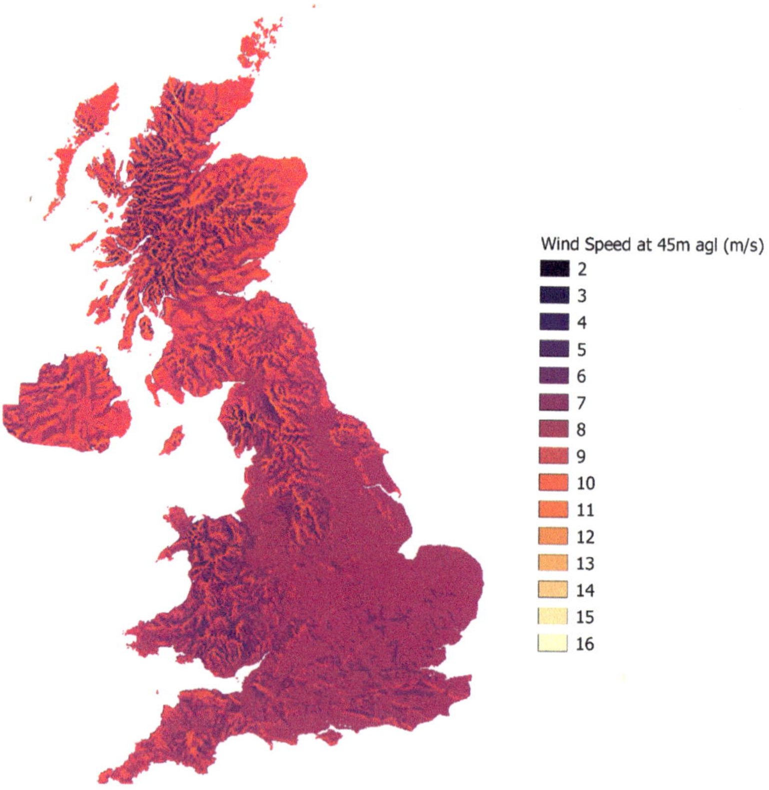

Figure 5.3 Map of wind speeds at 45 m above ground level for the UK calculated using the NOABL model.

Some well-known mass consistent models include:

- MATHEW (Sherman, 1978): one of the first models to use the α parameter as a means to adjust for stability.
- NOABL (Traci et al., 1978): capable of using a stair step terrain or TFC grid.
- AIOLOS (Ratto, 1996): an enhancement of NOABL to take account of surface roughness, to use more realistic wind profiles and to include Coriolis turning with height.
- WINDS: an enhancement of AIOLOS with more accurate parameterisation of the wind profiles under different stability conditions.
- ATMOS1 and ATMOS2 (Davis et al., 1984): uses an expanding grid in the vertical.
- NUATMOS (Ross et al., 1988): based on the ATMOS models but incorporates a more logical parameterisation of α based on stability.

5.2.2 Linearised Models

This class of models solves the continuity and the conservation of momentum equations (Eqs. 5.11–5.13). As the addition of the momentum equations adds significant additional complexity to any iterative solution, significant computing resource is required to solve these equations. Indeed, the turbulent stress transport terms (in square brackets in Eqs. 5.11–5.13) comprise additional variance and covariance terms, which means that there are more variables than equations creating a problem which is not closed (the *closure problem*). This is one of the fundamental problems of numerical fluid dynamics modelling. The way around this is to devise additional equations based on measurable properties of the flow to model the turbulence. The additional equations provide a closed set of equations, though at the expense of accuracy.

During the 1970s, computing resources were limited to the point that even providing an iterative solution to Navier–Stokes equations was not feasible. A simplifying approach was made to linearise the momentum equations and, along with a relatively simple turbulence model, to allow an analytical solution to be computed. This formed the basis of work by Jackson and Hunt (JH), who originally applied their theory to 2D hills (Jackson and Hunt, 1975), which was later updated by Mason and Sykes for 3D hills (Mason and Sykes, 1979). The first step in this theory is to divide the flow over a hill into an inner and an outer region, as shown in Figure 5.4. The figure shows the flow over a hill (slope not exceeding ~30%) with a characteristic height h and characteristic width L_W which is often

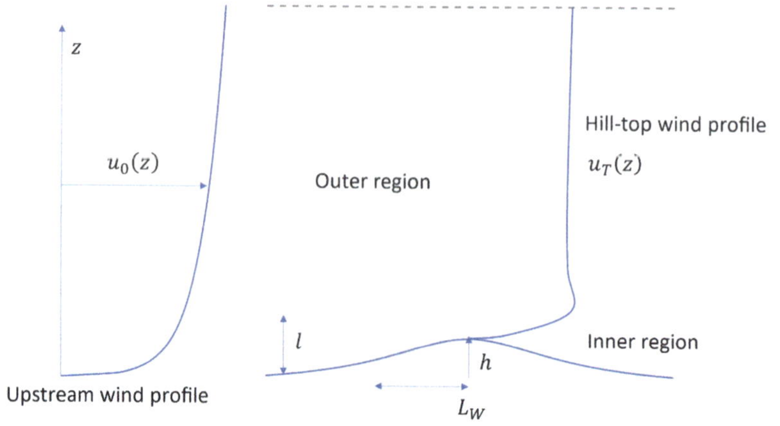

Figure 5.4 Flow over a low hill: division of flow regime into two layers.

taken as the half-width at half-height. The inner layer has a vertical extent l and, in this layer, viscous forces and Reynolds shear stresses are significant. Above this, in the outer layer, these forces are proportionately much smaller and the flow is dominated by a pressure gradient. This is sometimes known as the *potential flow region* where the effect of viscosity is negligible (i.e. the flow is *inviscid*).

A detailed explanation of the Jackson and Hunt theory is beyond the scope of this book, but an outline of the salient features shows how the flow solution is derived.

The basic premise of the linearisation process for the momentum equations is to treat the effect of the hill as a perturbation so that the upstream wind speed profile seen in Figure 5.4, once above the hill, experiences a small perturbation to its mean value. If only the flow in the streamwise direction is considered, this means that

$$u_T = u_0 + u'' \tag{5.26}$$

$$w_T = w'' \tag{5.27}$$

where u_T is the hilltop streamwise wind speed (at a certain height z), u_0 is the upstream wind speed in the streamwise direction, w_T is the hill-top vertical wind speed, and u'' and w'' are the streamwise and vertical perturbations to the wind speed, respectively, due to the hill. For the linear assumptions to be valid, $u''/u_0 << 1$, $w''/u_0 << 1$, and $h/L_w << 1$. The roughness length of the surface is assumed to be constant with $z_0/L_w << 1$. The linearisation process allows the nonlinear terms in the momentum equation to be replaced by constants or constants whose variation as a function of the independent variable in the equations is already known. In the outer inviscid region, the solution is a relatively straightforward solution of a potential flow equation. In the inner region, there remains the problem of parameterising the Reynolds stress terms. This requires a turbulence closure scheme. The scheme used in this case is the relatively simple *mixing length* as first proposed by Prandtl (1925).

In mixing length theory, a parcel of air perturbed vertically by a distance z' experiences a perturbation in streamwise wind speed of u' relative to the local wind speed gradient at the new height, as seen in Figure 5.5, whereby

$$u' = -\left(\frac{\partial u(z)}{\partial z}\right)z' \tag{5.28}$$

If it is assumed that the vertical perturbation in speed w' is proportional to the magnitude of u'

$$w' = c\left|u'\right| \tag{5.29}$$

$$w' = -c\left|\left(\frac{\partial u(z)}{\partial z}\right)\right|z' \tag{5.30}$$

The quantity $\overline{z'^2}$ is the variance of the parcel displacement distance, so the square root of the quantity is a measure of the average distance moved by a parcel in the turbulent mixing process. A mixing length l_m can be defined in terms of this value

$$l_m^2 = c\overline{z'^2} \tag{5.31}$$

Combining Eqs. (5.28), (5.30), and (5.31) gives

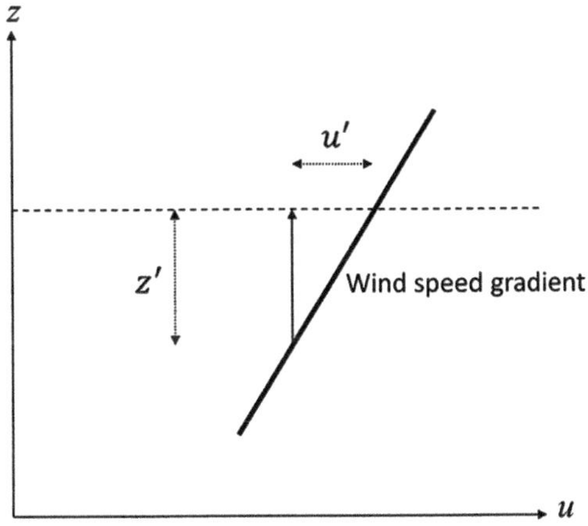

Figure 5.5 Perturbation of a parcel of air and the effect on its streamwise wind speed.

$$\overline{u'w'} = -u_*^2 = -l_m^2 \left| \left(\frac{\partial u(z)}{\partial z} \right) \right| \frac{\partial u(z)}{\partial z} \tag{5.32}$$

The mixing length can have various forms but in this case is assumed to be equal to κz. This allows the Reynolds stress term $\overline{u'w'}$ in Eq. (5.11) to be calculated based on the local wind speed gradient so that the x and z momentum equations can be solved. Note that as the flow is considered approximately homogeneous in the streamwise direction, the $\partial / \partial y$ and $\partial / \partial y$ derivatives are neglected.

By assuming that in the inner layer the shear stress divergence terms are of the same magnitude as the inertial terms along with the mixing layer turbulence scheme, the inner layer height l can be estimated using

$$\frac{l}{L} \ln \left(\frac{l}{z_0} \right) = 2\kappa^2 \tag{5.33}$$

although subsequent researchers have adapted the original JH theory and used different equations to calculate the inner layer height.

Finally, to solve the equations in the inner and outer layers, the linearised equations are Fourier transformed to provide analytical solutions which are then reverse transformed to give the final wind speed values.

The linearised theory as developed by JH is the basis for several codes which have been used extensively in wind resource assessment, including:

- MS3DJH/3R (Walmsley et al., 1986): a 3D implementation of the theory which uses a range of length scales and incorporates changes in surface roughness.
- FLOWSTAR (Finardi et al., 1993): based on a later version of the JH theory where the outer layer is further subdivided into two layers, where the upper layer is influenced by stratification.
- MS-Micro: a version of MS3DJH/3R developed for a personal computer.
- MSFD (Beljaars et al., 1987): a version of MS-Micro which incorporates a more sophisticated turbulence closure model ($k-\epsilon$, which is described in Section 5.2.4) requiring an iterative finite difference solution.
- WAsP (Wind Atlas Analysis and Application Programme) (Troen, 1990): developed as part of the European Wind Atlas project. Instead of a cartesian grid, this model is based on a zooming polar coordinate grid giving greater accuracy at the point of interest. A roughness change model and shelter model (for obstacles) are also incorporated.

The usual way of using a linearised model is to calculate the terrain-induced wind speed perturbations by direction sector relative to a given location (reference site) where the wind speed is known (e.g. where a meteorological mast is located). The absolute wind speed at another site of interest can then be determined from the wind speed measured at the reference site adjusted by the perturbation factor, in the form of a speed-up factor which can be positive or negative. Many onshore wind farm sites, if they are large, may have multiple masts and it is common to produce a weighted combination of the speed-up factors dependent on the distance away of the point of interest from the masts.

5.2.2.1 Simple Guidelines for Calculating Speed-up over Topographic Features

Based on the JH theory, a set of simple equations has been devised for calculating the speed-up effect of a simple topographic feature such as a hill or ridge (Taylor and Lee, 1984; Walmsley et al., 1989). In addition, the effect of a roughness change close to the feature was also considered. The linearised form of the momentum equations and the solutions in Fourier space lead to the conclusion that the speed-up over a hill depends, in magnitude, on the slope of the hill. This speed-up reduces exponentially with height above the hill in proportion to the scale of the hill. This holds above l in the potential flow region where turbulent stresses can be neglected. For most practical wind resource assessment purposes, the height of interest is well above this.

Figure 5.6 shows a schematic flow situation where the guidelines can be applied. To the left of the figure, a measurement of the upstream wind speed u_0 is made at a height Δz above the ground. An estimate is required of the wind speed on top of a nearby hill u_T at the same height above the ground. As in Figure 5.4, the hill is parameterised by a height h and half-width at half-height L_W. Close to the upstream anemometer, there is an instantaneous change in the surface roughness from z_{0u} upstream to z_0 downstream. From this point onwards, an *internal boundary layer* (IBL) develops and the wind profile adapts to the change in surface conditions whereby

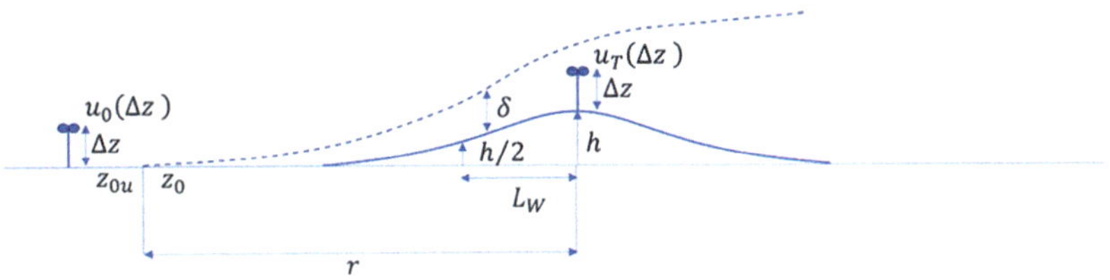

Figure 5.6 Schematic of the wind flow situation to which the guideline equations can be applied. *Source:* adapted from Taylor and Lee (1984).

below the height of this IBL it has adapted to the downstream roughness z_0, but above the layer it still feels the influence of the upstream roughness z_{0u}. In practice, the interface between these two flow regimes is not instantaneous and there is a thin layer over which the profile is adjusting. The distance from the start of this roughness length to the measurement point on top of the hill is r.

The premise of the guidelines is that:

- The change in wind speed due to the topographic feature can be treated independently from the change in surface roughness.
- The change in wind speed due to the topographic feature and the roughness change can be parameterised as multiplying or speed-up factors relative to the upstream wind speed.

Therefore

$$u_T(\Delta z) = u_0(\Delta z) + \Delta u_T + \Delta u_R \tag{5.34}$$

where Δu_T is the wind speed change due to the topographic feature and Δu_R is the wind speed change due to a roughness change. Note that Δu_T is normally positive though in principle can be negative if the topographic feature is a valley, and Δu_R is positive for a rough-to-smooth surface transition and negative vice versa. The topographically induced increase is given by

$$\Delta u_T = \Delta S u_0(\Delta z) \tag{5.35}$$

where the speed-up factor ΔS is

$$\Delta S = \Delta S_{max} \exp(-A\Delta z / L_W) \tag{5.36}$$

and the maximum speed-up ΔS_{max} (at l above the ground) is

$$\Delta S_{max} = Bh / L_W \tag{5.37}$$

where A and B are empirical constants which depend on the topographic feature and account for the fact that not all of the pressure gradient will usefully drive the flow over the top of the feature.

The values of A and B from observations and scale model experiments are given in Table 5.1. Note that 'rolling terrain' is parameterised from the average of hill-top to valley floor over the site of interest.

In the case of the wind speed change due to the roughness change

$$\Delta u_R = \left\{ \left(\frac{\ln[\Delta z / z_0]}{\ln[\delta / z_0]} \right) \bullet \left(\frac{\ln[\delta / z_{0u}]}{\ln[\Delta z / z_{0u}]} \right) - 1 \right\} u_0(\Delta z) \tag{5.38}$$

Table 5.1 Values of A and B for different topographical features.

Terrain type	A	B
2D hills (ridges)	3.0	2.0
3D hills	4.0	1.6
2D escarpments	2.5	0.8
2D rolling terrain	3.5	1.55
3D rolling terrain	4.4	1.1

where the IBL height δ is given by (Elliott, 1958)

$$\delta / z_0 = A_1 \, (r / z_0)^{0.8} \tag{5.39}$$

where

$$A_1 = 0.75 + 0.03M \tag{5.40}$$

and

$$M = \ln z_{0u} - \ln z_0 \tag{5.41}$$

It has been shown that the effect of M is quite small (Walmsley, 1989) and can be set to 0 with little change in accuracy.

The assumptions made in applying the guidelines are:

- Neutral surface layer stability.
- Horizontal length scales $\lesssim$ 1 km.
- Gentle slopes ($\lesssim$ 0.35).

The guidelines provide a useful practical tool that can be used in a spreadsheet to estimate the effect of simple topographic features and surface coverage on the approaching wind speed, though in many cases a more detailed model is required for all but the simplest terrain.

5.2.3 Linearised Models in Steep Terrain

Linearised models have been widely used for wind flow modelling, particularly where terrain slopes are not excessive (less than 30–35%). Even for larger slopes, there have been attempts to make empirical corrections to WAsP model predictions based on a ruggedness index (RIX) number (Bowen and Mortensen, 1996). The RIX number is defined as the fraction of terrain within a certain distance of a site which exceeds a certain critical slope when flow separation begins to occur (explained in Section 5.2.4). Figure 5.7 illustrates how the RIX number is calculated based on an 8 km × 8 km tile of contour data. The calculation is based on several radii extending out from the site of interest. Two examples are shown in Figure 5.7 with (a) 12 radii and (b) 72 radii. The calculation is made by dividing the tile into 250 m × 250 m cells. For each radius line, the fraction of the terrain with a slope greater than 30% is calculated and then the average is calculated based on all radii to give the overall RIX value of the point of interest.

When the WAsP model is used to predict the wind speed at one site based on the value at another site, the RIX number is calculated at both sites and the difference Δ RIX is calculated. An empirical correction factor to the predicted wind speed u_p based on the Δ RIX value can then be used when terrain slopes are greater than around 30% to estimate the true measured value u_m

$$u_m = \frac{1}{\exp\left(\alpha_{\mathrm{RIX}} \Delta \mathrm{RIX}\right)} u_p \tag{5.42}$$

where α_{RIX} is an empirical parameter which, based on values from several complex terrain sites, was found to be around 1.5 (Mortensen et al., 2008). This has been shown to produce reasonable results for the WAsP linearised model, though its general applicability for other linearised models has not been demonstrated.

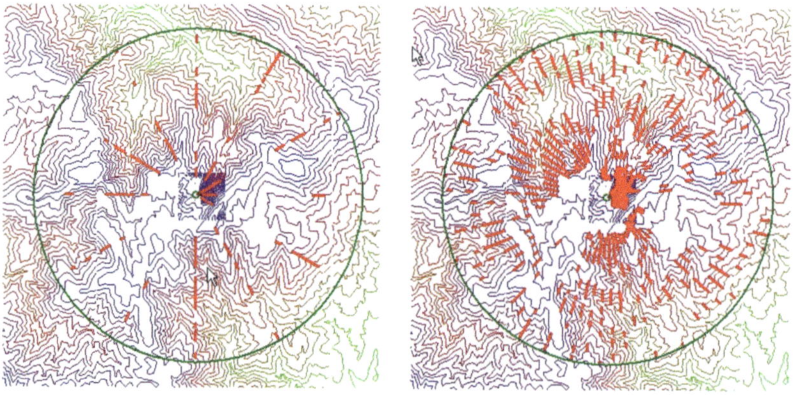

Figure 5.7 Illustration of the calculation of the RIX number using (a) 12 radii and (b) 72 radii extending from a site of interest based on an orographic contour tile of 8 km × 8 km. The red line segments represent terrain steeper than 0.3. The green circle shows the limit of the area for which the RIX number is calculated. *Source:* Bowen and Mortensen (2004) / DTU Wind Energy.

5.2.4 Navier–Stokes Solvers and Computational Fluid Dynamics

In general, linearised models perform poorly in areas of complex terrain with steep slopes as the assumptions that underly the linearisation of the momentum equations break down. When the wind flow passes over a slope beyond a certain critical value, the adverse (negative) pressure gradient acting to decelerate the flow combined with the viscous drag become so great that the flow stops or indeed reverses close to the surface causing the flow to *separate*. This is illustrated in Figure 5.8.

It can be seen that in Figure 5.8(a) the stream lines follow the contours of the hill, whereas in Figure 5.8(b) they separate from the hill and a region of turbulent reverse flow is seen in the lee of the hill. This behaviour cannot be captured by a linearised model. To provide a more accurate prediction of the wind flow where terrain slopes are above the critical level to trigger flow separation, a more complete solution of Eqs. (5.11)–(5.13) is required. Such a model is often referred to as a *computational fluid dynamics* (CFD) or *Navier–Stokes model*, though this is just loose terminology for a flow solver which models the turbulence more accurately than a linearised model. In addition to the momentum equations, CFD models are sometimes used to solve the energy equation (Eq. 5.14), particularly if the effects of stability are to be accurately captured. Other scalars such as moisture (Eq. 5.15) or pollutants can be modelled using a CFD model, but this is not normally the case for wind resource assessment.

5.2.4.1 k–ε Turbulence Closure

As mentioned in Sections 5.1 and 5.2.2, the turbulent stress transport terms in the momentum equations mean that the set of equations is not closed. The approach that was described in Section 5.2.2, was to use a mixing length closure. Although this can be done even if the momentum equations are not linearised, normally a more sophisticated turbulence model is used. The most common of these is the $k - \varepsilon$ model (Launder and Spalding, 1974). There are many CFD textbooks which explain such models in detail (e.g. Versteeg and Malalasekera, 2007), so only the basic

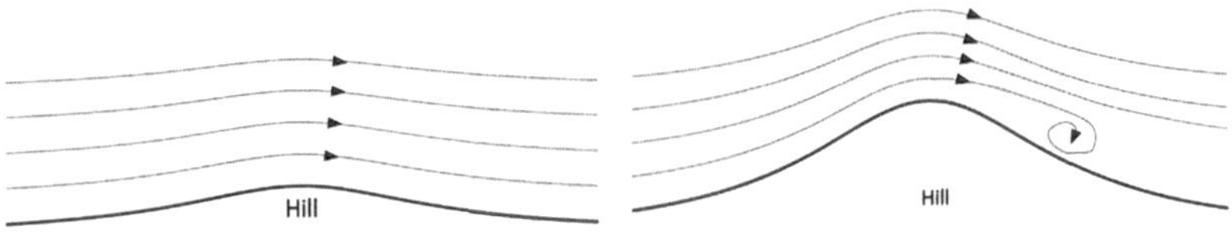

Figure 5.8 Wind flow over (a) a hill with a gentle slope and no separation and (b) a steeply sloping hill with separation and reverse flow in the lee.

principles are explained here. The basic principle of the $k-\varepsilon$ closure model is that turbulent kinetic energy production and its dissipation rate can be determined from known mean properties of the flow, specifically the mean rate of deformation.

First of all, we define the turbulent kinetic energy of the flow

$$k=\frac{1}{2}\left(\overline{u'^2}+\overline{v'^2}+\overline{w'^2}\right) \tag{5.43}$$

A more general version of Eq. 2.8 can be defined relating the stress tensor τ_{ij} to the rate of deformation e_{ij} of an element of fluid (the air in this case)

$$\tau_{ij}=\mu\left(\frac{\partial u_i}{\partial x_j}+\frac{\partial u_j}{\partial x_i}\right)=2\mu e_{ij} \tag{5.44}$$

where, for convenience, suffix notation is introduced such that i or $j=1$ refers to the x direction, i or $j=2$ refers to the y direction, and i or $j=3$ refers to the z direction. So, for example, using the nomenclature above, $u_1=u$, $u_2=v$, etc.

The rate of deformation tensor (analogous to a strain rate for a material) can also be Reynolds decomposed into a mean and fluctuating part

$$e_{ij}=\bar{e}_{ij}+e'_{ij} \tag{5.45}$$

which can be also written as

$$e_{ij}=\frac{1}{2}\left(\frac{\partial \overline{u_i}}{\partial x_i}+\frac{\partial \overline{u_j}}{\partial x_j}\right)+\frac{1}{2}\left(\frac{\partial u'_i}{\partial x_i}+\frac{\partial u'_j}{\partial x_j}\right) \tag{5.46}$$

As kinetic energy is produced in the flow, it is also dissipated by the viscous stresses and so a rate of dissipation ε is introduced

$$\varepsilon=2\nu\overline{e'_{ij}.e'_{ij}} \tag{5.47}$$

where $\nu=\mu/\rho$ is the kinematic viscosity.

There are two scales defined within the model for large-scale turbulence: a *velocity scale* ϑ_k and a *length scale* l_k

$$\vartheta_k=k^{\frac{1}{2}} \tag{5.48}$$

$$l_k=\frac{k^{\frac{3}{2}}}{\varepsilon} \tag{5.49}$$

This is based on the (reasonable) assumption that there is a balance between the energy extracted from the flow by the largest turbulence structures and that cascaded down to the smallest eddies. Based on these scales, an eddy viscosity μ_t is defined

$$\mu_t = C\rho\vartheta_k l_k = \rho C_\mu \frac{k^2}{\varepsilon} \tag{5.50}$$

where C_μ is a dimensionless constant. Note that this is equivalent to K_M defined in Section 2.2 and reflects the use of the parameter in two different communities: those researchers working in fluid dynamics and those working in atmospheric flows.

To close the set of equations, two additional transport equations are introduced for k and ε

$$\frac{\partial(\rho k)}{\partial t} + \nabla.(\rho k\mathbf{u}) = \nabla.\left(\frac{\mu_t}{\sigma_k}\nabla k\right) + 2\mu_t \overline{e_{ij}}.\overline{e_{ij}} - \rho\varepsilon + S_k \tag{5.51}$$

$$\frac{\partial(\rho\varepsilon)}{\partial t} + \nabla\cdot(\rho\varepsilon\mathbf{u}) = \nabla\cdot\left(\frac{\mu_t}{\sigma_\varepsilon}\nabla\varepsilon\right) + C_{1\varepsilon}\frac{\varepsilon}{k}2\mu_t\overline{e_{ij}}\cdot\overline{e_{ij}} - C_{2\varepsilon}\rho\frac{\varepsilon^2}{k} + s_\varepsilon \tag{5.52}$$

where S_k and S_ε are source terms. The values of the constants normally used are

$$C_\mu = 0.09,\ \sigma_k = 1.00,\ \sigma_\varepsilon = 1.30,\ C_{1\varepsilon} = 1.44,\ C_{2\varepsilon} = 1.92$$

As these constants have been empirically determined from constrained rather than atmospheric flow, some researchers have adjusted some of these constants for wind resource assessment, though the justification for this has been questioned by others.

To determine the Reynolds stress terms in Eqs. (5.11)–(5.13), the extended Boussinesq approximation is used

$$-\rho\overline{u_i'u_j'} = \mu_t\left(\frac{\partial\overline{u_i}}{\partial x_i} + \frac{\partial\overline{u_j}}{\partial x_j}\right) - \frac{2}{3}\rho k\delta_{ij} = 2\mu_t\overline{e_{ij}} - \frac{2}{3}\rho k\delta_{ij} \tag{5.53}$$

where δ_{ij} is the Kronecker delta function which equals 1 if $i = j$ or 0 if $i \neq j$.

5.2.4.2 *k*–*ω* Shear Stress Turbulence Closure

An alternative model which is better able to predict flow separation is the $k-\omega$ shear stress turbulence (SST) model (Menter, 1994). This uses a specific dissipation which defines the mean scale of the turbulence

$$\omega = \frac{1}{\beta^*}\frac{\varepsilon}{k} \tag{5.54}$$

where β^* is normally set equal to C_μ. A transport equation like those for k and ε is solved for ω. Close to the ground k and ω transport equations are used, and away from the ground the k and ε transport equations are solved.

5.2.4.2.1 Simplifying the Equations in Atmospheric Flow

The Reynolds number defines the balance between inertial and viscous forces and is defined as

$$Re = \frac{u_c l_c}{\nu} \tag{5.55}$$

where ν is the kinematic viscosity, u_c is the characteristic speed of the flow, and l_c is the characteristic scale of the flow. Depending on the value of Re, the characteristics of the flow may change. For atmospheric flows, the value of u_c corresponds to the wind speed at the height of interest. The value l_c is less well defined, and different definitions may be used depending on the situation of interest. For example, if the flow is over a hill, it may be appropriate to use the height or width of the hill. For some applications, the height of the atmospheric boundary layer might be an appropriate scale. However, typically, for atmospheric flows, $\nu = 1.5 \times 10^{-5}\,\text{m}^2\text{s}^{-1}$, $u_c \sim 10\,\text{ms}^{-1}$, and $l_c \sim$ 100s of metres so that $Re \sim 10^7 - 10^8$ which means that inertial forces completely dominate over viscous forces and at these very high values the flow is, to all intents and purposes, Reynolds number independent. This means that the viscous terms $\nabla \cdot (\mu \nabla \bar{u}), \nabla \cdot (\mu \nabla \bar{v})$, and $\nabla \cdot (\mu \nabla w)$ are small compared with the divergence of the Reynolds stress terms and can be neglected. If the wind flow is then assumed to be horizontal, homogeneous, and steady state then this allows significant simplification of the momentum equations, so that Eqs. (5.11) and (5.12) become

$$0 = -\frac{\partial p}{\partial x} - \frac{\partial\left(\rho \overline{u'w'}\right)}{\partial z} + \rho f \bar{u} \tag{5.56}$$

$$0 = -\frac{\partial p}{\partial y} - \frac{\partial\left(\rho \overline{v'w'}\right)}{\partial z} - \rho f \bar{v} \tag{5.57}$$

From the definition of the geostrophic wind in Eqs. 2.30 and 2.31, these then become

$$0 = -\frac{\partial\left(\rho \overline{u'w'}\right)}{\partial z} + \rho f (\bar{u} - u_g) \tag{5.58}$$

$$0 = -\frac{\partial\left(\rho\overline{v'w'}\right)}{\partial z} - \rho f(\overline{v} - v_g) \tag{5.59}$$

Recalling Eq. 2.8, which is repeated here for convenience, equating the Reynolds stresses to the mean wind shear

$$\tau_x = -\rho\overline{u'w'} = \rho K_m \frac{\partial \overline{u}}{\partial z} \tag{5.60}$$

there is an equivalent for the stress in the y direction

$$\tau_y = -\rho\overline{v'w'} = \rho K_m \frac{\partial \overline{v}}{\partial z} \tag{5.61}$$

Combining Eqs. (5.56), (5.57), (5.60), and (5.61) and assuming the geostrophic wind is aligned with the x direction ($v_g = 0$) gives (Garratt, 1994)

$$f\overline{v} = -K_M \frac{\partial^2 \overline{u}}{\partial z^2} \tag{5.62}$$

$$f\left(\overline{u} - u_g\right) = -K_M \frac{\partial^2 \overline{v}}{\partial z^2} \tag{5.63}$$

Note that the Coriolis parameter has both a magnitude and a sign, which can be written as

$$f = |f|\,\mathrm{sgn}(f) \tag{5.64}$$

where $\mathrm{sgn} f = +1$ in the northern hemisphere and -1 in the southern hemisphere.

When Eqs. (5.62) and (5.63) are solved with the boundary conditions that $\overline{u} = \overline{v} = 0$ when $z = 0$ and $\overline{u} \to u_g$, $\overline{v} \to 0$ when $z \to \infty$, this results in the solution

$$\overline{u} = u_g\left[1 - \exp(-a_0 z)\cos(a_0 z)\right] \tag{5.65}$$

$$\overline{v} = \mathrm{sgn}(f) u_g\left[\exp(-a_0 z)\sin(a_0 z)\right] \tag{5.66}$$

where

$$a_0^2 = \frac{|f|}{2K_M} \tag{5.67}$$

The solution is known as an *Ekman spiral* as the wind direction turns from being aligned with the geostrophic wind above the atmospheric boundary layer (ABL) to rotating anticlockwise 45° (theoretically) to the geostrophic wind at the surface in the northern hemisphere for a neutral atmosphere.

The turning is clockwise in the southern hemisphere. In practice, the maximum turning angle is less than this and it is influenced by the surface roughness and the stability (i.e. it reduces under unstable conditions and smoother surfaces, e.g. it can be less than 5° over the sea under very unstable conditions).

5.2.4.3 Unsteady Reynolds Averaged Navier–Stokes Models

For most wind resource applications where the atmosphere is assumed neutral throughout, Eqs. (5.1) and (5.11)–(5.13) can be solved ignoring the time derivate and assuming a steady state solution (i.e. that once the solver has found a converged solution, it will not change in time). However, there are certain flow situations where this assumption is no longer valid, for example where stability above or within the atmospheric boundary layer, especially in the surface layer, is no longer neutral. A non-neutral surface layer will result if the surface is warmer or cooler than the air above or if it is a source or sink of heat flux resulting in a time evolving situation. In this situation, the Navier–Stokes equations need to be solved in time. If the Reynolds decomposed equations are used, then the model is referred to as an *unsteady Reynolds averaged Navier–Stokes (URANS) model.*

If a URANS model is used to model a wind flow situation which incorporates stability or more generally *potential temperature stratification* then some modifications and additional source terms need to be added into the momentum and turbulence equations. Moreover, to properly capture the effects of heat flux to or from the surface, the energy equation (Eq. 5.14) also needs to be solved. This results in the modified momentum equations (Montavon, 1998a, 1998b), using the extended Boussinesq approximation (reminder: where changes in density are assumed negligible apart from where they may affect a buoyancy force) to replace the Reynolds stress terms

$$\frac{\partial(\rho\bar{u})}{\partial t} + \boldsymbol{\nabla}.(\rho\bar{u}\mathbf{u}) = -\frac{\partial p'}{\partial x} + \boldsymbol{\nabla}.(\mu_{eff}\nabla\bar{u}) \tag{5.68}$$

$$\frac{\partial(\rho\bar{v})}{\partial t} + \boldsymbol{\nabla}.(\rho\bar{v}\mathbf{u}) = -\frac{\partial p'}{\partial y} + \boldsymbol{\nabla}.(\mu_{eff}\nabla\bar{v}) \tag{5.69}$$

$$\frac{\partial(\rho\bar{w})}{\partial t} + \boldsymbol{\nabla}.(\rho\bar{w}\mathbf{u}) = -\frac{\partial p'}{\partial z} + \boldsymbol{\nabla}.(\mu_{eff}\nabla\bar{w}) + g\frac{\rho_h}{\theta_h}(\theta - \theta_0)' \tag{5.70}$$

where p' is the modified pressure

$$p' = p + \frac{2}{3}\rho_h k - p_h \tag{5.71}$$

and p_h and ρ_h are the pressure and density, respectively, for a reference atmosphere which is in hydrostatic equilibrium.

The modified turbulence equations are

$$\frac{\partial(\rho k)}{\partial t}+\boldsymbol{\nabla}.(\rho k\mathbf{u})=\boldsymbol{\nabla}.\left(\frac{\mu_t}{\sigma_k}\nabla k\right)+2\mu_t\overline{e_{ij}}.\overline{e_{ij}}-\rho\varepsilon+G \tag{5.72}$$

$$\frac{\partial(\rho\varepsilon)}{\partial t}+\boldsymbol{\nabla}.(\rho\varepsilon\mathbf{u})=\boldsymbol{\nabla}.\left(\frac{\mu_t}{\sigma_\varepsilon}\nabla\varepsilon\right)+C_{1\varepsilon}\frac{\varepsilon}{k}2\mu_t\overline{e_{ij}}.\overline{e_{ij}}-C_{2\varepsilon}\rho\frac{\varepsilon^2}{k}+C_{1\varepsilon}\frac{\varepsilon}{k}\max(G,0) \tag{5.73}$$

$$G=-\frac{\mu_{eff}}{\sigma_H}\frac{g}{\theta_h}\frac{\partial\theta}{\partial z} \tag{5.74}$$

and the energy equation (using θ instead of T) is

$$\frac{\partial(\rho\bar{\theta})}{\partial t}+\boldsymbol{\nabla}.(\rho\bar{\theta}\mathbf{u})=\boldsymbol{\nabla}.\left(\left(\frac{k_T}{C_p}+\frac{\mu_t}{\sigma_\theta}\right)\nabla\bar{\theta}\right) \tag{5.75}$$

where θ_h is the potential temperature for a reference atmosphere which is in hydrostatic equilibrium, $\mu_{eff}=\mu+\mu_t$ is the effective viscosity, and $\sigma_H=\sigma_\theta=0.9$.

Incorporating the effect of stability can have a profound effect on the wind flow as is expanded upon in Section 5.6.

5.2.4.4 Large Eddy Simulation Models

A yet more sophisticated solution to the Navier–Stokes equations uses a large eddy simulation (LES) model. This type of model explicitly resolves scales of turbulence above a certain size, Δ. Below that size, the turbulence is modelled in an analogous way to RANS-equations-based models. To do this, a filter is applied so that the velocity components are split into two parts

$$u_i=\tilde{u}_i+u_i' \tag{5.76}$$

where $\tilde{u}_i$ is the resolved part and u_i' is unresolved (modelled). Now the basic momentum equation becomes

$$\frac{\partial(\rho\tilde{u}_i)}{\partial t}+\frac{\partial(\rho\tilde{u}_i\tilde{u}_j)}{\partial x_j}=\frac{\partial\tilde{p}}{\partial x_i}+\frac{\partial}{\partial x_i}\left[\mu\left(\frac{\partial\tilde{u}_i}{\partial x_j}+\frac{\partial\tilde{u}_j}{\partial x_i}\right)\right]-\frac{\partial\tau_{ij}}{\partial x_j} \tag{5.77}$$

where τ_{ij} is the *sub-grid-scale stress tensor*

$$\tau_{ij} = \widetilde{u_i u_j} - \tilde{u}_i \tilde{u}_j \tag{5.78}$$

This tensor needs to be modelled using a sub-grid-scale (SGS) model. A model frequently used is that by Smagorinsky (1963).

Although LES models are more accurate than RANS-equations-based models and can resolve time-varying turbulence, they are much more difficult to set up, require precursor simulations to be run, and need much greater computing resources. For this reason, they are still restricted to the research domain for wind resource assessment.

5.2.4.5 The Use of CFD Models in Wind Resource Assessment

In contrast to linearised models, the solutions to the nonlinear RANS equations are iterative, requiring significant computing resources depending on the scale of the problem. Until relatively recently, the use of such models was restricted to the research domain, but as personal computers have become more powerful, their use has started to become more widespread in the wind resource assessment industry. The setting-up of a CFD model with the correct grid and boundary conditions is a nontrivial task and can give rise to erroneous solutions if not done correctly. To maximise the use of computing resources, the model is normally run with a grid or mesh whose resolution becomes finer closer to the region of interest. This can be done using a structured (regular) or unstructured (irregular) grid which does not necessarily use a terrain following system, such as that described in Section 5.2.1. Terrain and roughness length data are digitised and used to provide surface boundary conditions to the model. An example of digitised terrain and a structured mesh for modelling Bolund hill in Denmark is shown in Figure 5.9. Mean and turbulence quantities are prescribed at the inlet to the model domain. Where stratification is incorporated in the model, temperature profile, heat flux, or surface temperature offset data are required. These data can be provided from direct observations (e.g. radiosonde data, mast data, etc.) or from numerical model output such as reanalysis datasets. An example of a prescribed wind speed and potential temperature profile is shown in Figure 5.10. The wind speed profile corresponds to a Zilitinkevich profile (Zilitinkevich et al., 1998; Zilitinkevich and Mironov, 1996) with modified model coefficients, and the potential temperature profile prescribes an initially neutral profile up to an atmospheric boundary layer height of 1700 m with a stable inversion layer above. In this example, the ground is 1 K cooler than the air directly above which will cause a stable surface layer to develop during the simulation. Several companies have developed software products to make the setting up of a CFD model for wind resource assessment more straightforward; nonetheless, care is required in using such models.

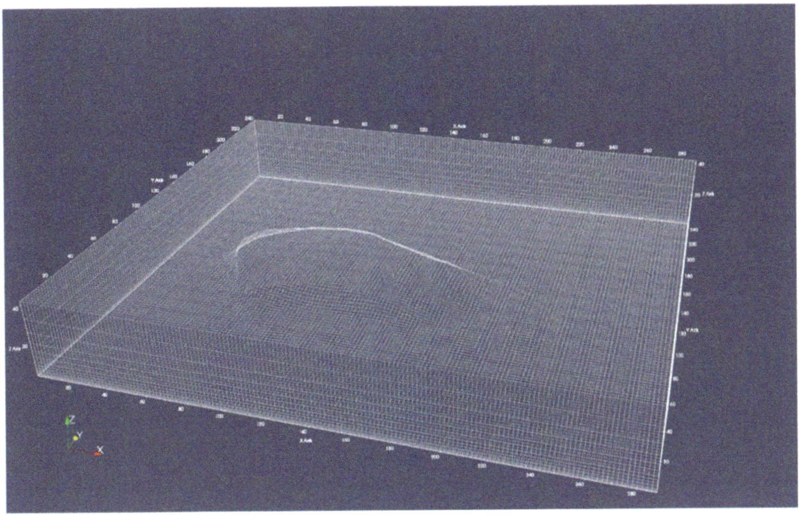

Figure 5.9 A structured mesh above digitised terrain height data for Bolund hill.

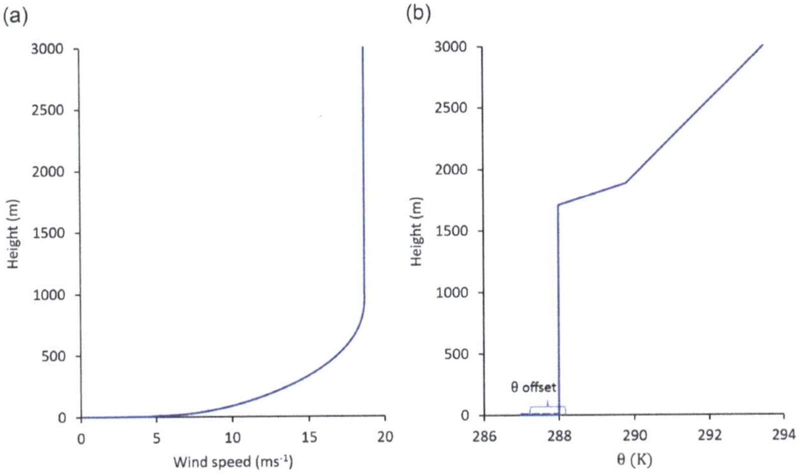

Figure 5.10 Example of (a) mean wind speed and (b) potential temperature profiles used as inlet conditions for the domain of a CFD model.

Popular CFD codes that are used for wind resource assessment include:

- WindSim (Gravdahl, 1998): one of first CFD codes developed specifically for wind resource assessment.
- Ansys CFX and Fluent with Windmodeller (Ansys, n.d.): the Windmodeller macros have been developed to help set up the commercial CFX or Fluent codes (formerly independent industrial CFD codes but now both part of Ansys, Inc) to carry out a wind resource assessment for a site.

- Meteodyn WT (https://meteodyn.com): another CFD code specifically developed for wind resource assessment. This uses a RANS solver with a $k-L$ turbulence model (Dimonte and Tipton, 2006) which incorporates a transport equation for the eddy scale.
- Simcenter Star-CCM+/Environmental Wizard (Siemens, n.d.): another popular commercial CFD code with some basic front-end software for setting up terrain modelling.
- OpenFOAM/ZephyScience (www.openfoam.com; www.zephy-science.com): OpenFOAM (Open-source Field Operations and Manipulations) is probably the most-well-known open-source CFD code. ZephyScience provide a suite of tools that can be used with OpenFOAM for wind resource and wind farm yield assessment.
- SOWFA (Simulator for Wind Farm Applications) (National Renewable Energy Laboratory, 2020): this is based on OpenFOAM and was developed by the National Renewable Energy Laboratory (NREL) specifically to simulate wind flows in and around wind farms.
- WAsP CFD: this uses much of the WAsP suite of software, but the wind flow modelling, rather than using a linearised solver, is based on the EllipSys code developed at the Danish Technical University (Sørensen, 1995).

5.3 Mesoscale Models

The models described in Section 5.2 are generally applied over the scale of tens of kilometres and are primarily concerned with resolving the wind flow, ignoring most other physical processes in the atmosphere aside from the effect of heat flux at the surface and its impact on stability. In general, this gives a good approximation of the wind conditions on the microscale. However, there are certain wind flow situations such as sea breezes and valley winds that are only properly captured by modelling over larger areas (i.e. hundreds of kilometres) that require a much more complete representation of all the physical processes in the atmosphere. This is the realm of *mesoscale* (medium-scale) models. These models solve the continuity, momentum, and energy equations but also other scalar flux quantities, such as moisture, and include several other models which can account for the effects, for example of infrared radiation, soil physics, evaporation, precipitation, etc. Another significant difference compared with microscale models is that mesoscale models are normally nonhydrostatic (i.e. the vertical momentum equation is explicitly solved allowing the modelling of processes such as vertical convection). The compressible form of the momentum equations is often used. Mesoscale models are frequently used to solve the governing equations in time to provide forecasts, with their boundary conditions provided by the output of coarser resolution global models (described in Section 5.4). Vertical coordinates are terrain following and normally parameterised using hydrostatic pressure (rather than normalised height) levels. As well as providing forecasts (prognosis), they can provide a *downscaled* (finer resolution) assessment of the state of the atmosphere based on an initialisation using a reanalysis dataset. In this way, mesoscale models can be used in resource assessment, to provide a wind atlas, for example. In principle, to build up the wind resource for a location, it is possible to run a mesoscale model in time over a period of many years to capture all climatological conditions. This has the advantage of producing a time series of values and captures diurnal and seasonal cycles. However, it also requires a large amount of computing resource, so an alternative approach is to run the model for a sufficiently representative range of climatological situations to build up an average wind resource weighted by the occurrence of these situations. In order that the model does not drift too far from the large-scale circulation patterns of the atmosphere when running a time domain simulation, it can be nudged back in several ways:

- Using observational data at given locations which nudge the solution at given time intervals to the actual measured value weighted by distance from the location (observational nudging).
- Using an analysis of observational values (e.g. a reanalysis dataset) to periodically nudge all grid points in the model domain (analysis nudging).
- Using a variation of analysis nudging where only the larger scales are nudged using some sort of spectral decomposition of the atmospheric fields (i.e. spectral nudging). Nudging may only be carried out on grid cells above a certain height.

To capture all the required physical processes requires a significant amount of computing power, and to resolve features down to the microscale is not feasible. For this reason, such models are normally run with the smallest grid size of the order of kilometres, or in some cases hundreds of metres. Moreover, to conserve computer resources the models are set up with a nested grid arrangement, as shown in Figure 5.11. Using three nests is common (but more may be used) whereby the model resolution is coarser in the outer domain where less resolution is required. The difference in resolution between one nest and the next should not be too large or the details at one scale cannot be properly fed to the next scale. A factor of three is quite common. The coarser outer domain should be large enough to capture the details of the important large-scale weather systems which

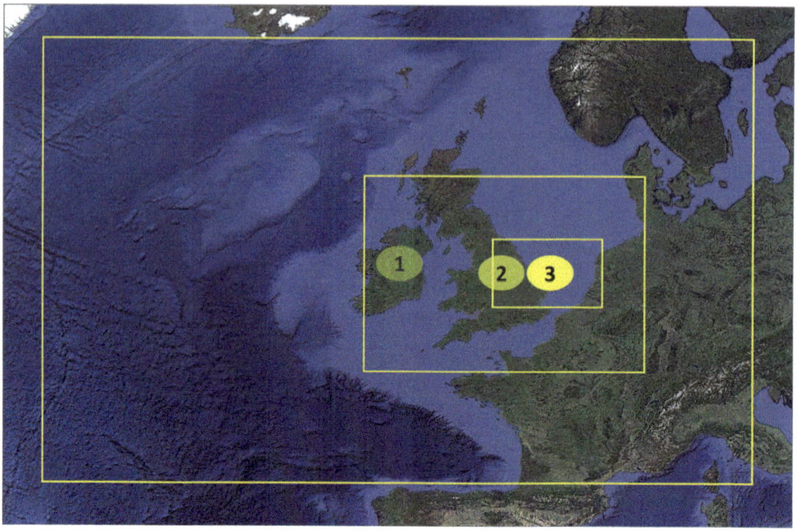

Figure 5.11 An example of a mesoscale model nested grid configuration. In this case, the coarsest-scale outer domain (centred at 1) has a horizontal grid point resolution of 18 km. The output resolved at this scale is used as the boundary conditions for the intermediate-scale domain, which has a horizontal grid point resolution of 6 km (centred at 2). The output from this scale in turn provides the boundary conditions to the finest-scale domain, which has a grid point resolution of 2 km (centred at 3).

drive the climatology at the site of interest. The output at this level acts as the boundary conditions for a finer resolution domain whose output in turn acts as the boundary conditions for yet a finer domain. In this way, the resolution of the physical atmospheric processes is concentrated near the site of interest. Normally, the flow of information is only from coarse to finer resolution (one-way coupling). In principle, the finer resolution grid could feed model information back to the coarser resolution (two-way coupling) and the solution could iterate back and forth to provide a converged solution. In practice this tends not to be done due to the extra computational requirement and the relatively small difference made to the modelled values in the finest grid domain. It should be noted that a mesoscale model can typically only resolve phenomena which are around 5–7 times the grid spacing (ESMAP, 2018).

In a similar fashion to RANS CFD models, turbulent processes are frequently modelled rather than resolved. The closure schemes in this case are called *planetary boundary layer* (PBL) schemes. There are multiple schemes available which have different strengths and weaknesses depending on the atmospheric conditions and the meteorological phenomena that are being studied. Some of the common PBL schemes are listed in Table 5.2. They are classified under two main headings: whether they are local or nonlocal and their closure order. Each PBL scheme is also used in combination with a surface layer scheme to resolve the surface layer fluxes (e.g. u_*). Such surface layer schemes are normally a variation on Monin–Obukhov similarity described in Section 2.2.

It can be seen from Eqs. (5.11)–(5.13) that the Reynolds decomposition has related quantities of first-order moment (mean values) to quantities of a second-order moment (the Reynolds stress variance and covariance terms). It is possible to express these second-order quantities in terms of

Table 5.2 Commonly used PBL closure schemes in mesoscale modelling.

PBL scheme	Order	Closure	Reference(s)
MRF (medium-range forecast model)	1	Nonlocal	(Hong and Pan, 1996)
YSU (Yonsei University)	1	Nonlocal	(Hong et al., 2006)
MYJ (Mellor–Yamada–Janjić)	1.5	Local	(Janjić, 1994, 1990; Mellor and Yamada, 1982, 1974)
QNSE (quasi-normal scale elimination)	1.5	Local	(Sukoriansky et al., 2005)
MYNN2/ MYNN3 (Mellor–Yamada–Nakanishi–Niino)	1.5/2	Local	(Nakanishi and Niino, 2004, 2006)
BouLac (Bougeault and Lacarrère)	1.5	Local	(Bougeault and Lacarrère, 1989)
GBM (Grenier–Bretherton–McCaa)	1.5	Local	(Grenier and Bretherton, 2001)
UWMT (University of Washington moist turbulence)	1.5	Local	(Bretherton and Park, 2009)
ACM/ACM2 (asymmetric convective model)	1	Nonlocal/ local*	(Pleim, 2007)
TEMF (total energy–mass flux)	1.5	Nonlocal/ local*	(Angevine et al., 2010)

*Upward fluxes nonlocal, downward fluxes local.
Source: adapted from Cohen et al. (2015).

third-order quantities, and so on, but as mentioned, this leads to an open set of equations. In the case of the $k - \varepsilon$ model described in Section 5.2.4, the closure is made at first order such that the second-order terms are related to known first-order terms. PBL schemes are formulated in similar terms. Such models are normally first, 1.5, or second order. In the case of the 1.5-order closure, some second-order terms (namely *turbulent kinetic energy*) are predicted explicitly, and some are modelled as functions of first-order terms. In second-order models, all second-order terms are predicted explicitly.

Local closure refers to where the PBL scheme only considers the effect of adjacent vertical model levels on properties at a given point. *Nonlocal closure*, on the other hand, considers the effect of several vertical levels in the vicinity of the point. The advantages and disadvantages of different PBL schemes are described in Cohen et al. (2015). No one scheme stands out as best under all situations when predicting wind resource, and their performance can differ under different atmospheric conditions (e.g. stability), with some models better during stable conditions and some better under convective conditions. Although there is not really consensus about which scheme should be used when, Draxl et al. (2014) suggest that the MYJ scheme performs best under stable and very stable conditions, the ACM2 performs best for neutral and near stable conditions, and the YSU scheme performs best under unstable conditions. The QNSE scheme was seen to give the best representation of the wind profile overall but was the worst for specific stability conditions.

As with CFD models, it is possible with a mesoscale model to resolve turbulent quantities explicitly down to a certain scale. This is termed *mesoscale LES modelling*. Using this method allows the simulation of turbulent processes down to the scale of 50–100 m compared with kilometres when using a PBL closure model, although it is clearly more computationally intensive. This can present its own challenges, as shown in Figure 5.12. This figure shows the turbulence spectrum as a function of wavenumber (reciprocal of wavelength) around the region of $1/l$ where l is the length scale of the most energetic, turbulent structures (eddies) in the atmosphere. Mesoscale models are designed to resolve turbulent structures beyond a given size. A filter of size Δ_{meso} is applied to filter out turbulent structures below a certain size which are modelled using a PBL scheme. In this case $\Delta_{\text{meso}} \gg l$. Microscale LES models (as described in Section 5.2.4.4) are designed to resolve rather smaller turbulent eddies in the turbulence spectrum inertial subrange. In this case, the cut-off filter between resolved and modelled turbulence is denoted in Figure 5.12 as

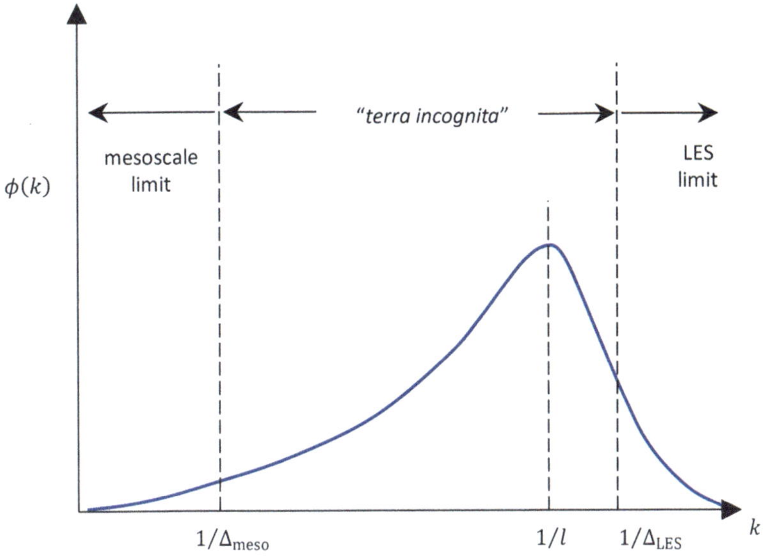

Figure 5.12 The turbulence spectrum ϕ as a function of wavenumber k (= 1/wavelength) and the so-called terra incognita which lies between the realm of mesoscale models and microscale LES models. *Source:* adapted from Wyngaard (2004).

Δ_{LES} where $\Delta_{LES} \ll 1$. Between wavenumbers l/Δ_{meso} and l/Δ_{LES} lies the region of the turbulence spectrum known as *terra incognita* (Wyngaard, 2004) for which neither mesoscale models nor microscale LES models are designed. As already mentioned, mesoscale LES models use increased resolution grids to resolve structures down to 50–100 m. In this case, the filter Δ_{meso} separates the scales of motion resolved from those which are modelled using a sub-filter-scale (SFS) parameterisation scheme (as opposed to a PBL scheme). In such cases, $\Delta_{meso} \sim l$, which is within the terra incognita for which the turbulence parameterisation is not designed. This is known as *grey zone modelling* and requires a modified approach to prevent inaccurate resource estimates.

The most commonly used mesoscale models for wind resource assessment are:

- WRF (Weather Research and Forecasting) (Skamarock et al., 2008): developed by the National Center for Atmospheric Research (NCAR) and the National Oceanic and Atmospheric Administration (NOAA). This is by far the most popular model when assessing wind resource. It is open source and has two solvers: ARW (Advance Research WRF) and NMM (Nonhydrostatic Mesoscale Model), with the former aimed at research and latter at operational weather forecasting.

- MM5 (Fifth-Generation Penn State/NCAR Mesoscale Model) (Anthes and Warner, 1978): developed by Penn State University and NCAR. One of thc earliest research mesoscale models and the predecessor of the WRF model (by which it is now superseded).
- KAMM (Karlsruhe Atmospheric Mesoscale Model) (Adrian and Fiedler, 1991): developed by the Institute for Meteorology and Climate Research, Karlsruhe Institute of Technology (KIT). The model has been used with the WAsP model (KAMM/WAsP) to produce regional wind atlases.

5.4 Global Models

Global models, as the name suggests, produce forecasts or analyses on a global scale. These types of model also solve the governing equations for fluid flow as well as a host of other equations which capture all the physical processes important to predict the earth's weather (in the short term) and climate (in the long term). As such models require a vast amount of terrestrial and remote sensing data as well as significant supercomputing resource, they are only run by large meteorological organisations and their outputs are then used to drive mesoscale models which can be used in a model chain with a microscale model to provide a wind resource assessment. For that reason, only brief details of the major global models whose output is relevant for wind resource assessment are given here. Their use in the production of global reanalyses is described in Chapter 9. Table 5.3 lists some of the principal models and gives details such as their temporal and spatial resolution.

Each of the models listed provides a range of products depending on application. The products can be either operational forecasts or reanalyses. In general, the longer ahead that the models are run, the lower the spatial and temporal resolution of the data which are output.

5.5 Model Chains

Because the models described in Sections 5.2–5.4 are applied on different scales then their application depends on what is required in terms of a resource assessment. If a general large-scale, long-term climatology is required then the output of a global model may be sufficient. If an

Table 5.3 Some commonly used global model output for wind resource assessment.

Name	Source	Temporal resolution	Highest spatial resolution (distances for midlatitudes)
CFS: Climate Forecast System	National Centers for Environmental Prediction (NCEP)	Hourly to monthly	$0.5° \times 0.5°$
GEM: Global Environmental Multiscale Model	Canadian Meteorological Centre (CMC), Recherche en Prévision Numérique (RPN), Meteorological Research Branch (MRB)	Six-hourly	~22 km
GFS: Global Forecast System	National Centers for Environmental Prediction (NCEP)	Three-hourly to twelve-hourly	~23 km
ICON: Icosahedral Nonhydrostatic	Deutscher Wetterdienst (DWD)	Three-hourly	~13 km
IFS: Integrated Forecast System	European Centre for Medium-Range Weather Forecasts (ECMWF)	Hourly to monthly	HRES model ~9 km
UM: Unified Model	United Kingdom Meteorological Office (UKMO)	Six-hourly up to seasonal	~10 km

understanding of diurnal phenomena such as valley winds, low-level jets, or sea breezes is required then a mesoscale model may be appropriate. If the details of very local scale flow around a hill is required for micrositing then a microscale model should be used. Sometimes it is necessary to combine the flow phenomena over a range of scales, for example, to model site conditions in complex terrain where time-dependent valley winds are prevalent. To do this requires coupling models together as a chain using the output from a larger-scale model to provide the boundary conditions for a finer-scale model, as illustrated in Figure 5.13. This may not be as straightforward as it sounds given the different physics captured by each type of model, the approximations made, particularly in how turbulent eddies are resolved, and the sudden change in resolution from one model to another (i.e. the problem of resolving the terra incognita as described in Section 5.3), especially when going from mesoscale to microscale.

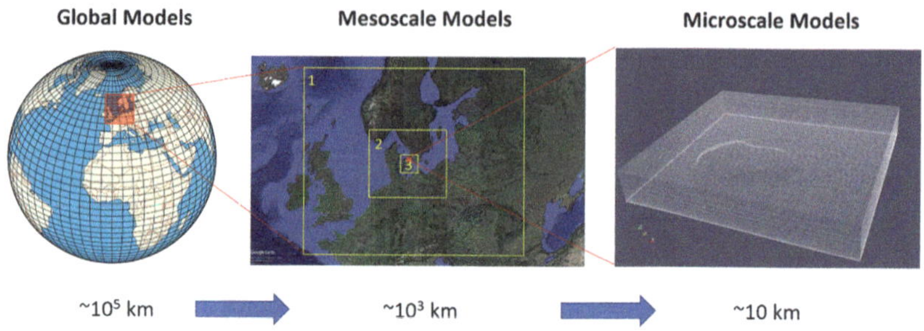

Figure 5.13 Model chains from global to microscale. *Source:* Wikimedia Commons.

In the production of a fine-detail wind atlas, it is common to couple the output from a mesoscale model to a microscale model. However, it may not be appropriate to simply take the output at, say, 1 km scale and input these data directly as boundary conditions to the microscale model which is then run at a finer resolution to produce the final result. The reason for this is that the mesoscale model has already incorporated the effects of terrain and surface roughness but at a lower resolution than is to be simulated by the microscale model. What is normally required is a generalised climatology which is representative of the area to be modelled. This can then be used as input to the microscale model. This was the approach used in the development of the European Wind Atlas (Troen and Lundtang Petersen, 1989) shown in Figure 5.14(a), except in this case actual site measurements were converted into a generalised climatology. The approach is used by the WAsP linearised model to translate wind speed measurements at one location to an estimate of the local wind resource at another location, as shown in Figure 5.14(b). This approach can also be used to downscale mesoscale model data in the production of a wind atlas. Namely, the microscale model is used to remove the terrain and roughness change effects at the lower resolution used by the mesoscale model to create the generalised climatology. This climatology is then used as input to the microscale model which can then be rerun using the finer-scale terrain and roughness length data required to calculate the detailed wind resource at the site of interest.

If site-specific time-evolving phenomena are to be downscaled then a more sophisticated coupling approach is required (Haupt et al., 2019). When using a nested mesoscale model such as WRF, it is possible to use coarse outer domains which parameterise turbulence using a PBL scheme (WRF-mesoscale) and then have finer-scale inner domains which resolve turbulence (WRF-LES) as long as the jump in scales between the domains is

(a)

(b)

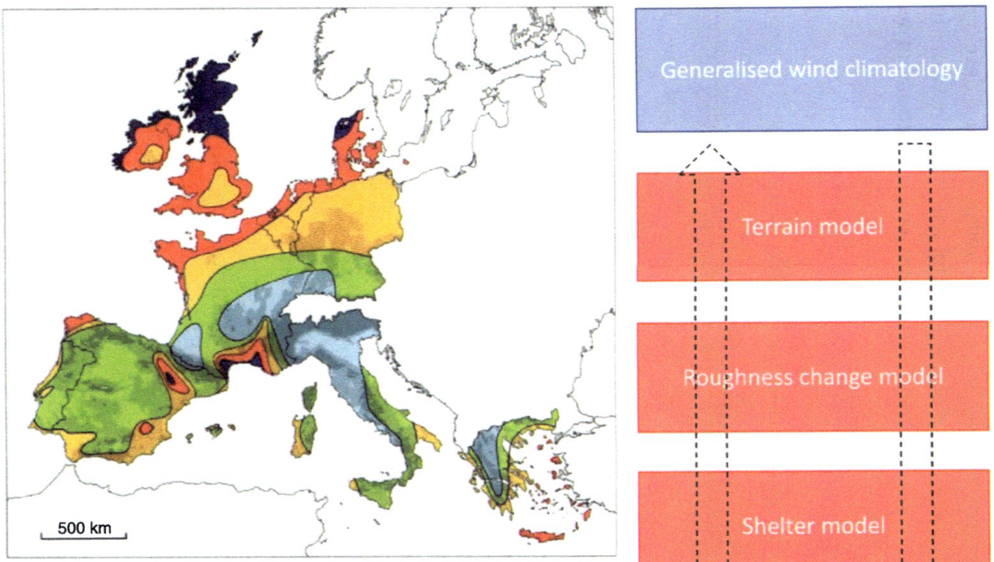

wind resources at 50m above ground level for five different topographic conditions

sheltered terrain ms^{-1}	sheltered terrain Wms^{-2}	sheltered terrain ms^{-1}	sheltered terrain Wms^{-2}	sheltered terrain ms^{-1}	sheltered terrain Wms^{-2}	sheltered terrain ms^{-1}	sheltered terrain Wms^{-2}	sheltered terrain ms^{-1}	sheltered terrain Wms^{-2}
>6.0	>250	>7.5	>500	>8.5	>700	>9.0	>800	>11.5	>800
5.0–6.0	150–250	6.5–7.5	300–500	7.0–8.5	400–700	8.0–9.0	600–800	10.0–11.5	1200–1800
4.5–5.0	100–150	5.5–6.5	200–300	6.0–7.0	250–400	7.5–8.0	400–600	8.5–10.0	700–1200
3.5–4.5	50–100	4.5–5.5	100–200	5.0–6.0	150–250	5.5–7.5	200–400	7.5–8.5	400–700
<3.5	<50	<4.5	<100	<5.0	<150	<5.5	<7.0	<7.0	<400

Figure 5.14 (a) The European Wind Atlas (Troen and Lundtang Petersen, 1989) (*Source:* courtesy of the Danish Technical University); (b) the methodology used to produce the European Wind Atlas and used by the WAsP model.

chosen carefully (i.e. a domain cell resolution is not within the terra incognita range). Alternatively, the output from a mesoscale model, extracted either from a limited vertical column of cells or at the boundaries of a microscale model can provide forcing source terms which can be incorporated into the microscale model momentum and energy equations in the time domain. An additional method is then required to develop the correct levels of turbulence within the microscale model. In all cases, global model data such as a reanalysis are used to drive the mesoscale model. It is not generally considered appropriate to drive a microscale model directly with reanalysis data as energy contained in the larger turbulent structures is lost, leading to inaccurate results.

The coupling of mesoscale and microscale models is still a developing area, but with the advent of ever-more-powerful computers new methods are being developed and more accurate downscaled predictions of wind resource are becoming available.

5.6 The Importance of Atmospheric Stability

In recent years, the importance of incorporating the effects of atmospheric stability in the atmospheric boundary layer and the free atmosphere above has become apparent in wind resource assessment. As is shown in Figure 2.2, surface layer stability has a large impact on the wind shear, which needs to be considered when extrapolating wind speeds from one height to another. Stratification can also have a profound effect on the flow, particularly in complex terrain. For example, whether the surface layer is stable or unstable affects the tendency of the flow to go above or around an orographic feature and whether the flow separates in the lee of the feature.

This is illustrated in Figure 5.15 in terms of the Froude number

$$Fr = \frac{U}{NL_W} \tag{5.79}$$

where U is the characteristic wind speed in the atmospheric boundary layer above the hill and N is the Brunt–Väisälä frequency defined in Eq. 2.44. If $Fr << 1$ then the conditions are very stable. This is the case for the first flow situation in Figure 5.15. Under this condition, there is some degree of blockage of the flow upstream of the hill leading to reduced wind speeds. Vertical motions are inhibited, creating a tendency for wind to flow

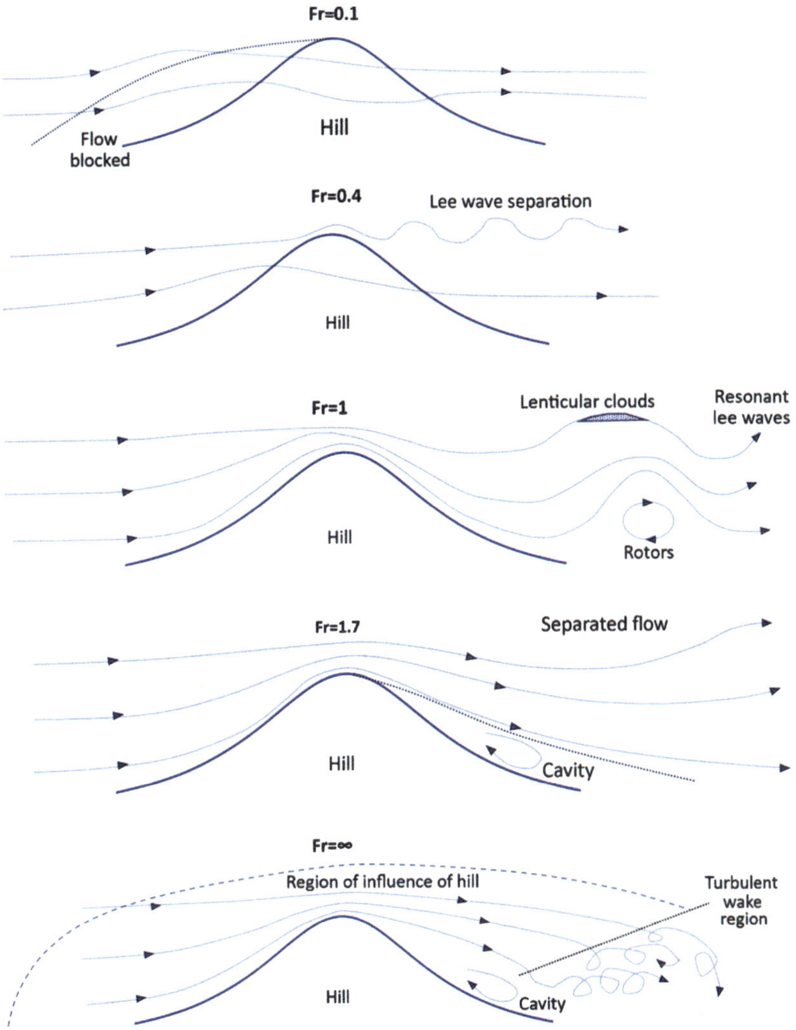

Figure 5.15 Flow above and around a hill under different stability regimes (Froude number). *Source:* adapted from Stull (1988).

around the hill. As Fr is increased to 0.4, the stability is weaker. Here some of the flow is around the hill, but lee waves of a wavelength significantly smaller than the dimensions of the hill develop. At a value of $Fr = 1$, then resonant conditions occur, and strong vertical oscillations of the flow are seen downstream of the hill with a wavelength of the order of the dimensions of the hill. Strong rotors can also develop and the peaks in the lee waves (a particular type of mountain wave) are often visually apparent due to the condensation of water vapour brought from a lower level to a higher, cooler layer giving rise to lenticular (lens-like) clouds. Multiple waves can be seen downstream where there is a mountain chain creating rows of parallel clouds. Oscillations in the wind speed (and wind power) have been seen in wind farms downstream of mountain chains (Draxl et al., 2021; Wilczak et al., 2019), an example of which is shown in Figure 5.16. As conditions become even less stable ($Fr = 1.7$), the wavelength of perturbation over the hill increases and the flow starts to separate downstream of the hill (depending on the steepness of the leeward side of the hill) with reverse flow in the separation region (cavity). When conditions are neutral ($Fr = \infty$), the flow separation becomes more pronounced, and the lee of hill develops a turbulent wake. The hill gives rise to a region of influence above which the flow is not influenced by the orographic feature. Within this region above the hill, the flow is accelerated, as discussed in Section 5.2.2, though the maximum speed-up observed will be affected by the degree of flow separation in the leeside of the hill if the slope is above the critical value for separation to occur.

A further illustration of how stability affects the flow over terrain is shown in Figure 5.17, where a CFD simulation of the wind flow over two mountains (Chasseral and Mont Crosin) in north-western Switzerland is

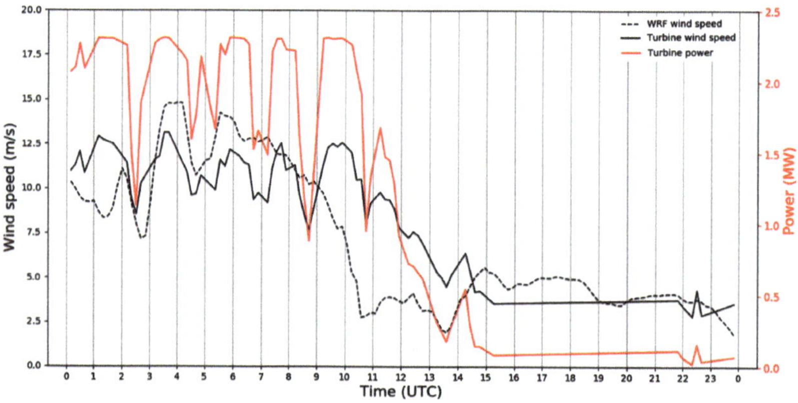

Figure 5.16 The effect of mountain waves observed in the wind speed and wind power from a wind turbine in a wind farm close to the Cascade Mountains near the Washington/Oregon border in the US. The dotted lines represent the wind speed at turbine hub height (80 m) simulated by the WRF mesoscale model, the solid line is the wind speed measured at the turbine hub height, and the red line in the corresponding turbine power output (Draxl et al., 2021). *Source:* Draxl et al. 2021 / Copernicus Publications / CC BY 4.0.

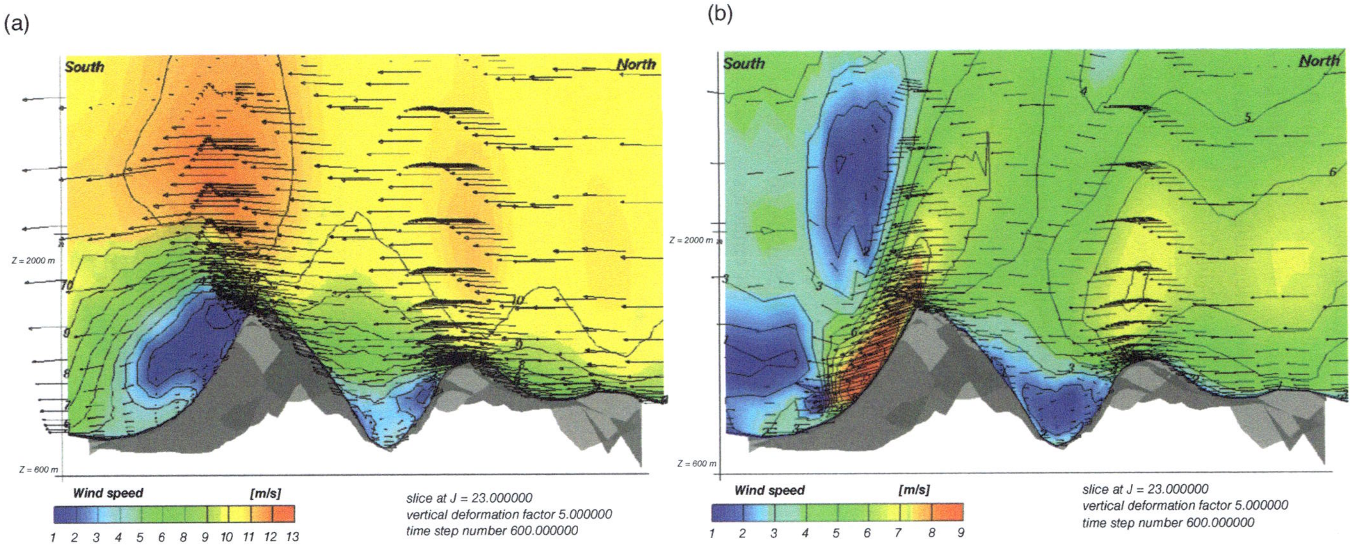

Figure 5.17 A CFD simulation of the wind flow over Chasseral (left mountain) and Mont Crosin (right mountain) in north-western Switzerland under (a) neutral and (b) stable conditions (Montavon, 1998b). *Source:* produced with permission of Christiane Montavon.

illustrated (Montavon, 1998b). The flow is from right to left. In Figure 5.17(a), neutral conditions are simulated. It can be seen that the flow has separated and the wake regions downwind of Chasseral and Mont Crosin are clearly shown. In Figure 5.17(b), conditions are stable with a Froude number close to a value of one. This gives rise to a strong mountain wave resonance causing a significant downdraft on the leeside of Chasseral. Indeed, the wind speed in this downdraft is around twice the value of the geostrophic wind speed used in the simulation. This clearly illustrates that, in complex orography, stability can have a profound effect on the wind flow and this needs to be factored into any simulation of the wind resource.

References

Adrian, G., Fiedler, F., 1991. Simulation of unstationary wind and temperature fields over complex terrain and comparison with observations. *Beitr. Zur Phys. D. Atmosphaere* 64, 27–48.

Angevine, W.M., Jiang, H., Mauritsen, T., 2010. Performance of an eddy diffusivity–mass flux scheme for shallow cumulus boundary layers. *Mon. Weather Rev.* 138, 2895–2912. https://doi.org/10.1175/2010MWR3142.1.

Ansys, n.d. *Computational fluid dynamics (CFD) simulation.* https://www.ansys.com/products/fluids (accessed 25 September 2020).

Anthes, R.A., Warner, T.T., 1978. Development of hydrodynamic models suitable for air pollution and other mesometerological studies. *Mon. Weather Rev.* 106, 1045–1078. https://doi.org/10.1175/1520-0493(1978)106<1045:DOHMSF>2.0.CO;2.

Beljaars, A.C.M., Walmsley, J.L., Taylor, P.A., 1987. A mixed spectral finite-difference model for neutrally stratified boundary-layer flow over roughness changes and topography. *Boundary-Layer Meteorol.* 38, 273–303. https://doi.org/10.1007/BF00122448.

Bougeault, P., Lacarrère, P., 1989. Parameterization of orography-induced turbulence in a mesobeta scale model. *Mon. Weather Rev.* 117, 1872–1890. https://doi.org/10.1175/1520-0493(1989)117<1872:POOITI>2.0.CO;2.

Bowen, A.J., Mortensen, N.G., 1996. Exploring the limits of WAsP the Wind Atlas Analysis and Application program, in: *Proceedings of the 1996 European Union Wind Energy Conference, 20–24 May 1996.* Gothenburg, Sweden. pp. 584–587.

Bowen, A.J., Mortensen, N.G., 2004. *WAsP prediction errors due to site orography,* Risoe-R No. 995(EN). https://orbit.dtu.dk/files/7711496/ris_r_995.pdf (accessed 4 October 2022).

Bretherton, C.S., Park, S., 2009. A new moist turbulence parameterization in the community atmosphere model. *J. Clim.* 22, 3422–3448. https://doi.org/10.1175/2008JCLI2556.1.

Burch, S.F., Ravenscroft, F., 1992. Computer modelling of the UK wind energy resource: final overview report. *Technical report ETSU-WN-7055.*

Businger, J.A., 1984. Equations and concepts, in: F.T.M. Nieuwstadt and H. van Doop (eds.) *Atmospheric Turbulence and Air Pollution Modelling.* Springer, Dordrecht, Netherlands, pp. 1–36. https://doi.org/10.1007/978-94-010-9112-1_1.

Cohen, A.E., Cavallo, S.M., Coniglio, M.C., Brooks, H.E., 2015. A review of planetary boundary layer parameterization schemes and their sensitivity in simulating southeastern US: cold season severe weather environments. *Weather Forecast.* 30, 591–612. https://doi.org/10.1175/WAF-D-14-00105.1.

Davis, C.G., Bunker, S.S., Mutschlecner, J.P., 1984. Atmospheric transport models for complex terrain. *J. Clim. Appl. Meteorol.* 23, 235–238. https://doi.org/10.1175/1520-0450(1984)023<0235:ATMFCT>2.0.CO;2.

Dimonte, G., Tipton, R., 2006. K–L turbulence model for the self-similar growth of the Rayleigh–Taylor and Richtmyer–Meshkov instabilities. *Phys. Fluids* 18(8), 085101. https://doi.org/10.1063/1.2219768.

Draxl, C., Hahmann, A.N., Peña, A., Giebel, G., 2014. Evaluating winds and vertical wind shear from Weather Research and Forecasting model forecasts using seven planetary boundary layer schemes. *Wind Energy* 17, 39–55. https://doi.org/10.1002/we.1555.

Draxl, C., Worsnop, R.P., Xia, G., Pichugina, Y., Chand, D., Lundquist, J.K., Sharp, J., Wedam, G., Wilczak, J.M., Berg, L.K., 2021. Mountain waves can impact wind power generation. *Wind Energy Sci.* 6, 45–60. https://doi.org/10.5194/WES-6-45-2021.

Elliott, W.P., 1958. The growth of the atmospheric internal boundary layer. *Eos, Trans. Am. Geophys. Union* 39, 1048–1054. https://doi.org/10.1029/TR039i006p01048.

ESMAP, 2018. *Guidance on mesoscale wind mapping.* ESMAP working paper. Washington, DC. World Bank.

Finardi, S., Brusasca, G., Morselli, M.G., Trombetti, F., Tampieri, F., 1993. Boundary-layer flow over analytical two-dimensional hills: a systematic comparison of different models with wind tunnel data. *Boundary-Layer Meteorol.* 63, 259–291. https://doi.org/10.1007/BF00710462.

Garratt, J.R., 1994. *The Atmospheric Boundary Layer.* Cambridge University Press.

Gravdahl, A., 1998. Meso scale modeling with a Reynolds averaged Navier–Stokes solver: assessment of wind resources along the Norwegian coast, in: *31st IEA Experts Meeting State of the Art on Wind Resource Estimation, Risoe, Denmark, October 1998.* IEA, Roskilde, Denmark. p. 16.

Grenier, H., Bretherton, C.S., 2001. A moist PBL parameterization for large-scale models and its application to subtropical cloud-topped marine boundary layers. *Mon. Weather Rev.* 129, 357–377. https://doi.org/10.1175/1520-0493(2001)129<0357:AMPPFL>2.0.CO;2.

Guo, X., Palutikof, J.P., 1990. A study of two mass-consistent models: problems and possible solutions. *Boundary-Layer Meteorol.* 53, 303–332. https://doi.org/10.1007/BF02186092.

Haupt, S.E., Kosovic, B., Shaw, W., Berg, L.K., Churchfield, M., Cline, J., Draxl, C., Ennis, B., Koo, E., Kotamarthi, R., Mazzaro, L., Mirocha, J., Moriarty, P., Muñoz-Esparza, D., Quon, E., Rai, R.K., Robinson, M., Sever, G., 2019. On bridging a modeling scale gap: mesoscale to microscale coupling for wind energy. *Bull. Am. Meteorol. Soc.* 100, 2533–2550. https://doi.org/10.1175/BAMS-D-18-0033.1.

Hong, S.-Y., Noh, Y., Dudhia, J., 2006. A new vertical diffusion package with an explicit treatment of entrainment processes. *Mon. Weather Rev.* 134, 2318–2341. https://doi.org/10.1175/MWR3199.1.

Hong, S.-Y., Pan, H.-L., 1996. Nonlocal boundary layer vertical diffusion in a medium-range forecast model. *Mon. Weather Rev.* 124, 2322–2339. https://doi.org/10.1175/1520-0493(1996)124<2322:NBLVDI>2.0.CO;2.

Jackson, P.S., Hunt, J.C.R., 1975. Turbulent wind flow over a low hill. *Q. J. R. Meteorol. Soc.* 101, 929–955. https://doi.org/10.1002/qj.49710143015.

Janjić, Z.I., 1990. The step-mountain coordinate: physical package. *Mon. Weather Rev.* 118, 1429–1443. https://doi.org/10.1175/1520-0493(1990)118<1429:TSMCPP>2.0.CO;2.

Janjić, Z.I., 1994. The step-mountain eta coordinate model: further developments of the convection, viscous sublayer, and turbulence closure schemes. *Mon. Weather Rev.* 122, 927–945. https://doi.org/10.1175/1520-0493(1994)122<0927:TSMECM>2.0.CO;2.

Kitada, T., Kaki, A., Ueda, H., Peters, L.K., 1983. Estimation of vertical air motion from limited horizontal wind data: a numerical experiment. *Atmos. Environ.* 17, 2181–2192. https://doi.org/10.1016/0004-6981(83)90214-7.

Launder, B.E., Spalding, D.B., 1974. The numerical computation of turbulent flows. *Comput. Methods Appl. Mech. Eng.* 3, 269–289. https://doi.org/10.1016/0045-7825(74)90029-2.

Mason, P.J., Sykes, R.I., 1979. Flow over an isolated hill of moderate slope. *Q. J. R. Meteorol. Soc.* 105, 383–395. https://doi.org/10.1002/qj.49710544405.

Mathur, R., Peters, L.K., 1990. Adjustment of wind fields for application in air pollution modeling. *Atmos. Environ. Part A. Gen. Top.* 24, 1095–1106. https://doi.org/10.1016/0960-1686(90).

Mellor, G.L., Yamada, T., 1974. A hierarchy of turbulence closure models for planetary boundary layers. *J. Atmos. Sci.* 31, 1791–1806. https://doi.org/10.1175/1520-0469(1974)031<1791:.

Mellor, G.L., Yamada, T., 1982. Development of a turbulence closure model for geophysical fluid problems. *Rev. Geophys.* 20, 851. https://doi.org/10.1029/RG020i004p00851.

Menter, F.R., 1994. Two-equation eddy-viscosity turbulence models for engineering applications. *AIAA J.* 32, 1598–1605. https://doi.org/10.2514/3.12149.

Montavon, C., 1998a. Validation of a non-hydrostatic numerical model to simulate stratified wind fields over complex topography. *J. Wind Eng. Ind. Aerodyn.* 74–76, 273–282. https://doi.org/10.1016/S0167-6105(98)00024-5.

Montavon, C., 1998b. Simulation of atmospheric flows over complex terrain for wind power potential assessment. *EPFL.* https://doi.org/10.5075/EPFL-THESIS-1855.

Mortensen, N.G., Tindal, A., Landberg, L., 2008. Field validation of the ΔRIX performance indicator for flow in complex terrain, in: *Proceedings of the European Wind Energy Conference and Exhibition, Brussels, Belgium, 31 Mar 2008–3 Apr 2008.* European Wind Energy Association, Brussels. p. 10.

Nakanishi, M., Niino, H., 2004. An improved Mellor–Yamada level-3 model with condensation physics: its design and verification. *Boundary-Layer Meteorol.* 112, 1–31. https://doi.org/10.1023/B:BOUN.0000020164.04146.98.

Nakanishi, M., Niino, H., 2006. An improved Mellor–Yamada level-3 model: its numerical stability and application to a regional prediction of advection fog. *Boundary-Layer Meteorol.* 119, 397–407. https://doi.org/10.1007/s10546-005-9030-8.

Phillips, G.T., 1979. *Preliminary User's Guide for the NOABL Objective Analysis Code: Special Report, 15 June 1977–15 June 1978. U.S. Department of Energy.* Richland, WA. https://doi.org/10.2172/7080538.

Pleim, J.E., 2007. A combined local and nonlocal closure model for the atmospheric boundary layer: part I: model description and testing. *J. Appl. Meteorol. Climatol.* 46, 1383–1395. https://doi.org/10.1175/JAM2539.1.

Prandtl, L. (1925). Bericht über Untersuchungen zur ausgebildeten Turbulenz. *Zeitschrift für Angewandte Mathematik und Mechanik* 5: 136–139. https://doi.org/10.1002/ZAMM.19250050212.

Ratto, C.F., 1996. The AIOLOS and WINDS codes, in: Lalas, D.P. *Modelling of Atmospheric Flow Fields.* World Scientific, pp. 421–431. https://doi.org/10.1142/9789814447164_0017.

Ratto, C.F., Festa, R., Romeo, C., Frumento, O.A., Galluzzi, M., 1994. Mass-consistent models for wind fields over complex terrain: the state of the art. *Environ. Softw.* 9, 247–268. https://doi.org/10.1016/0266-9838(94)90023-X.

Ross, D.G., Smith, I.N., Manins, P.C., Fox, D.G., 1988. Diagnostic wind field modeling for complex terrain: model development and testing. *J. Appl. Meteorol.* 27, 785–796. https://doi.org/10.1175/1520-0450(1988)027<0785:DWFMFC>2.0.CO;2.

Sherman, C.A., 1978. A mass-consistent model for wind fields over complex terrain. *J. Appl. Meteorol.* 17, 312–319. https://doi.org/10.1175/1520-0450(1978)017<0312:AMCMFW>2.0.CO;2.

Siemens, n.d. *STAR-CCM+: engineer innovation with multiphysics computational fluid dynamics (CFD) simulation.* https://www.plm.automation.siemens.com/global/en/products/simcenter/STAR-CCM.html (accessed 25 September 2020).

Skamarock, W.A., Klemp, J.B., Dudhia, J., Gill, D.O., Barker, D.M., Duda, M.G., Huang, X.-Y., Wang, W., Power, J.G., 2008. A description of the advanced research WRF version 3. *NCAR Technical Note NCAR/TN-475+STR*. University Corporation for Atmospheric Research. doi:10.5065/D68S4MVH. Boulder, CO.

Smagorinsky, J., 1963. General circulation experiments with the primitive equations. *Mon. Weather Rev.* 91, 99–164. https://doi.org/10.1175/1520-0493(1963)091<0099:GCEWTP>2.3.CO;2.

Sørensen, N.N., 1995. General purpose flow solver applied to flow over hills. *Technical Report Risø-R-827(EN).* Risø National Laboratory. Roskilde, Denmark.

National Renewable Energy Laboratory, 2020. *SOWFA: Simulator for Wind Farm Applications.* https://www.nrel.gov/wind/nwtc/sowfa.html (accessed 8 October 2020).

Stull, R., 1988. *An Introduction to Boundary Layer Meteorology*. Springer, Dordrecht, Netherlands. https://doi.org/10.1007/978-94-009-3027-8.

Sukoriansky, S., Galperin, B., Perov, V., 2005. Application of a new spectral theory of stably stratified turbulence to the atmospheric boundary layer over sea ice. *Boundary-Layer Meteorol.* 117, 231–257. https://doi.org/10.1007/s10546-004-6848-4.

Taylor, P.A., Lee, R.J., 1984. Simple guidelines for estimating wind speed variations due to small-scale topographic features. *Clim. Bull.* 18, 3–32.

Traci, R.M., Phillips, G.T., Patnaik, P.C., 1978. *Developing a Site Selection Methodology for Wind Energy Conversion Systems: Final Report, 15 June 1977–15 September 1978. U.S. Department of Energy*. Richland, WA. https://doi.org/10.2172/5049443.

Troen, I., 1990. A high resolution spectral model for flow in complex terrain, in: *Ninth Symposium on Turbulence and Diffusion, April 30–May 3, 1990.* Risoe National Laboratory, Roskilde, Denmark, pp. 417–420.

Troen, I., Lundtang Petersen, E., 1989. *European Wind Atlas.* Risø National Laboratory, Roskilde, Denmark.

Versteeg, H.K., Malalasekera, W., 2007. *An Introduction to Computational Fluid Dynamics: The Finite Volume Method.* Pearson Education.

Walmsley, J.L., 1989. Internal boundary-layer height formulae: a comparison with atmospheric data. *Boundary-Layer Meteorol.* 47, 251–262. https://doi.org/10.1007/BF00122332.

Walmsley, J.L., Taylor, P.A., Keith, T., 1986. A simple model of neutrally stratified boundary-layer flow over complex terrain with surface roughness modulations (MS3DJH/3R). *Boundary-Layer Meteorol.* 36, 157–186. https://doi.org/10.1007/BF00117466.

Walmsley, J.L., Taylor, P.A., Salmon, J.R., 1989. Simple guidelines for estimating wind speed variations due to small-scale topographic features: an update. *Climatol. Bull.* 23, 3–14.

Wilczak, J.M., Stoelinga, M., Berg, L.K., Sharp, J., Draxl, C., McCaffrey, K., Banta, R.M., Bianco, L., Djalalova, I., Lundquist, J.K., Muradyan, P., Choukulkar, A., Leo, L., Bonin, T., Pichugina, Y., Eckman, R., Long, C.N., Lantz, K., Worsnop, R.P., Bickford, J., Bodini, N., Chand, D., Clifton, A., Cline, J., Cook, D.R., Fernando, H.J.S., Friedrich, K., Krishnamurthy, R., Marquis, M., McCaa, J., Olson, J.B., Otarola-Bustos, S., Scott, G., Shaw, W.J., Wharton, S., White, A.B., 2019. The Second Wind Forecast Improvement Project (WFIP2): observational field campaign. *Bull. Am. Meteorol. Soc.* 100, 1701–1723. https://doi.org/10.1175/BAMS-D-18-0035.1.

Wyngaard, J.C., 2004. Toward numerical modeling in the 'terra incognita'. *J. Atmos. Sci.* 61, 1816–1826. https://doi.org/10.1175/1520-0469(2004)061 <1816:TNMITT>2.0.CO;2.

Zilitinkevich, S., Johansson, P.-E., Mironov, D., Baklanov, A., 1998. A similarity-theory model for wind profile and resistance law in stably stratified planetary boundary layers. *J. Wind Eng. Ind. Aerodyn.* 74–76, 209–218. https://doi.org/10.1016/S0167-6105(98)00018-X.

Zilitinkevich, S., Mironov, D.V., 1996. A multi-limit formulation for the equilibrium depth of a stably stratified boundary layer. *Boundary-Layer Meteorol.* 81, 325–351. https://doi.org/10.1007/BF02430334.

6

Wind Resource Estimation in Complex Terrain, Offshore, and in Urban Areas

Complex terrain is defined as terrain which has irregular orography, such as hills and mountains, particularly where the slopes are steep enough to cause flow separation. It also encompasses significant changes in land coverage and land/sea transitions. All these factors pose challenges when trying to predict wind resource, as rapid spatial changes in wind speed often result that are sensitive to small changes in topographic conditions and stratification. Determining the wind resource in urban areas represents a challenge due to the complex flow around buildings. This chapter looks at some empirical or semi-empirical models that can be used to estimate the resource in these situations.

6.1 Coastal and Offshore

The siting of wind farms in coastal areas and particularly offshore is becoming more widespread with the multiple attraction of higher average wind speeds, less variability, and lower levels of turbulence. However, predicting the coastal or offshore resource has particular challenges. Although the offshore environment is flat, the coast may well have complex terrain associated with it, and at the very least wind flow in coastal or offshore regions will involve the adjustment of the wind profile from a surface of one terrain coverage to that of another giving rise to an internal boundary layer (IBL). In Eq. (5.39), an expression is given for the IBL height following a roughness change which is dependent on distance and the up- and downwind roughness length. This formula can also be used in conditions of neutral stability to estimate the IBL height δ for a coastal transition. Figure 6.1 shows the growth of such an IBL with a wind blowing from onshore to offshore (an offshore wind) with an illustration of the wind profile at around 5 km from the coast. At this distance, the height of the IBL is around 150 m, which is within the vertical span of a typical modern wind turbine rotor disc. In principle, it is possible to estimate the form of the logarithmic profile above and below the IBL by

Handbook of Wind Resource Assessment, First Edition. Simon Watson.

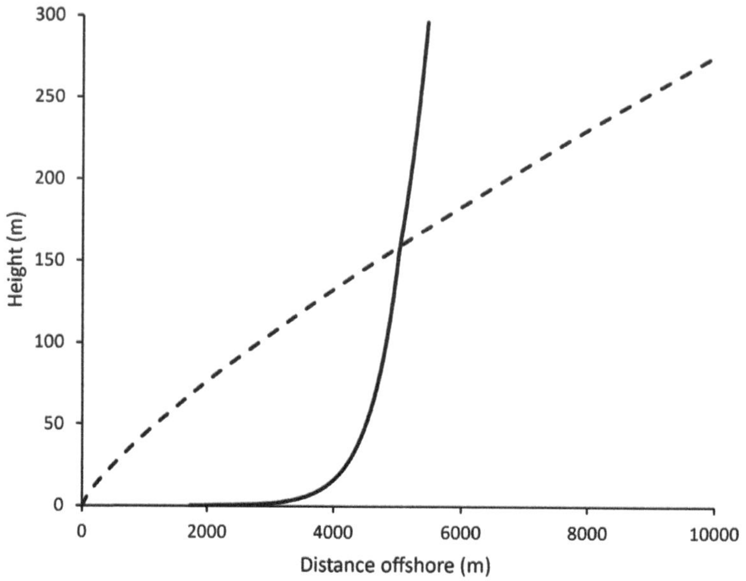

Figure 6.1 Growth of an IBL as wind blows offshore with an illustration of the wind speed profile at around 5 km offshore.

equating the friction velocity above the IBL, u_{*a} with that below the IBL, u_{*b} (Jensen, 1978)

$$\frac{u_{*a}}{u_{*b}} = \frac{\ln\left(\frac{\delta}{z_0}\right)}{\ln\left(\frac{\delta}{z_{0u}}\right)} \tag{6.1}$$

In which above the IBL

$$u(z) = \frac{u_{*a}}{\kappa}\ln\left(\frac{z}{z_{0u}}\right) \tag{6.2}$$

and below

$$u(z) = \frac{u_{*b}}{\kappa}\ln\left(\frac{z}{z_0}\right) \tag{6.3}$$

The drawback of this approach is that the estimation of the vertical profile is only as good as the estimate of δ, which will not be precise and can lead to a significant error in the shape of the profile. Such an approach should thus be used with caution when attempting to extrapolate wind speed measurements in the surface layer at distances from the coast where the height of the IBL is likely to be in the range of interest.

The development of the IBL is also dependent on the surface layer stability, which offshore, even at moderately high wind speeds, is frequently non-neutral, as mentioned in Section 2.2, and there have been attempts to parameterise this. One such empirical expression (Bergström et al., 1988) based on the Obukhov length L followed a numerical study and measurements close to the coast for an onshore wind

$$\delta_{\mathrm{IBL}} = 0.2x^{(0.78-3.63/L)} \tag{6.4}$$

where x is the distance from the coast. It should be noted that this expression has only been validated for an inland fetch of 1.5 km and there was a lot of scatter in the measured data from which this equation was inferred. For neutral conditions, the exponent term tends to 0.78 which is close to that in Eq. (5.39). It might therefore be assumed that a more general form of Eq. (6.4) for an arbitrary roughness change is (neglecting any stability dependence of the factor multiplying the power of x)

$$\delta_{\mathrm{IBL}} = z_0^{0.22} A_1 x^{(0.78-3.63/L)} \tag{6.5}$$

which for an offshore wind gives

$$\delta_{\mathrm{IBL}} = 0.15x^{(0.78-3.63/L)} \tag{6.6}$$

However, it should be noted that this has not been validated for offshore winds.

For very stable offshore conditions (e.g. $L = 50$ m) this would give an IBL height of around 150 m at 10 km from the coast (i.e. the same IBL height at a distance twice that of the neutral stability example above). This highlights the importance of measuring stability offshore in the assessment of wind resource and the challenges of extrapolating measurements from below hub (or rotor tip) heights, particularly for near coastal sites.

It can be expected that eventually the IBL height will reach an equilibrium at a certain distance from the coast. A more complicated expression for the IBL – which incorporates downstream roughness, downstream stability, and the height of the atmospheric boundary layer (ABL) h_{ABL} – is given by (Van Wijk et al., 1990)

$$\delta_{\mathrm{IBL}}\left[\ln(\delta_{\mathrm{IBL}}/z_0)-\psi_m(\delta_{\mathrm{IBL}}/L)-1\right]$$
$$\phi_m(\delta_{\mathrm{IBL}}/4L)(1-\delta_{\mathrm{IBL}}/h_{\mathrm{ABL}})^{-1}=2\kappa^2 x \tag{6.7}$$

where ϕ_m and ψ_m are defined in Eqs. (2.13)–(2.16).

The height of the ABL under neutral or stable conditions can be assumed to be (Nieuwstadt, 1981)

$$\frac{h_{\mathrm{ABL}}}{L}=\frac{0.3\left(u_*/(fL)\right)}{1+1.9h_{\mathrm{ABL}}/L} \tag{6.8}$$

The estimation of the ABL height under unstable (convective) conditions is not so straightforward, but under these conditions the ABL height is high enough to have little influence on Eq. (6.7) and the neutral form of Eq. (6.8) ($L \to \infty$) gives acceptable values.

A comparison between Eqs. (6.4) and (6.7) shows relatively good agreement for an onshore wind.

In the same way that the friction velocity above and below an IBL can be determined in neutral conditions (see Eq. 6.1), similarly, it can be determined in the case of an IBL under non-neutral conditions (with similar caveats as before in relation to the accuracy of the IBL height)

$$\frac{u_{*a}}{u_{*b}}=\frac{\ln\left(\frac{\delta_{\mathrm{IBL}}}{z_0}\right)-\psi_m(\delta_{\mathrm{IBL}}/L)}{\ln\left(\frac{\delta_{\mathrm{IBL}}}{z_{0u}}\right)-\psi_m(\delta_{\mathrm{IBL}}/L)} \tag{6.9}$$

So that, above the IBL

$$u(z)=\frac{u_{*a}}{\kappa}\left[\ln\left(\frac{z}{z_{0u}}\right)-\psi_m(z/L)\right] \tag{6.10}$$

and below

$$u(z)=\frac{u_{*b}}{\kappa}\left[\ln\left(\frac{z}{z_0}\right)-\psi_m(z/L)\right] \tag{6.11}$$

This presupposes that the potential temperature profile is in equilibrium above and below the IBL, which strictly speaking is not the case. In Van Wijk et al. (1990), downwind equilibrium values below the IBL were used for L and the wind profile determined showed reasonably good agreement with measurements.

Expressions have been developed for the height of the potential temperature IBL, δ_θ. This is analogous to the IBL defined in Eq. (5.39), which is strictly a momentum IBL and is referred to as a *thermal internal boundary layer* (TIBL). Just as the wind speed profile adjusts to a change in surface roughness, the TIBL adjusts to a change in surface temperature. This is illustrated in Figure 6.2. In (Figure 6.2a), an offshore wind from cold land to warm sea is illustrated giving rise to a TIBL and an evolving temperature profile. (Figure 6.2b) shows the case where the adjustment is from warm land to cold sea.

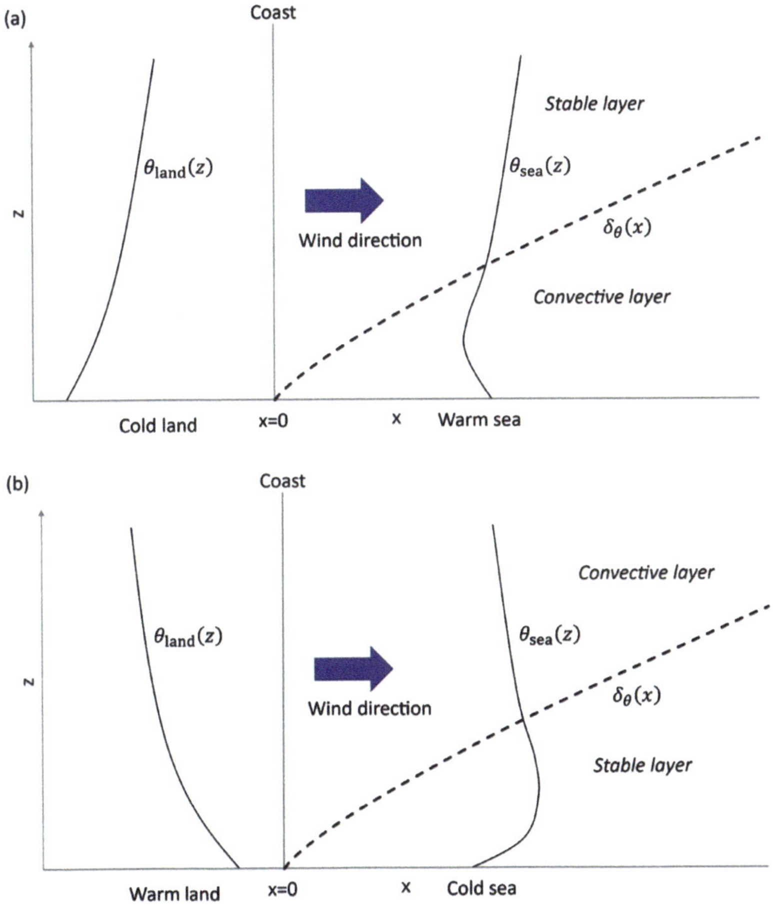

Figure 6.2 A TIBL and associated potential temperature profile for an offshore wind flowing from (a) cold land to warm sea and (b) warm land to cold sea.

Although the momentum IBL (δ_{IBL}) and the thermal internal boundary (δ_{TIBL}) are created by different mechanisms, they are related (Kaimal and Finnigan, 1994)

$$\frac{\delta_{\text{IBL}}}{\delta_{\text{TIBL}}} = (Pr_t)^{0.8} \tag{6.12}$$

where Pr_t is the turbulent Prandtl number (= 0.95). Thus, the ratio of the two IBLs is 0.96. For this reason, they can be treated as being practically the same, and expressions for the TIBL can be used in estimating the wind speed profile in the same way as for the IBL expressions.

One such expression for the TIBL at distances greater than a few kilometres from the coast for a *cold to warm* surface transition, resulting in a *convective internal boundary layer* (CIBL), is (Stunder and Sethuraman, 1985)

$$\delta_{\text{CIBL}} = \left[\frac{u_{*b}}{\bar{u}_b}\frac{\left(T_s - T_{su}\right)}{\partial T_u / \partial z}x\right]^{0.5} \tag{6.13}$$

where $\bar{u}_b$ is the average velocity in the convective layer below the CIBL height (assumed to be fairly constant with height as the layer is well mixed), T_s is the surface temperature downstream of the coastal discontinuity, T_{su} is surface temperature upstream of the coastal discontinuity, and $\partial T_u / \partial z$ is the upstream temperature profile.

An expression for a *warm to cold* surface transition, resulting in a *stable internal boundary layer* (SIBL) is (Mulhearn, 1981)

$$\delta_{\text{SIBL}} = 0.0146\left(\frac{g'}{\bar{u}_a^2}\right)^{-0.47} x^{0.53} \tag{6.14}$$

where

$$g' = g\frac{\Delta\rho}{\rho} \approx \frac{\theta_{su} - T_s}{\theta_s} \tag{6.15}$$

and where θ_{su} is the upstream surface potential temperature, θ_s is the downstream surface potential temperature, and $\bar{u}_a$ is the average wind speed in the well-mixed convective layer above the SIBL.

To estimate the profiles above and below a TIBL (Eqs. 6.10 and 6.11, respectively), some estimate of the potential temperature is required in order to estimate the stability term in the diabatic profile. As can be seen in

Figure 6.2, the shape of the profile above and below the TIBL is not straightforward to estimate, although there have been attempts to parameterise it. The best approach is to make temperature measurements at two heights both onshore and offshore or, failing that, to rely on the measurements at two heights at the site of interest and use the same value of L for both profiles following the approach of (Van Wijk et al., 1990).

Although it has been shown that it is possible to estimate the shape of the wind speed profile at a coastal or offshore site based on wind speed and temperature measurements, any estimate will carry a degree of uncertainty. The best that can be said is that if an estimate of the local IBL height lies close to the hub height of a potential wind turbine that is to be sited nearby then extrapolation of the wind profile in order to estimate the wind resource should be done with caution. Numerical modelling using a combination of mesoscale or microscale models may well give a better insight into the wind resource close to a coastal transition and measurements of the wind profile should, if possible, be made over a height range which would encompass the span of any prospective wind turbine rotor to be located at the site.

6.2 Forested Terrain

In many parts of the world, wind farms are sited in hilly or mountainous terrain to benefit from higher average wind speeds. Such terrain is also frequently forested, so prediction of the wind resource in and around forest canopies is important. One way to treat the effect of forestry on the wind resource is to treat it as a roughness length. Far away from the canopy this may yield satisfactory results but, given that wind farms are often placed in or close to areas of forest, this approach is not always appropriate. Above the canopy, a common approach is to treat the forest as an effective displacement to the wind profile as it flows over the terrain, so that the logarithmic profile over the canopy becomes

$$u(z) = \frac{u_*}{\kappa}\left[\ln\left(\frac{z-d}{z_0}\right) - \psi_m((z-d)/L)\right] \tag{6.16}$$

where d is an effective displacement height. This has a value somewhat less than the actual height of the canopy and is usually around 70% of the canopy height, h_c. As the canopy acts as a large thermal mass and provides a buffer between the air and the ground, this can lead to significant temperature stratification above the canopy and thus non-neutral conditions. It should be noted that the stability function ψ_m is less certain than over nonforested terrain below about $3h_c$ and is significantly affected by the canopy type. This suggests caution in the use of this profile function for non-neutral conditions.

A relatively simple estimate of d and z_0 can be made based on h_c and a canopy area index, Λ (Raupach, 1994)

$$d = h_c\left(1 - \left(\frac{1-\exp\left(-\sqrt{c_{d1}\Lambda}\right)}{\sqrt{c_{d1}\Lambda}}\right)\right) \tag{6.17}$$

$$z_0 = (h_c - d)\exp(-\kappa\gamma - \psi_h) \tag{6.18}$$

where

$$\gamma^{-1} = \frac{u_*}{u(h_c)} = \min\left[(C_S + C_R\Lambda/2)^{0.5}, (u_*/u(h_c)_{\max})\right] \tag{6.19}$$

The parameter ψ_h is the roughness-sublayer influence function, C_S and C_R are drag coefficients which relate the ground and the roughness elements of the canopy, and c_{d1} is a constant. Reasonably accurate values for d and z_0 are obtained when $c_{d1} = 7.5$, $C_S = 0.003$, $C_R = 0.3$, $(u_*/u(h_c)_{\max}) = 0.3$, and $\psi_h = 0.193$.

The canopy area index, assuming isotropic distribution of the roughness elements, can be related to a *frontal area index* λ, which is defined as the frontal area of the roughness elements of interest (i.e. trees) in the prevailing wind direction divided by the unit area of the roughness elements (i.e. the forest)

$$\Lambda = 2\lambda \tag{6.20}$$

To a reasonable approximation, when trees are coniferous or in full leaf, Λ can also be related to A_I, the *leaf area index* (LAI), which is defined as the total leaf area in the direction of interest divided by the floor area of the canopy and is a metric commonly used to parameterise canopies.

A later study (Nakai et al., 2008) proposes a model which is found to be more accurate in predicting d and z_0 based on h_c, A_I, and the *stand density* ρ_s defined as the number of trees per hectare

$$d = h_c\left[1 - \frac{\left(1-\exp\left(-\alpha\rho_s\right)\right)\left(1-\exp\left(-\beta A_I\right)\right)}{\alpha\beta\rho_s A_I}\right] \tag{6.21}$$

$$z_0 = 0.264 h_c \left(1 - \frac{d}{h_c}\right) \tag{6.22}$$

where $\alpha = 0.000724$ and $\beta = 0.273$. The formulas have been validated for stand densities between about 800 and 4000 trees per hectare, LAIs between about 0.5 and 6.25 and tree heights between around 9 m to 25 m.

In areas close to, though not in the forest canopy, the effect of the forest on the wind flow becomes more complex and a numerical model is required to properly characterise the wind flow. Because the forest acts as a momentum sink for the wind profile impinging on the canopy, a common approach in numerical modelling is to incorporate source terms in the momentum and turbulence equations using a Reynolds averaged Navier–Stokes (RANS) computational fluid dynamics (CFD) solver with, for example, $k-\varepsilon$ or $k-\omega$ closure models. The general form of these terms is:

In the momentum equations (a canopy drag force)

$$S_{M_i} = -\rho C_d A_D(z) \left|\bar{u}\right| \bar{u}_i \tag{6.23}$$

where C_d is the drag coefficient of the canopy, $A_D(z)$ is the *leaf area density* (LAD), $\left|\bar{u}\right|$ is magnitude of the wind speed and $\bar{u}_i$ is the component of the wind speed $(i = x, y, z)$. The canopy drag coefficient depends on the wind speed and based on wind tunnel measurements of simulated spruce trees (Stacey et al., 1994) was found to have the form

$$C_d = 0.04257 \exp(-0.0009779 \bar{u}^2) \tag{6.24}$$

The LAD is related to the LAI through the equation

$$A_I = \int_0^{h_c} A_D(z) dz \tag{6.25}$$

The LAD reflects the vertical variation in the density of the foliage from the ground to the top of the canopy. If a measurement of the canopy height is available, for example from airborne lidar measurements,[1] LAD profiles can be estimated based on tree type. One such empirical form of the LAD uses two parameters p and q to determine the LAD (Bonan et al., 2018)

1 Note that in contrast to the use of lidar for measuring wind speed, reflections off the different elements of the forest canopy can be used to determine canopy height.

$$A_D(z) = \frac{A_I}{z} \frac{(z/h_c)^{p-1}(1-z/h_c)^{q-1}}{\mathrm{B}(p,q)} \tag{6.26}$$

where B is the beta function. Two example profiles are shown in Figure 6.3. The profiles shown tend to LAD values of zero close to the ground. It is common to add some form of stem area density below the main part area of the canopy to simulate the blockage effect of the trunks.

In the turbulent kinetic energy equation

$$S_k = \rho C_d A_D(z)|\bar{u}|(\beta_p|\bar{u}|^2 - \beta_d k) \tag{6.27}$$

In the turbulent dissipation rate equation

$$S_\varepsilon = \rho C_d A_D(z)|\bar{u}|\varepsilon\left(\frac{C_{\varepsilon 4}\beta_p|\bar{u}|^2}{k} - C_{\varepsilon 5}\beta_d\right) \tag{6.28}$$

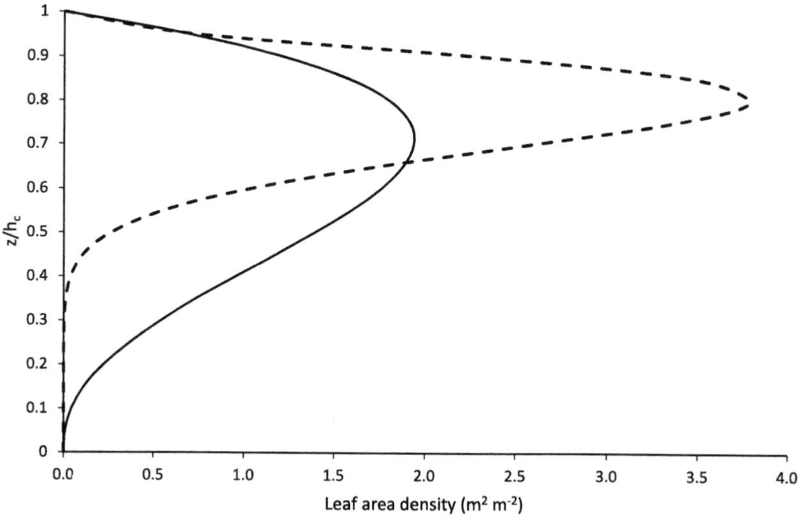

Figure 6.3 An empirical form of the LAD based on Eq. 6.26 as a function of height normalised by canopy height. The solid line is for deciduous and spruce trees ($p = 3.5$, $q = 2.0$) and the dashed line is for pine trees ($p = 11.5$, $q = 3.5$). The values are based on a value of $A_I/z = 1$. *Source:* adapted from Bonan et al. (2018).

Table 6.1 Model constants used by different researchers for the canopy model implemented in a RANS CFD with a $k-\varepsilon$ or $k-\omega$ closure model.

Model	β_p	β_d	$C_{\varepsilon 4}$	$C_{\varepsilon 5}$	Reference
Svensson and Häggkvist	1.0	0.0	1.95	0.0	(Svensson and Häggkvist, 1990)
Lopes da Costa	0.17	3.37	0.9	0.9	(Lopes da Costa, 2007)
Katul et al.	1.0	5.1	0.9	0.9	(Katul et al., 2004)

In the turbulent eddy frequency equation

$$S_\omega = \rho C_d A_D(z)|\bar{u}|\omega\left(\frac{(C_{\varepsilon 4}-1)\beta_p|\bar{u}|^2}{k}-(C_{\varepsilon 5}-1)\beta_d\right) \tag{6.29}$$

where β_p, β_d, $C_{\varepsilon 4}$, and $C_{\varepsilon 5}$ are constants. Different researchers have used different model constants based on experimental data. These are listed in Table 6.1.

Figure 6.4 shows an example where a RANS CFD solver using the shear stress turbulence (SST) closure model with source terms described

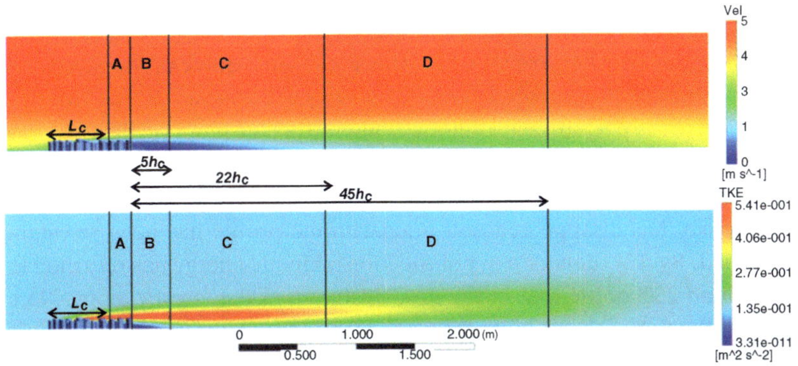

Figure 6.4 Contour plots of streamwise mean wind speed (top) and turbulent kinetic energy (bottom) for wind flowing from left to right on a plane through the centre of a forest canopy above and downstream of the canopy using a RANS CFD model with the SST turbulence closure model under conditions of neutral stability. Trees are represented as cylinders of several different tree sizes and LAD. The canopy extends $8.9h_c$ in both the streamwise and lateral directions. *Source:* reproduced with permission of Cian Desmond (Desmond, 2014).

in Eqs. (6.27–6.29) is used to simulate the 3D wind flow above and downstream of a square forest canopy consisting of individual trees of different sizes and LAD (Desmond, 2014). The simulated flow was under conditions of neutral stability with surface roughness length upstream and downstream of the canopy equivalent to $z_0 = 0.2$ m. For the first part of the flow over the canopy, the wind adjusts until there is a balance between the downward flow of momentum, and that which is removed due to canopy drag. This is known as the *canopy adjustment length* L_c which is inversely proportional to the canopy drag coefficient and average LAD. The flow is then divided into a series of zones (Desmond, 2014; Lee, 2000):

A) Developed canopy profile: the flow has fully developed to the underlying canopy conditions. In some situations where the tree trunk spacing is sparse, a subcanopy jet of wind speed higher than in the canopy region can be observed. Above the canopy, high shear and high levels of turbulence are observed.
B) Quiet zone: the flow is in the immediate shelter zone of the forest having low speed below canopy height and high shear above, creating an expanding jet of turbulent flow at canopy height. The flow changes little in this zone as the flow has not yet begun to adjust to the change in downstream surface roughness. In this simulation the zone extends $5h_c$ but can persist between $4h_c$ and $7h_c$ (Lee, 2000). In some cases, the subcanopy jet can be visible in this zone.
C) Mixing zone: the strong wind shear causes significant mixing and spreading of the turbulent kinetic energy jet. The wind profile starts to adjust to the new surface roughness, but the profile is not yet logarithmic. This zone persists to around $22h_c$ downstream of the forest.
D) Re-equilibrium zone: any structure in the profile due to the canopy morphology has disappeared and a logarithmic profile has been re-established. By $45h_c$ downstream of the forest, kinetic energy has returned to around 25% of the that in the flow upstream of the forest canopy. Note that the flow has not quite fully adjusted to the downstream roughness by the time it has reached the end of the simulation domain (~$65h_c$ or >7 horizontal canopy lengths).

It should be stressed that the distances over which the different flow regimes persist depend on the downstream surface roughness. They will also be influenced by atmospheric stability. Conditions where the downstream surface roughness is relatively low under stable conditions will mean that the zones will persist far longer than where the surface roughness is high and conditions unstable. Nonetheless, the observations in the list give a good idea of how the forest canopy affects the downstream flow and the importance of measurements and numerical modelling to determine the wind resource.

6.3 Urban Areas

From a wind energy perspective, urban areas are representative of only a limited area for development. Such areas, by their very nature, are low wind speed areas and buildings create high levels of turbulence, neither of which is an ideal condition for wind turbines. Nonetheless, there is some interest in siting small turbines on top of or next to, or incorporating into, buildings.

The impact of isolated buildings may also be important, particularly when siting small machines to provide energy for industrial units or farms, for example. Figure 6.5 shows an example of a simulation of the flow over a simple square pitched roof building using a RANS CFD model. A cross-section of the wind blowing at 45° orientation relative to the building is shown. In the lee of the building is an area of reverse circulation as the flow separates at the apex of the roof. The wind flow does not fully reattach to the ground until around $5h_o$ downstream, where h_o is the height of the

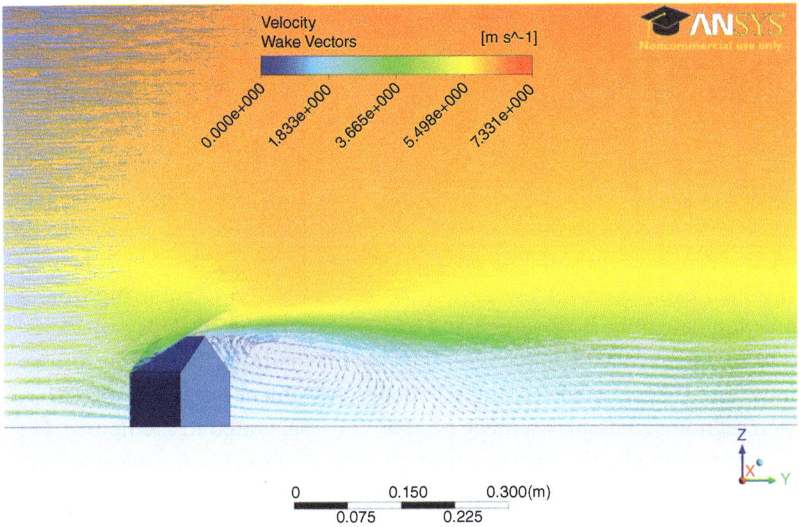

Figure 6.5 RANS CFD simulation of the wind flow over an isolated square pitched roof building with the wind approaching at an angle of 45° to the building.

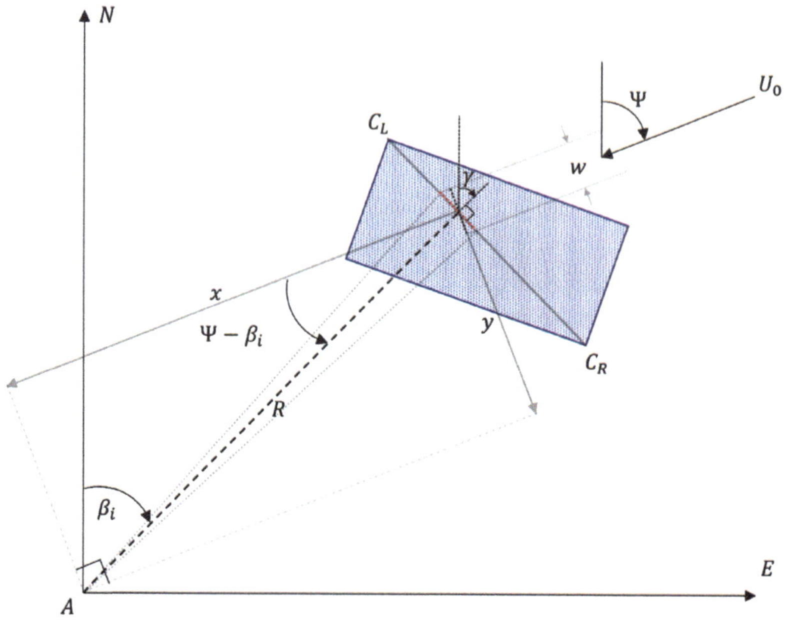

Figure 6.6 Plan view of the sheltering effect of a segment (in red) of an obstacle shaded in blue as seen at location A with a wind direction ψ. The segment is assumed to act as a windbreak of width *w* perpendicular to the wind direction. *Source:* adapted from Taylor and Salmon (1993).

obstacle (from ground to roof apex in this case). This point onwards is referred to as the *far wake*. There have been several studies to develop empirical models of the wind profile in this far wake region. One such study (Taylor and Salmon, 1993) considers a Gaussian wake behind an obstacle extending an empirical model based on the sheltering effect of porous 2D fences (Perera, 1981). Based on the geometry shown in Figure 6.6, an empirical expression for the reduction in wind speed U_o blowing from a direction ψ can be determined at a location A. The obstacle is considered as M segments (one of which is marked in red) spanning its extreme dimensions from C_L to C_R. The central point of each segment i is at a bearing of β_i. Each segment subtends an angle of 1° at A (exaggerated in the figure). For any given segment, the distance from the segment to A is R projected onto the x and y axes where x is aligned to the wind direction and y is cross-wind. Each segment is assumed to act as a windbreak of width w perpendicular to the wind direction.

The velocity reduction $u_{n,i}$ normalised to the upstream wind speed at the height of the obstacle $U_0(h_o)$ due to segment i for a given obstacle n is then given by

$$\frac{u_{n,i}}{U_0(h_o)} = \Gamma \tilde{C}_h \left(\frac{w}{h_o}\right)\left(\frac{x}{h_o}\right)^{-1.5} F(\eta)G(\zeta) \tag{6.30}$$

where

$$\eta = (y/h_o)(x/h_o)^{-\frac{1}{2}} \tag{6.31}$$

$$\zeta = (z/h_o)(x/h_o)^{-\frac{1}{2}} \tag{6.32}$$

$$F(\eta) = \frac{1}{(2\pi)^{\frac{1}{2}} a_f} \exp\left(\frac{-\eta^2}{2a_f^2}\right) \tag{6.33}$$

$$G(\zeta) = c_a \zeta \exp(-a_g \zeta^{1.5}) \tag{6.34}$$

$$c_a = \left\{\frac{\ln\left[(h_o + z_0)/z_0\right]}{2\kappa^2}\right\}^{\frac{1}{2}} \tag{6.35}$$

$$a_g = 0.67 c_a^{1.5} \tag{6.36}$$

Γ and a_f are empirical constants (set to 12.2 and 0.5, respectively) and $\tilde{C}_h$ depends on obstacle type ranging in value from 0.2 to 0.35 where the lower end is more appropriate for pitched roof buildings. To estimate the relative wind speed reduction based on an upstream wind speed at an arbitrary height z_A then

$$s_{n,i} = \frac{u_{n,i}}{U_0(z_A)} = \frac{u_{n,i}}{U_0(h_o)} \frac{\ln\left[(h_o + z_0)/z_0\right]}{\ln\left[(z_A + z_0)/z_0\right]} \tag{6.37}$$

To calculate the overall relative wind speed reduction seen at A caused by the obstacle n when the wind is blowing from the direction ψ then the calculation is repeated and summed up over all M segments

$$S_n = \sum_{i=1}^{M} s_{n,i} \tag{6.38}$$

This obstacle model has been validated against wind tunnel and full-scale measurements and was found to produce good results for $x \geq 3h_o$. In principle, the shelter effect of multiple upstream obstacles can be determined by a linear superposition of the relative wind speed reductions caused by each obstacle. However, this may lead to an overestimate drawing on the experience of combining the effect of multiple wind turbine wakes, which has been widely studied and where combinations of relative wake loss deficits are first squared, then summed and the square root of the total calculated (Archer et al., 2018).

Where a wind turbine is to be located on top of a building, an estimate of the wind resource needs to be made to take account of the flow over the roof, which can result in flow acceleration in some cases. Often, the building in question is within an urban area with other surrounding buildings. This requires, first, that an estimate is made of the urban wind profile (Macdonald et al., 1998). Above the mean building height H, this is a form of the logarithmic profile adjusted for a displacement height d

$$u(z) = \frac{u_*}{\kappa} \ln\left(\frac{z-d}{z_0}\right) \tag{6.39}$$

Below the mean building height, individual buildings will have a significant local effect on the wind profile. However, the mean urban wind speed profile has an approximately exponential profile

$$u(z) = u(H)\exp(a(z/H-1)) \tag{6.40}$$

where a depends on the building morphology and can be approximated by $a = 9.6\lambda_f$, where λ_f is the frontal area density defined as the total frontal surface area of the buildings (as seen by the wind) divided by the total land area, and is analogous to λ defined in Section 6.1.2 for forest canopies. Equations (6.39) and (6.40) should match at $z = H$.

The urban surface roughness length and displacement height can be estimated from the building morphology

$$\frac{z_0}{H} = \left(1 - \frac{d}{H}\right) \exp\left[-\left(0.5\beta \frac{C_D}{\kappa^2}\left(1 - \frac{d}{H}\right)\lambda_f\right)^{-0.5}\right] \tag{6.41}$$

$$\frac{d}{H} = 1 + A^{-\lambda_p}(\lambda_p - 1) \tag{6.42}$$

where λ_p is the *plan area density* (defined as the ratio of the total plan area of the buildings to the total overall land area covered by the buildings), C_D is the theoretical drag coefficient of the buildings, A is a constant which depends on the building layout, and β is a building drag correction factor based on realistic conditions. Based on Macdonald et al. (1998), values of $A = 4.4$, $\beta = 0.55$, and $C_D = 1.2$ are suggested.

Typically, an estimate is required of the friction velocity or more likely a reference wind speed. This might be done using a point wind speed value upwind of the urban area, u_u at a height z_u taken from a wind atlas or a reanalysis dataset which is representative of the extra-urban area. To adjust for the effect of the urban area, it is assumed that an urban IBL develops following the approach taken in Mertens (2006) treating the edge of an urban area as a roughness change. The reference wind speed can then be used to estimate the wind speed above the mean building height at an arbitrary distance downstream of the urban boundary modifying the approach in Taylor and Lee (1984) to take into account the displacement height

$$u(z) = \left(\frac{\ln\frac{(z-d)}{z_0}}{\ln\frac{(\delta-d)}{z_0}}\right) \bullet \left(\frac{\ln\frac{\delta}{z_{0u}}}{\ln\frac{z_u}{z_{0u}}}\right) u_u(z_u) \tag{6.43}$$

where δ is the urban IBL height defined in Eq. (5.39) and z_{0u} is the upwind surface roughness length.

The final step in determining the wind speed just above the roof level of a building is to modify the urban profile for the local effect of the building. It is not straightforward to do this. However, the approach in Heath et al. (2007) is to apply a correction factor based on CFD simulations of the flow around buildings. The correction factor $l_{m,d}$ is then multiplied by the urban wind profile value to give an estimate of the final wind speed value $u_{m,d}$ for the potential roof mounting point m of the wind turbine and the prevailing wind direction d of interest

$$u_{m,d}(z) = l_{m,d}\, u(z) \tag{6.44}$$

An example of the $l_{m,d}$ values for the optimum roof mounting points for different wind directions for a small wind turbine installed on a pitched roof building embedded within a series of similar buildings (e.g. a residential house) is shown in Table 6.2. Mounting positions are illustrated in Figure 6.7.

Table 6.2 Optimum roof mounting points for a small wind turbine installed at three different heights on a pitched roof building and the associated wind speed correction factors to the urban profile, $l_{m,d}$. Optimum mounting positions are given for different wind directions and for a situation where the wind distribution is equal from all directions.

	Mounting height, z/H		
Wind direction	**0.8**	**1.0**	**1.3**
North	South corners $l_n = 0.6$	South corners $l_n = 1.0$	South gable or centre $l_n = 1.3$
North-east	SE corner $l_{ne} = 0.6$	SE corner $l_{ne} = 0.9$	SE corner $l_{ne} = 1.1$
East	Gable ends $l_e = 0.3$	East front $l_e = 0.5$	East front or centre $l_e = 0.9$
South-east	NE corner $l_{se} = 0.6$	NE corner $l_{se} = 0.8$	NE corner $l_{se} = 1.1$
South	North corners $l_s = 0.6$	North corners $l_s = 1.0$	North gable or centre $l_s = 1.3$
South-west	NW corner $l_{sw} = 0.6$	NW corner $l_{sw} = 0.8$	NW corner $l_{sw} = 1.1$
West	Gable ends $l_w = 0.3$	West front $l_w = 0.5$	West front or centre $l_w = 0.9$
North-west	SW corner $l_{nw} = 0.6$	SW corner $l_{nw} = 0.8$	SW corner $l_{nw} = 1.1$
Equal distribution	Corners $l_{mean} = 0.5$	Corners $l_{mean} = 0.6$	East or west front $l_{mean} = 1.0$

Source: adapted from Heath et al. (2007).

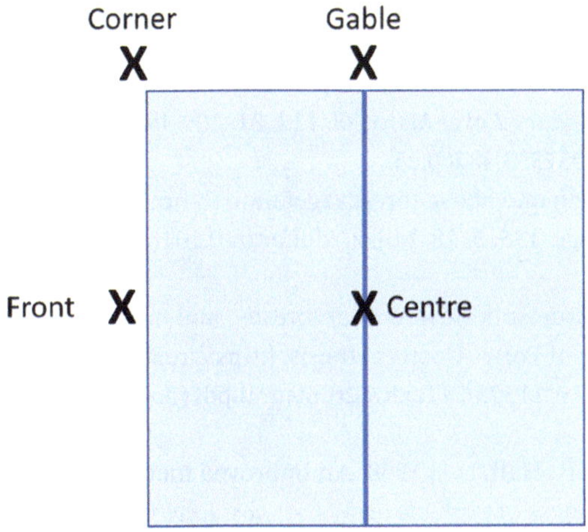

Figure 6.7 Plan view of mounting positions used to assess the wind resource on a pitched roof house.

References

Archer, C.L., Vasel-Be-Hagh, A., Yan, C., Wu, S., Pan, Y., Brodie, J.F., Maguire, A.E., 2018. Review and evaluation of wake loss models for wind energy applications. *Appl. Energy.* 226, 1187–1207. https://doi.org/10.1016/j.apenergy.2018.05.085.

Bergström, H., Johansson, P.-E., Smedman, A.-S., 1988. A study of wind speed modification and internal boundary-layer heights in a coastal region. *Boundary-Layer Meteorol.* 42, 313–335. https://doi.org/10.1007/BF00121589.

Bonan, G.B., Patton, E.G., Harman, I.N., Oleson, K.W., Finnigan, J.J., Lu, Y., Burakowski, E.A., 2018. Modeling canopy-induced turbulence in the earth system: a unified parameterization of turbulent exchange within plant canopies and the roughness sublayer (CLM-ml v0). *Geosci. Model Dev.* 11, 1467–1496. https://doi.org/10.5194/gmd-11-1467-2018.

Desmond, C.J., 2014. The consideration of forestry effects in wind energy resource assessment. Thesis. Loughborough University.

Heath, M.A., Walshe, J.D., Watson, S.J., 2007. Estimating the potential yield of small building-mounted wind turbines. *Wind Energy* 10, 271–287. https://doi.org/10.1002/we.222.

Jensen, N.O., 1978. Change of surface roughness and the planetary boundary layer. *Q. J. R. Meteorol. Soc.* 104, 351–356. https://doi.org/10.1002/qj.49710444009.

Kaimal, J.C., Finnigan, J.J., 1994. *Atmospheric Boundary Layer Flows: Their Structure and Measurement.* Oxford University Press.

Katul, G.G., Mahrt, L., Poggi, D., Sanz, C., 2004. One- and two-equation models for canopy turbulence. *Boundary-Layer Meteorol.* 113, 81–109. https://doi.org/10.1023/B:BOUN.0000037333.48760.e5.

Lee, X., 2000. Air motion within and above forest vegetation in non-ideal conditions. *For. Ecol. Manage.* 135, 3–18. https://doi.org/10.1016/S0378-1127(00)00294-2.

Lopes da Costa, J.C.P., 2007. Atmospheric flow over forested and non-forested complex terrain. University of Porto. Doctoral thesis. https://repositorio-aberto.up.pt/bitstream/10216/11375/2/Texto%20integral.pdf (accessed 4 October 2022).

Macdonald, R.W., Griffiths, R.F., Hall, D.J., 1998. An improved method for the estimation of surface roughness of obstacle arrays. *Atmos. Environ.* 32, 1857–1864. https://doi.org/10.1016/S1352-2310(97)00403-2.

Mertens, S., 2006. *Wind Energy in the Built Environment.* Multi-Science.

Mulhearn, P.J., 1981. On the formation of a stably stratified internal boundary-layer by advection of warm air over a cooler sea. *Boundary-Layer Meteorol.* 21, 247–254. https://doi.org/10.1007/BF02033942.

Nakai, T., Sumida, A., Daikoku, K., Matsumoto, K., van der Molen, M.K., Kodama, Y., Kononov, A.V., Maximov, T.C., Dolman, A.J., Yabuki, H., Hara, T., Ohta, T., 2008. Parameterisation of aerodynamic roughness over boreal, cool- and warm-temperate forests. *Agric. For. Meteorol.* 148, 1916–1925. https://doi.org/10.1016/j.agrformet.2008.03.009.

Nieuwstadt, F.T.M., 1981. The steady-state height and resistance laws of the nocturnal boundary layer: theory compared with cabauw observations. *Boundary-Layer Meteorol.* 20, 3–17. https://doi.org/10.1007/BF00119920.

Perera, M.D.A.E.S., 1981. Shelter behind two-dimensional solid and porous fences. *J. Wind Eng. Ind. Aerodyn.* 8, 93–104. https://doi.org/10.1016/0167-6105(81)90010-6.

Raupach, M.R., 1994. Simplified expressions for vegetation roughness length and zero-plane displacement as functions of canopy height and area index. *Boundary-Layer Meteorol.* 71, 211–216. https://doi.org/10.1007/BF00709229.

Stacey, G.R., Belcher, R.E., Wood, C.J., Gardiner, B.A., 1994. Wind flows and forces in a model spruce forest. *Boundary-Layer Meteorol.* 69, 311–334. https://doi.org/10.1007/BF00708860.

Stunder, M., Sethuraman, S., 1985. A comparative evaluation of the coastal internal boundary-layer height equations. *Boundary-Layer Meteorol.* 32, 177–204. https://doi.org/10.1007/BF00120934.

Svensson, U., Häggkvist, K., 1990. A two-equation turbulence model for canopy flows. *J. Wind Eng. Ind. Aerodyn.* 35, 201–211. https://doi.org/10.1016/0167-6105(90)90216-Y.

Taylor, P.A., Lee, R.J., 1984. Simple guidelines for estimating wind speed variations due to small-scale topographic features. *Clim. Bull.* 18, 3–32.

Taylor, P.A., Salmon, J.R., 1993. A model for the correction of surface wind data for sheltering by upwind obstacles. *J. Appl. Meteorol.* 32, 1683–1694. https://doi.org/10.1175/1520-0450(1993)032<1683:AMFTCO>2.0.CO;2.

Van Wijk, A.J.M., Beljaars, A.C.M., Holtslag, A.A.M., Turkenburg, W.C., 1990. Diabatic wind speed profiles in coastal regions: comparison of an internal boundary layer (IBL) model with observations. *Boundary-Layer Meteorol.* 51, 49–75. https://doi.org/10.1007/BF00120461.

7

Orographic Test Cases

Steep terrain represents a particular challenge in terms of predicting wind resource, as is alluded to in the earlier sections of Chapter 5. Much research has been carried out to assess the ability of numerical models and scale models in wind tunnels to simulate wind flow over terrain with steep slopes where flow separation is likely to occur. To provide comprehensive data to assess the accuracy of these approaches, several test cases have been studied where extensive measurements have been made at locations situated on top of and surrounding relatively isolated terrain features. Four well-known test cases and two lesser-known test cases are described in this chapter.

7.1 Askervein Hill

One of the first campaigns to make comprehensive measurements of mean wind speed and turbulence over a hill was set-up on Askervein Hill on the Scottish island of South Uist in the Outer Hebrides (Taylor and Teunissen, 1987). Other smaller measurement campaigns had been taken place on other hills, but the campaign which took place on Askervein Hill was significantly more extensive. This international campaign was part of the International Energy Agency's (IEA) Wind Task VI to help the embryonic wind energy industry of the time develop and test siting tools such as wind flow models. The hill was chosen as it was reasonably symmetric (an approximately elliptical shape), isolated, and of uniform surface coverage (estimated to correspond to a roughness length of 0.03 m). It was also relatively accessible. Mean wind speed and turbulence measurements were made on a series of masts laid out along three lines across the hill. The contours of the hill and the position of the lines are shown in Figure 7.1 and the vertical profiles of the lines are shown in Figure 7.2.

Handbook of Wind Resource Assessment, First Edition. Simon Watson.

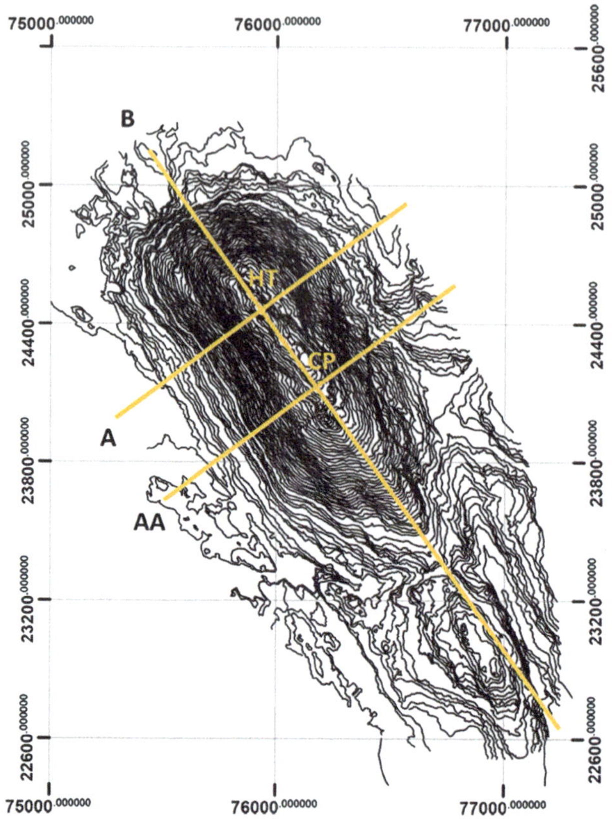

Figure 7.1 A contour map of Askervein Hill. Wind measurements were made along lines A, AA, and B. Lines A and B intersect at the hilltop (HT) (126 m above sea level). Lines AA and B intersect at the centre point of the hill (CP). Reference measurements were made at a mast some distance away in relatively flat terrain at a point labelled RS (not shown on this map). Horizontal units are in metres and the contours are every two metres (with some additional one-metre contours).

The first line (marked *B*) ran along the major axis of the hill passing through the highest point of the hill, 126 m above sea level (HT, or hilltop) and the centre point of the hill (CP). The second and third lines, denoted *A* and *AA*, ran parallel to the minor axis of the hill passing through HT and CP, respectively. Some distance from the hill, a mast was sited in relatively flat terrain to provide reference wind speed measurements (at RS, reference site). The majority of the towers were 10 m high with additional towers of 16 m, 30 m, and 50 m at key locations. Tethered aerodynamic lifting anemometer (TALA) kites were used to make some limited measurements up to 500 m above ground. Wind speed measurements were made using wind

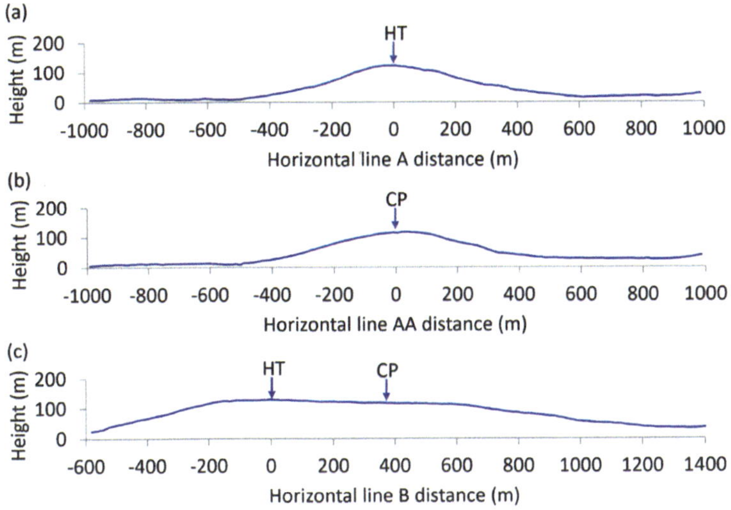

Figure 7.2 Askervein Hill profiles along (a) line A (south-west to north-east), (b) line AA (south-west to north-east), and (c) line B (north-west to south-east).

vanes and cup anemometers with a smaller number of sonic anemometers for turbulence measurements. Measurements of pressure, temperature, and humidity were also made. Two measurement campaigns were made: (i) 20 September to 3 October 1982 and (ii) 14 September to 18 October 1983. Based on radiosonde measurements, the majority of the wind speed measurements were made in neutral or near-neutral conditions.

The measurements made during these campaigns have since been used extensively to validate the accuracy of mass-consistent, linearised, and CFD models. Data have also been compared to scale model experiments made in wind tunnels. A comparison of three wind tunnel scale model measurements is shown in Figure 7.3 (Teunissen et al., 1987).

Measurements were made for three different scale models tested in two different wind tunnels. The scale models (of different sizes) were:

- AES: a smooth surface model from the Atmospheric Environment Service, Canada tested in their own tunnel.
- NZS: a smooth surface model from the University of Canterbury, New Zealand tested in the AES wind tunnel.
- NZR: as NZS but the model had a rough surface representative of that of the hill.
- OXR: a rough surface model from the University of Oxford, UK tested in their own wind tunnel.

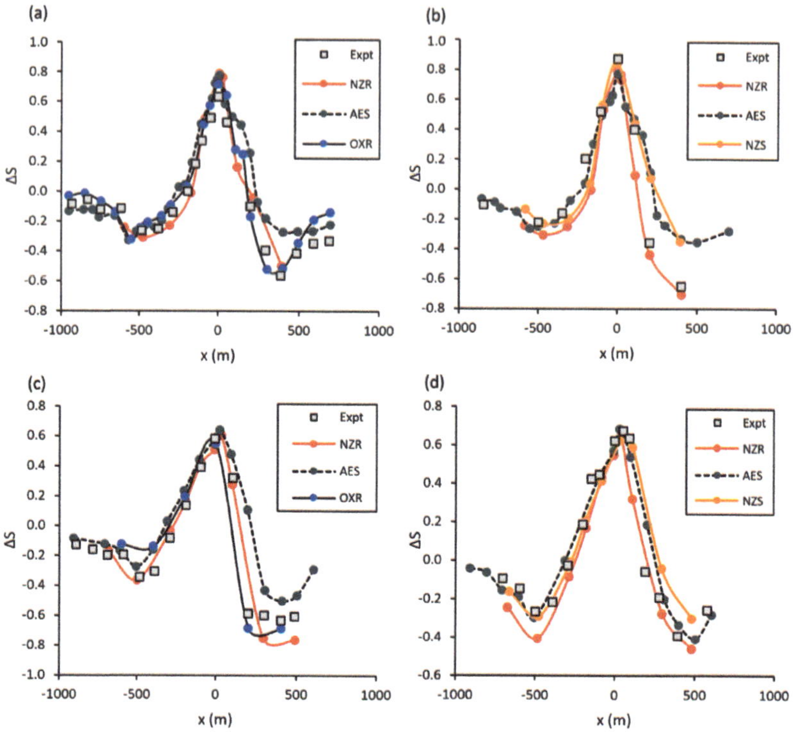

Figure 7.3 Comparisons between wind tunnel scale model measurements (labelled AES, NZR, NZS, OXR) and full-scale field measurements (marked *Expt*) over Askervein Hill at the full-scale equivalent of 10 m above ground level: (a) along line A for a wind direction at RS of 235°, (b) along line A for a wind direction at RS of 210°, (c) along line AA for a wind direction of 235°, (d) along line AA for a wind direction of 210°. *Source:* from Teunissen et al., (1987).

Wind tunnel measurements were compared with the full-scale field measurements along lines A and AA for two cases during the field measurement campaign: a wind direction at RS 210° and 235°. The measurements were converted into fractional speed-up values (ΔS) relative to the RS. In general, the different scale models showed good agreement with the models, except downstream of the hill in the smooth model cases. These cases overestimated the leeside speed-up, indicating the importance of accounting for surface roughness when attempting to predict wind speed changes over complex orography.

Figure 7.4 shows a comparison between measurements made along line A and wind flow models of varying fidelity. Upwind of the hill, all models perform reasonably well, though the LES model performs best. At the HT, the LES model predicts the maximum speed-up value well, and the mass-consistent

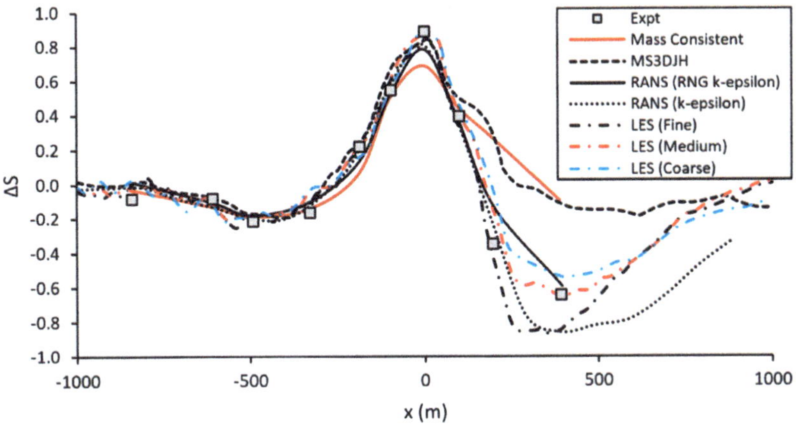

Figure 7.4 Comparison between speed-up values at 10 m above ground level measured on top of the hill ΔS along line A relative to the mast at RS when the wind direction was 210° (marked *Expt*) with several models: a mass consistent model (Forthofer et al., 2014); the linearised MS3DJH model (Beljaars et al., 1987); a RANS CFD model with RNG (renormalisation group) $k - \varepsilon$ turbulence closure (Forthofer et al., 2014); a RANS CFD model with the standard $k - \varepsilon$ turbulence closure (Castro et al., 2003); and an LES model run with three different grid resolutions marked *fine, medium,* and *coarse* (Silva Lopes et al., 2007). Point HT is at x = 0 m.

model significantly underpredicts this value. In the leeside of the hill, there is a significant divergence between the model results. The mass-consistent model and the linearised model MS3DJH significantly overpredict the speed-up. The Reynolds averaged Navier–Stokes (RANS) computational fluid dynamics CFD models provide a better prediction of the speed-up, and the LES model with the medium resolution mesh provides the best prediction overall.

The critical slope θ_{crit} at which flow separation in the lee of a hill is expected to occur under conditions of neutral stability can be estimated (Wood, 1995)

$$\theta_{\text{crit}} \approx \frac{\left(\ln\left(l / z_0\right)\right)^2}{\left(\ln\left(h_m / z_0\right)\right)^2 \left(1 + 4.2 / \ln\left(l / z_0\right)\right)} \tag{7.1}$$

where l was defined earlier in Eq. (5.33). and h_m is the height of an intermediate layer in an extended version of the original Jackson and Hunt (JH) theory where the outer layer is divided into two (Hunt et al., 1988)

$$h_m \approx L_W \left(\ln\left(L_W / z_0\right)\right)^{-\frac{1}{2}} \tag{7.2}$$

The maximum slope of Askervein Hill along line A is around 40%, which is just above the approximate critical value of around 35% for this hill. The results clearly show that the mass-consistent and linearised models fail to capture this separation, which is unsurprising given that both are only designed to be used for hills with a gentle slope. The RANS models better capture the separation but still show some divergence with the measured values. The LES model performs the best, though the degree to which it captures the flow separation is very dependent on the mesh resolution. It is interesting to note that the LES simulation with the finer-resolution grid provides, counterintuitively, a worse prediction of the leeside speed-up increasing the strength of the flow separation beyond that observed. This was assumed to be due to inaccurate capture of the intermittent nature of the separation occurring downwind of the hill (Silva Lopes et al., 2007).

The results show the challenge of modelling the wind resource in terrain with slopes where flow separation is likely to occur. It is also clear that the flow separation affects the speed-up which is seen at the HT, so mass-consistent and linearised models should be used with caution when trying to assess the maximum speed-up in such situations. Furthermore, although potentially the highest-fidelity LES models can provide the most accurate predictions, how they are applied, for example the mesh resolution used, can significantly affect their accuracy.

7.2 Bolund Hill

As mentioned in Section 7.1, the LES model results suggested that separation on the leeside of Askervein Hill was intermittent in nature most likely due to the fact that the slope only just fulfils that for separation to occur. A need for a further case study was apparent where the hill slope was steep enough to be certain to cause separation to provide data for the validation of nonlinear RANS and LES solvers. Bolund hill was chosen for this purpose. The site is a small hill attached to a peninsular jutting out into the Roskilde fjord near to the Danish Technical University campus at Risø and north of the town of Roskilde in Denmark. The hill is of uniform coverage (estimated to have a roughness length of 0.015 m) and has a relatively flat top and steep slopes in all directions, particularly on the western side. The fetch to the west is over water and provides ideal inflow conditions. A contour plot of the hill is shown in Figure 7.5. Because of its small size, variations in the wind flow due to atmospheric stability are considered negligible and neutral stability can be assumed. Ten small masts were erected to measure the flow in and around the hill. Eight were sited on two lines (A and B) on the hill

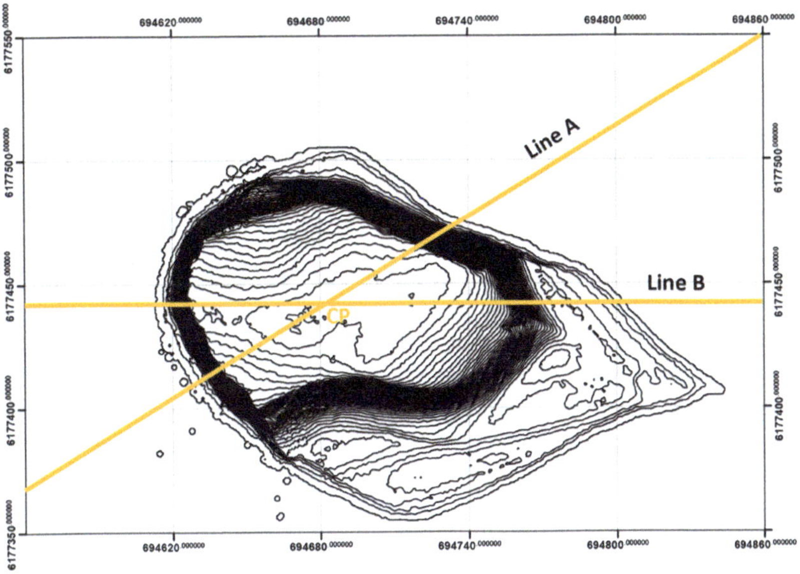

Figure 7.5 A contour map of Bolund hill. Wind measurements were made at several masts along lines A and B. The centre point of the hill (CP) is at the intersection of these two lines. Contours are every 0.25 m and the HT is 12 m above mean sea level.

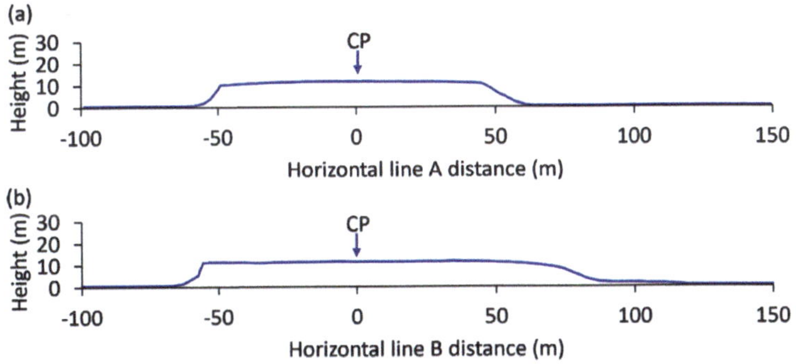

Figure 7.6 Bolund hill profiles along (a) line A (south-west to north-east) and (b) line B (west to east).

itself, one was placed in the fjord to measure the upstream flow west of the hill, and one was sited on a beach to the east. The profiles of the hill along lines A and B are shown in Figure 7.6. Measurements of wind speed and direction were made at several heights above ground level up to 10 m on most of the masts. Temperature measurements were also made to determine the stability conditions, which were found to be neutral or slightly stable for most of the time. The measurement campaign took place between 21 December 2007 and 28 February 2008. Details of the campaign are described in Berg et al. (2011).

The data collection campaign was used as the basis of a blind test comparison of microscale flow models (Bechmann et al., 2011), which is quite instructive to see how well the flow separation and flow recirculation can be captured both on the upwind and leeward sides of the hill. All model predictions were made on the basis of the upstream wind profile and the terrain details only so that no model tuning could be carried out. Over 50 model predictions were made by participants in the blind test, including a scale model wind tunnel test. Figure 7.7 shows a comparison between the speed-up factor (ΔS) relative to the upstream wind speed along line A for different model types as compared with the measurements. The models are coded by line type and colour, the wind tunnel model measurements are shown as 'x' symbols and the measurements on

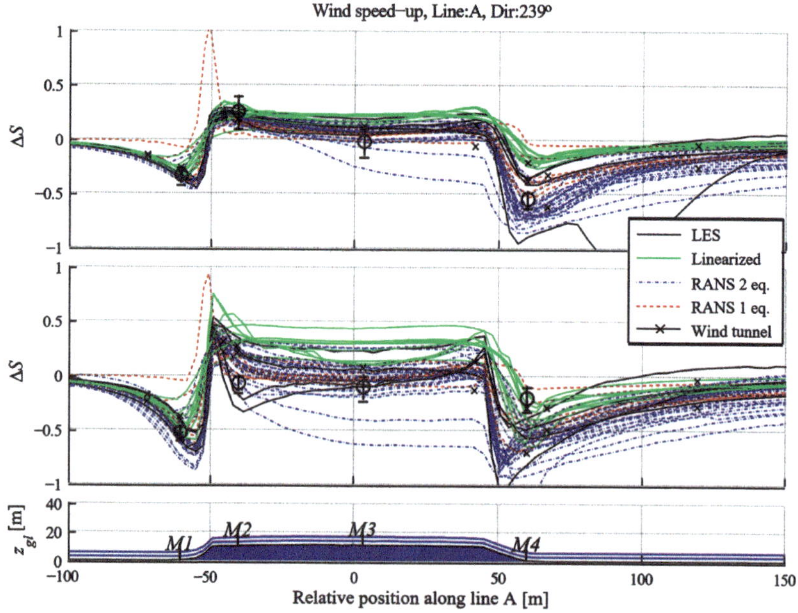

Figure 7.7 Bolund hill blind test comparison along line A with an upstream wind direction of 239°. Top plot is for measurements at 5 m above the ground, middle plot is at 2 m above the ground, and the bottom plot shows the position of the measurement masts along line A. *Source:* reproduced from Bechmann et al. (2011) / with permission of Springer Nature).

the masts are shown as open circles with error bars. The first thing to note is that there is a large degree of scatter in the results even for models of the same type. This shows that how a model is set up (mesh resolution, boundary conditions, etc.) is just as important as the type of model used. For the results at 5 m, all of the models capture the upstream blockage (and associated reduction in wind speed) on the upwind side of the hill. However, over the top of the hill, the model results start to deviate from the measurements. On the leeward side the deviation is even greater; indeed, even the wind tunnel measurements overestimate the speed-up. It can be seen that, as for Askervein Hill, the linearised models fail to capture the separation and overestimate the speed-up. The RANS models (particularly the two-equation turbulence models, e.g. $k-\varepsilon$) seem to give the best predictions of the speed-up in the separation region. Here, too, there is a significant variation in the LES model predictions, and one of these significantly overestimates the strength of the separation. The 2 m results show similar trends but with more scatter. The results reinforce the observations made for Askervein Hill, namely that, although the higher-fidelity LES models ought to provide the most accurate predictions, they are more difficult to set up compared with RANS models and should be used very carefully for wind resource assessment. It can also be seen that the RANS model predictions, although the closest to the measurements, are still in error to the tune of 10% on average, which is significant from the point of view of wind resource assessment.

7.3 Big Southern Butte (BSB) and Salmon River Canyon (SRC)

Although less well known than some other test cases, two sites were instrumented as part of a campaign to investigate complex flow phenomena in the presence of complex terrain features with thermally driven winds and large variations in surface layer stability (Butler et al., 2015). Wind speed and direction measurements sampled at 1 Hz and averaged every 30 s were made using a network of cup and vane anemometers at 3 m above ground level at each site. The measurements were supplemented by shorter-term sonic anemometer measurements, sodar wind profiles, and radiosonde ascents to give higher temporal resolution momentum and heat flux measurements.

This first of the two sites was an isolated volcanic cinder cone mountain situated above the upper Snake River Plain in south-eastern Idaho, known as *Big Southern Butte* (BSB). The mountain rises approximately 800 m above a flat plain with a long axis of around 4 km, as shown in Figure 7.8. The

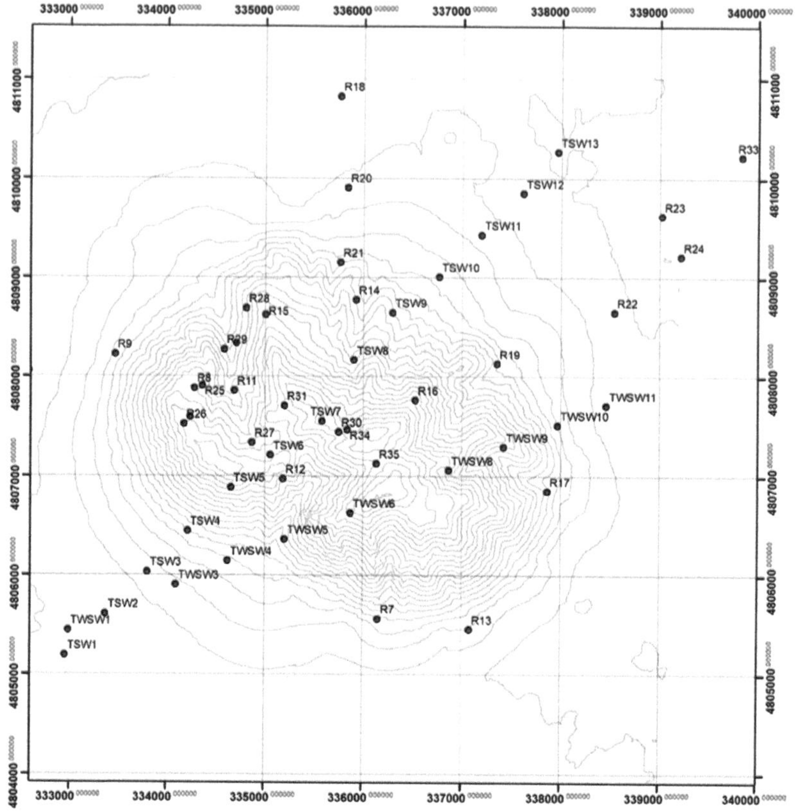

Figure 7.8 Contour map of BSB showing measurement points on and close to the mountain. NB: there are five additional measurement points (R1–R5) further away from the mountain clockwise from south-east to north-west which are not shown.

mountain is mainly grassy with steep slopes of average gradient 30–40%. Measurements were made at 53 locations across this site over the period 15 June to 9 September 2010.

The second site is a 5 km stretch of the *Salmon River Canyon* (SRC) which runs east to west and is situated about 300 km north-west of BSB in western Idaho. A network of 27 measurement locations was installed across the site from the bottom of the canyon at 550 m above mean sea level to the ridge tops above the canyon (~1600 m above mean sea level), as shown in Figure 7.9. Measurements were made over the period 14 July to 13 September 2011.

Both sites saw strong thermally driven diurnal winds giving quite different flow patterns between day and night. The strength of these patterns was

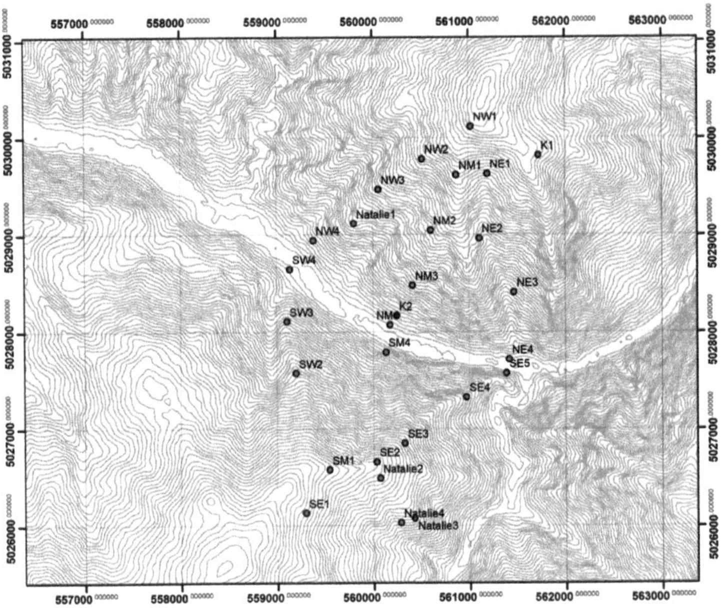

Figure 7.9 Contour map of SRC showing measurement points around the canyon.

also affected by the strength of the background synoptic winds. Figure 7.10 shows examples of these diurnal patterns in the absence of strong synoptic forcing (low background winds). The wind flow in these situations became almost completely decoupled from the large-scale synoptic forcing and gave rise to relatively complex flow patterns. Strong upslope winds were observed around late morning when surface heating was high and downslope winds could be observed during night-time. This type of behaviour is explained further in Section 10.1.2.

Conversely, during periods of strong synoptic forcing (higher background wind speeds), mechanical mixing once again coupled the surface conditions to the higher-level winds and flow associated with the terrain such as speed-up at the ridges and flow separation on the steep leeward side of terrain features was observed. These contrasting conditions represent a challenge to flow models. To properly capture the flow regimes requires mesoscale/microscale model coupling, high spatial resolution modelling, and an accurate parameterisation of the temperature stratification in the ABL. The data from these test cases is freely available for download.

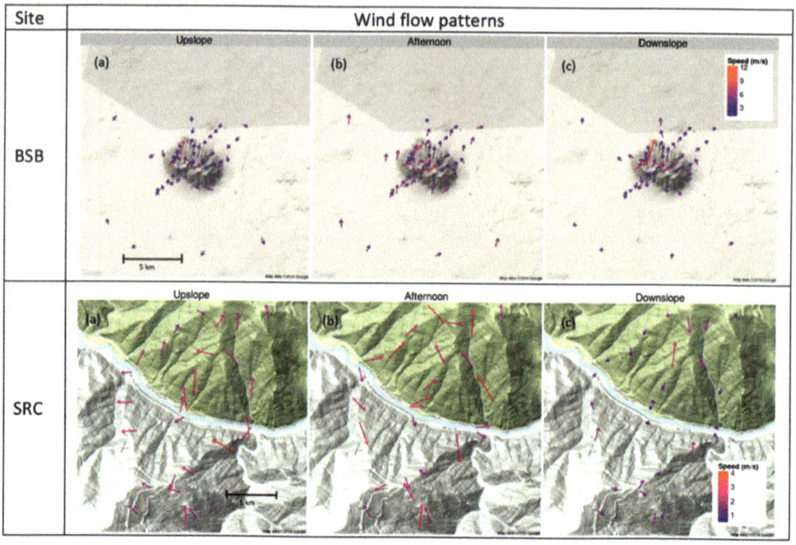

Figure 7.10 Wind flow averaged over several months at the measurement locations around BSB and SRC during (a) upslope (11:00 local time), (b) afternoon (16:00 local time), and (c) downslope (00:00 local time) regimes. Data are filtered for periods when the background synoptic winds were weak. *Source:* Butler et al. (2015) / European Geosciences Union, CC BY 4.0.

7.4 Perdigão

The next level in the validation of flow modelling in complex terrain came with the experimental measurement campaign at Perdigão, Vale do Cobrão, in eastern Portugal, 80 km south-east of Coimbra (https://perdigao.fe.up.pt). The site was particularly challenging from the point of view of wind resource assessment consisting of multiple hills with extensive canopy coverage and complex mesoscale circulations. The measurement campaign included multiple 100 m meteorological masts instrumented with sonic anemometers and other meteorological instruments, sodars, and scanning and profiling lidars installed along, between, and either side of two parallel ridges, 4 km long and 1.5 km apart with heights of 400–500 m above mean sea level running north-west to south-east, as shown in Figure 7.11. The height difference between valley bottom and ridge top is around 250 m. The campaign lasted between 15 December 2016 and 15 June 2017 (though there had been pilot measurements in the preceding period) producing

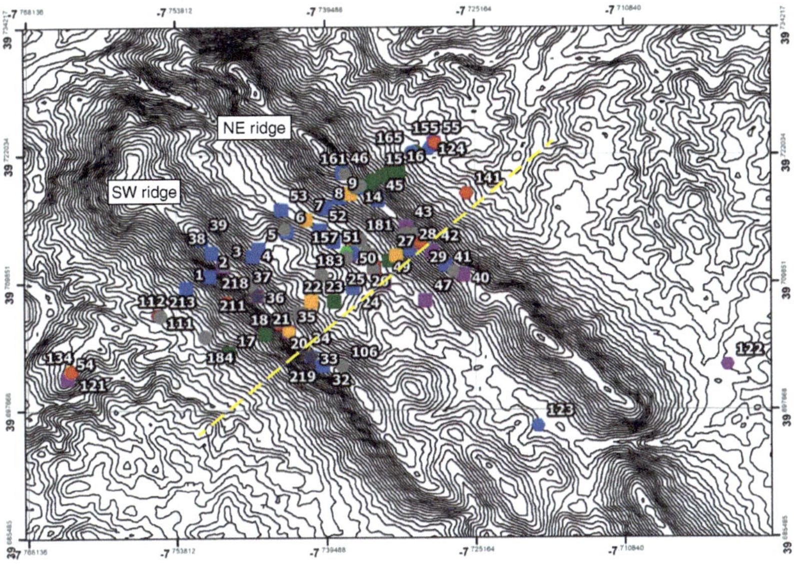

Figure 7.11 The Perdigão site. The different numbered measurement stations are marked by colour codes details of which can be found at Perdigão (2020). Contours are every 10 m. Where the instruments are located, the south-western (SW) ridge height is around 450–500 m above mean sea level (amsl) and the north-eastern (NE) ridge height is around 400–450 m amsl.

arguably the most extensive dataset available to date for the study of wind flow in complex terrain and the validation of both mesoscale and micro-scale models. There was also a wind turbine on the south-western ridge allowing the study of the structure of its wake under complex terrain and varying stability conditions.

Given the relatively recent nature of the campaign, there are as yet few model comparisons with the data, but to give an idea of the complexity of the flow, Figure 7.12 (Fernando et al., 2019) shows scanning lidar data from four instruments scanning along the south-eastern transect marked as a yellow dashed line in Figure 7.11, running perpendicular to the two ridges. The data show that there is a complex interaction between the large-scale synoptic flow driven by the prevailing pressure system and the local mesoscale circulation which depends on the orography and thermal stratification. In Figure 7.12(a) conditions are stable during the night resulting in a decoupling of the flow above the ridges and close to the surface. From associated radio-sonde data (not shown), the flow within the valley is south-easterly, changing

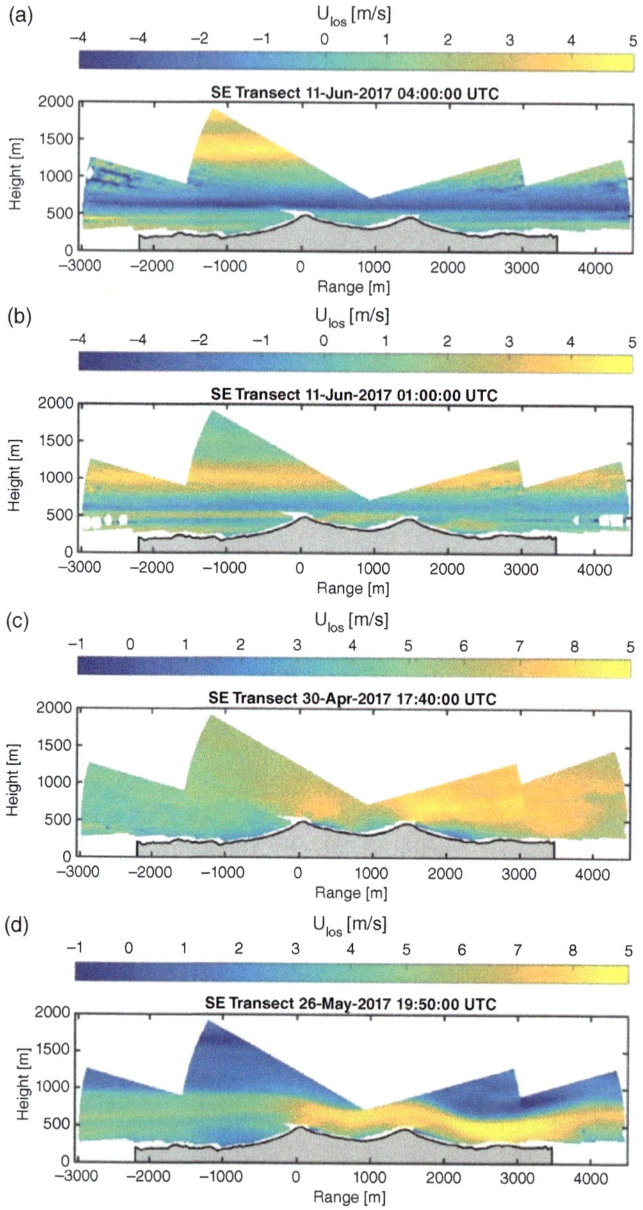

Figure 7.12 Scanning lidar measurements of wind speed along the SE transect perpendicular to the two ridges at the Perdigão site. Positive values of the line-of-sight wind speed (U_{LOS}) are from left to right (south-westerly). *Source:* Fernando et al. (2019) / with permission of American Meteorological Society.

to north-easterly immediately above and south-westerly further aloft. Figure 7.12(b) shows the situation three hours earlier with evidence of resonant lee waves. Figure 7.12(c) shows a different day during the late afternoon, where the flow is predominantly south-westerly and there is significant flow acceleration above the ridges, stronger for the second ridge. Flow separation is occurring in the lee of both ridges with reverse flow in the valleys. Figure 7.12(d) shows the situation during the early evening as conditions become more stable. Once again, there is a predominantly south-westerly flow.

There is a strong jet of accelerated wind above the ridges and shallow regions of flow separation and reverse flow in the valleys. The flow downstream of the second ridge is accelerated further and this is where the highest winds are observed. This flow situation is not unlike that simulated in Figure 5.17. Upstream of the first ridge the flow is blocked due to the stable conditions. Further aloft the wind direction becomes reversed. Clearly, the prediction of the wind resource in such situations represents a significant challenge and would require the use of coupled mesoscale and microscale models.

7.5 Alaiz

The Alaiz mountain range is in northern Spain, just south of Pamplona, and the topography is of a similar complexity to Perdigão, but an order of magnitude larger. This site was chosen to provide further data in the validation and testing of the coupling between mesoscale and microscale models (Cantero et al., 2019). The site is also in the vicinity of several wind farms making it an ideal location to study the complex flow patterns and how they translate to wind power production in steep mountainous terrain.

An orographic map of the site is shown in Figure 7.13. The Alaiz mountain range extends up to a height of 1100 m above mean sea level with the Elorz valley below descending steeply to 500 m. The valley is bounded to the north by the North Ridge (Tajonar), which reaches a height of 850 m. The mountains are partially forested, which increases the complexity of the flow. A measurement campaign to characterise the wind flow was carried out between July 2017 and July 2019. Ground-based instruments were sited across the valley to provide detailed surface flux measurements (turbulence and heat). This was achieved through several surface layer stations (SLSx) with instruments on small 2 m masts, and a surface energy budget (SEB) measurement site. A WindRASS sodar was located close to the SEB measurement site to provide simultaneous vertical profiles of wind speed and temperature. Several tall 80 m masts (Mx) were sited in a

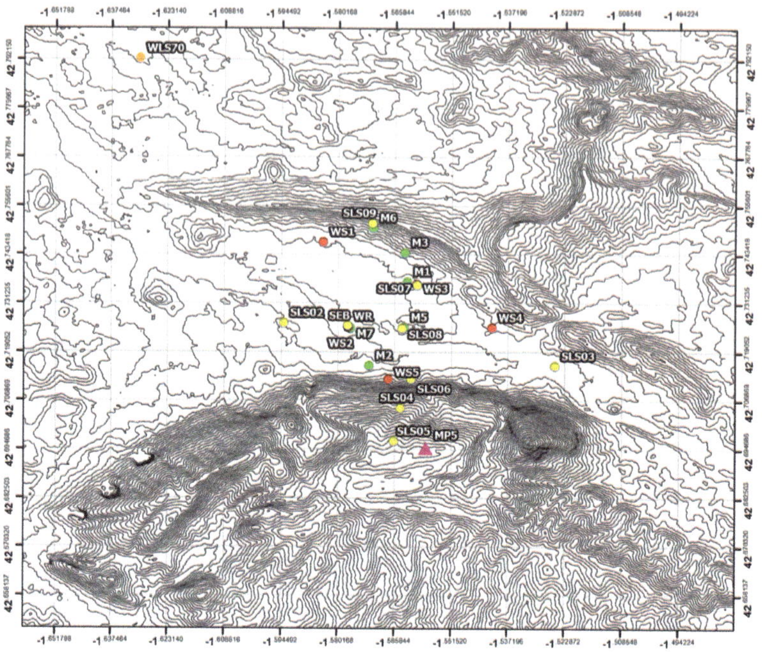

Figure 7.13 The Alaiz mountain range experimental site. The site was extensively instrumented with tall meteorological masts (Mx), wind scanning Doppler lidars (WSx), surface layer stations (SLSx), a surface energy budget measurement site (SEB), a WindRASS for simultaneous wind and temperature profiles (WR) and a wind profiling Doppler lidar (WLS70). A permanent tall meteorological mast is located at MP5.

valley and on the northern slope of the mountain ridge to measure wind speed, direction, and turbulent fluxes using wind vanes, cup anemometers, and, in the case of some of these masts, sonic anemometers. A further permanent 118 m mast was located on top of the ridge, similarly instrumented. Five scanning lidars (WindScanners, WSx) were sited on the valley floor to give detailed cross-sections of the wind flow in the valley and over the mountain ridge. A range of range height indicator (RHI) scan patterns was made where the WindScanners were operated synchronously in pairs or triplets to provide full 3D wind vectors to be constructed over the valley and ridge along various trajectories. Intersecting vertical scans were performed at various locations to provide virtual meteorological mast data. A profiling lidar (WLS70) was sited some distance away to the north-west of the ridge to determine the inflow conditions to the region.

Observations at the site revealed complex flow phenomena (Santos et al., 2020). For example, Figure 7.14 shows evidence of a so-called hydraulic jump. This phenomenon is observed where a fast-moving layer of fluid passes over an object to a region of slow-moving fluid, and as a consequence the energy dissipated causes a region of high turbulence and a change in the depth of this fast-moving layer immediately after the object. This is commonly observed when fast-moving water flows over a weir, for example. In Figure 7.14, the development of the hydraulic jump

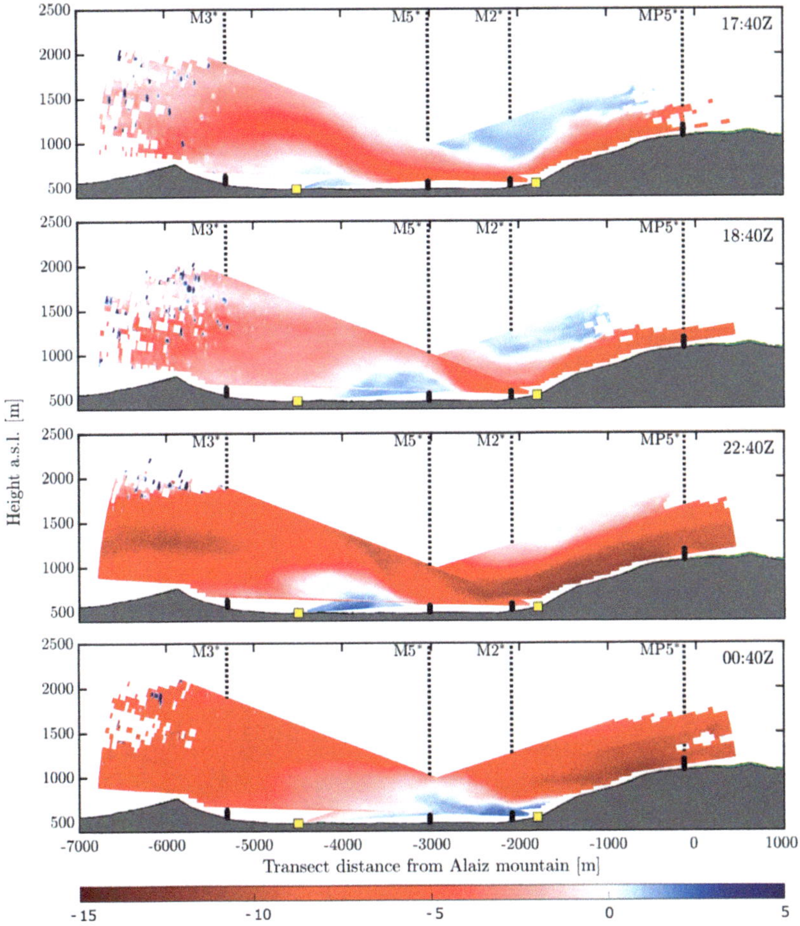

Figure 7.14 Evidence of a hydraulic jump downstream of the Alaiz mountain range (right) based on lidar RHI scans. The bottom scale gives the lidar radial wind speed in ms^{-1}. *Source:* reproduced from Santos et al. (2020). / Copernicus Publications / CC BY 4.0.

is shown over the period of a few hours in the evening into the night. The areas of red show the primarily southerly wind that accelerates down the mountain ridge under stable surface layer night-time conditions. The smaller areas of blue show the jump with a recirculating flow. The position of the jump is shown to vary during the period of these observations. Such events tend to occur when the height of the layer of faster-moving flow (dictated in this case by the height of the inversion layer) is around the same height as the mountain ridge and when the Froude number (mentioned in Section 5.6) transitions from a value greater than one above the ridge to less than one in the valley below. Figure 7.15 shows an example of a stable surface layer developing during the night causing stratification of the flow within the valley. Like Figure 7.12, this stratification causes a decoupling between the upper-level northerly wind flow (indicated by the blue regions of the RHI scans) and the flow within the valley where the red regions indicate a change in wind direction. Indeed, measurements from masts M2, M3, and M5 indicated a thermally driven westerly flow in contrast to the northerly flow on top of the ridge recorded at MP5. The two events show the importance of considering the effect of atmospheric stability on complex valley flows and further emphasise the requirement for coupled mesoscale and microscale modelling to fully understand the wind resource in mountainous terrain.

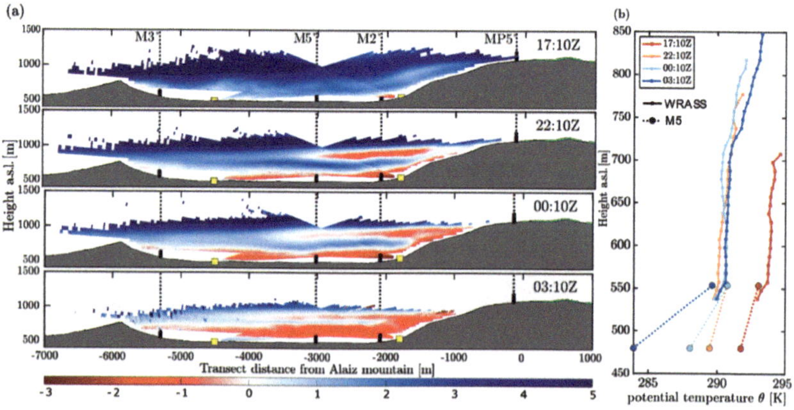

Figure 7.15 The left image (a) shows the development of stratified flow in the valley adjacent to the Alaiz mountain range (right) and bounded by the North Ridge (left). The image shows the measurements from the lidar RHI scans with the bottom scale giving the lidar radial wind speed in ms^{-1}. The right image (b) shows the vertical temperature profile measured on the mast M5 and by the WindRASS. *Source:* reproduced from Santos et al. (2020) / Copernicus Publications / CC BY 4.0.

References

Bechmann, A., Sørensen, N.N., Berg, J., Mann, J., Réthoré, P.E., 2011. The Bolund experiment: part II: blind comparison of microscale flow models. *Boundary-Layer Meteorol.* 141, 245–271. https://doi.org/10.1007/s10546-011-9637-x.

Beljaars, A.C.M., Walmsley, J.L., Taylor, P.A., 1987. A mixed spectral finite-difference model for neutrally stratified boundary-layer flow over roughness changes and topography. *Boundary-Layer Meteorol.* 38, 273–303. https://doi.org/10.1007/BF00122448.

Berg, J., Mann, J., Bechmann, A., Courtney, M.S., Jørgensen, H.E., 2011. The Bolund experiment: part I: flow over a steep, three-dimensional hill. *Boundary-Layer Meteorol.* 141, 219–243. https://doi.org/10.1007/S10546-011-9636-Y.

Butler, B.W., Wagenbrenner, N.S., Forthofer, J.M., Lamb, B.K., Shannon, K.S., Finn, D., Eckman, R.M., Clawson, K., Bradshaw, L., Sopko, P., Beard, S., Jimenez, D., Wold, C., Vosburgh, M., 2015. High-resolution observations of the near-surface wind field over an isolated mountain and in a steep river canyon. *Atmos. Chem. Phys.* 15, 3785–3801. https://doi.org/10.5194/ACP-15-3785-2015.

Cantero, E., Fernando, J., Pedro, J., Nikola, M., Daniel, B., Cuxart, J., 2019. *Alaiz experiment (ALEX17): campaign and data report, deliverable D2.21, New European Wind Atlas project.* https://doi.org/10.5281/ZENODO.3187482.

Castro, F.A., Palma, J.M.L.M., Lopes, A.S., 2003. Simulation of the Askervein flow: part 1: Reynolds averaged Navier–Stokes equations ($k\,\varepsilon$ turbulence model). *Boundary-Layer Meteorol.* 107, 501–530.

Fernando, H.J.S., Mann, J., Palma, J.M.L.M., Lundquist, J.K., Barthelmie, R.J., Belo-Pereira, M., Brown, W.O.J., Chow, F.K., Gerz, T., Hocut, C.M., Klein, P.M., Leo, L.S., Matos, J.C., Oncley, S.P., Pryor, S.C., Bariteau, L., Bell, T.M., Bodini, N., Carney, M.B., Courtney, M.S., Creegan, E.D., Dimitrova, R., Gomes, S., Hagen, M., Hyde, J.O., Kigle, S., Krishnamurthy, R., Lopes, J.C., Mazzaro, L., Neher, J.M.T., Menke, R., Murphy, P., Oswald, L., Otarola-Bustos, S., Pattantyus, A.K., Veiga Rodrigues, C., Schady, A., Sirin, N., Spuler, S., Svensson, E., Tomaszewski, J., Turner, D.D., Van Veen, L., Vasiljevic, N., Vassallo, D., Voss, S., Wildmann, N., Wang, Y., 2019. The Perdigão: peering into microscale details of mountain winds. *Bull. Am. Meteorol. Soc.* 100, 799–820. https://doi.org/10.1175/BAMS-D-17-0227.1.

Forthofer, J.M., Butler, B.W., Wagenbrenner, N.S., 2014. A comparison of three approaches for simulating fine-scale surface winds in support of wildland fire management: part I: model formulation and comparison against measurements. *Int. J. Wildl. Fire* 23, 969. https://doi.org/10.1071/WF12089.

Hunt, J.C.R., Leibovich, S., Richards, K.J., 1988. Turbulent shear flows over low hills. *Q. J. R. Meteorol. Soc.* 114, 1435–1470. https://doi.org/10.1002/qj.49711448405.

Santos, P., Mann, J., Vasiljević, N., Cantero, E., Rodrigo, J.S., Borbón, F., Martínez-Villagrasa, D., Martí, B., Cuxart, J., 2020 The Alaiz experiment: untangling multi-scale stratified flows over complex terrain. *Wind Energ. Sci.*, 5, 1793–1810. https://doi.org/10.5194/wes-2020-89.

Silva Lopes, A., Palma, J.M.L.M., Castro, F.A., 2007. Simulation of the Askervein flow: part II: large-eddy simulations. *Boundary-Layer Meteorol.* 125, 85–108. https://doi.org/10.1007/s10546-007-9195-4.

Taylor, P.A., Teunissen, H.W., 1987. The Askervein Hill project: overview and background data. *Boundary-Layer Meteorol.* 39, 15–39. https://doi.org/10.1007/BF00121863.

Teunissen, H.W., Shokr, M.E., Bowen, A.J., Wood, C.J., Green, D.W.R., 1987. The Askervein Hill project: wind-tunnel simulations at three length scales. *Boundary-Layer Meteorol.* 40, 1–29. https://doi.org/10.1007/BF00140067.

Wood, N., 1995. The onset of separation in neutral, turbulent flow over hills. *Boundary-Layer Meteorol.* 76, 137–164. https://doi.org/10.1007/BF00710894.

8

Statistical Methods

It has been shown in the previous chapters how wind speeds can vary significantly over relatively short distances, particularly where terrain conditions are complex. Several physical modelling approaches have been described in the previous chapters that attempt to predict this variation. These approaches vary in complexity and fidelity. If a relatively dense network of measurements is available, simpler statistically based interpolation techniques can be used to estimate the spatial variation in wind speeds without the need for a physical modelling approach. Even where physical models are required to capture complex spatial variation, they frequently require calibration using data from onsite measurements such as meteorological masts or lidars. However, onsite measurements are usually made over relatively short periods in climatological terms (e.g. 12–18 months). As is described in Section 4.3, significant fluctuations in annual mean wind speeds can be expected during the lifetime of a wind farm (20 years or more). Relying on only 1–2 years of data may give a very misleading picture of the expected long-term wind speed at a wind farm site. Some form of correction to the short-term data is required to provide a robust estimate of the long-term wind speed. The most common way to do this is to correlate the short-term onsite data with a source of data which is: (i) concurrent, (ii) available for a long-term period (at least 20 years), and (iii) shows a good statistical correlation with the short-term data. In this chapter, a review of statistical spatial interpolation and temporal extrapolation techniques is presented.

8.1 Spatial Coherence

Several factors influence the degree of correlation between wind speed measurements at one site and those at another, such as the complexity of the terrain, how similar the site locations are (e.g. both inland, both coastal), or

Handbook of Wind Resource Assessment, First Edition. Simon Watson.

the difference in elevation of the two sites. One of the most significant factors is the distance between the two sites. Sites which are relatively close together (~10–50 km) and in flat exposed terrain can be expected to be exposed to the same weather system at approximately the same time and display a high level of correlation on the timescale of ~1 h. At very short timescales (~1 s), small-scale turbulent structures will dominate and the degree of correlation (or *coherence* when considering spatial correlation) will be small. This can be seen when considering an example of longitudinal turbulence coherence between measurements at two masts at 18 m above ground level a distance of 62 m apart, as shown in Figure 8.1 (Schlez and Infield, 1998). The figure shows coherence (γ^2) as a function of a normalised frequency, $\tilde{n}$. When $\tilde{n}$ is around 3 Hz, coherence drops to almost zero. This corresponds to a timescale of around 15 s. However, for the purposes of wind resource assessment, we are less interested in such short-term correlation. On the other hand, if measurements are averaged at intervals of one hour, the correlation between sites decreases far more slowly. For example, Figure 8.2 shows the correlation between hourly wind speed measurements as a function of distance for pairwise combinations of 14 sites (Hill et al., 2012). The data have been de-trended to remove diurnal and

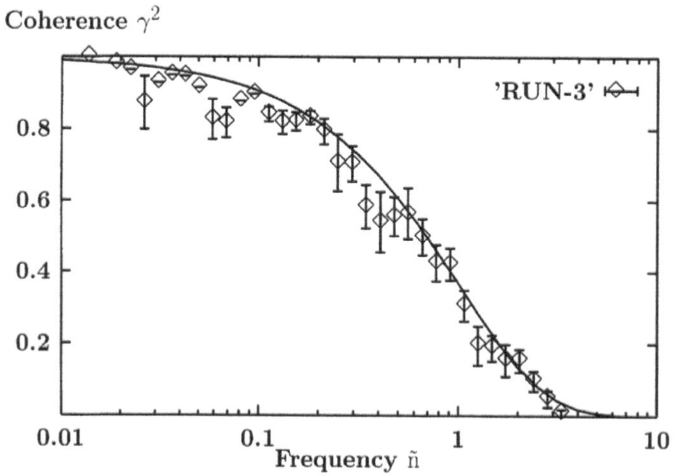

Figure 8.1 Longitudinal turbulence coherence (γ^2) between measurements at two masts 62 m apart at 18 m above ground level as a function of a normalised frequency parameter $\tilde{n}$. Measurements are shown as diamonds and an empirical function fitted to the data is shown as the solid line. *Source:* Schlez and Infield (1998) / with permission of Springer Nature.

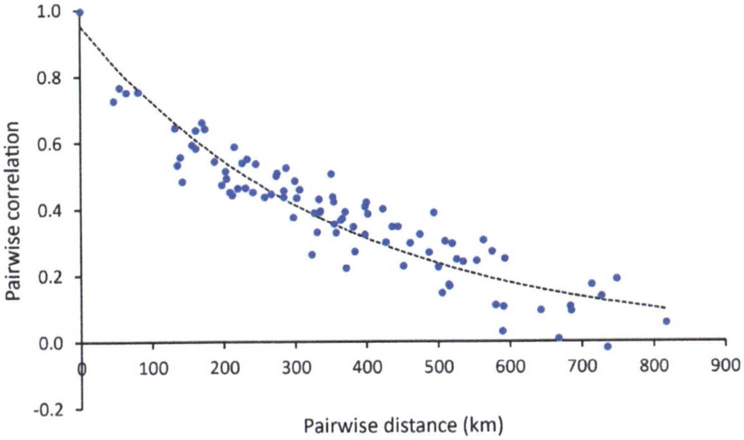

Figure 8.2 Correlation of wind speeds as a function of distance based on de-trended data from 14 sites in the UK. *Source:* Hill et al. (2012) / with permission of IEEE.

seasonal trends. Up to around 100 km, the wind speed measurements have a Pearson correlation coefficient in excess of 0.7. The correlation coefficient ρ_p appears to have an exponential tail-off with distance d and indeed several studies have found this trend (St. Martin et al., 2015). In the case of Figure 8.2, a fit of the form

$$\rho_p(d) = \alpha \exp\left(-\frac{d}{L_s}\right) \tag{8.1}$$

was found where $\alpha = 0.952$ and the length scale L_s = 357 km. Other studies have shown similar length scales, though differences have been noted between regions. Data that have not been de-trended tend to give less rapidly decaying correlations with length scales of around 500–700 km (Hill et al., 2012).

What this means is that if a reliable source of long-term data is available for a site within 100 km of the site of interest for which a long-term estimate of wind resource is required, it should be suitable to provide a climatological correction on an hourly timescale. This assumes that both the site with long-term data (the *reference site*) and the site of interest (the *target site*) are well exposed and situated in similar types of terrain. If one site, for example, is in a valley subject to very localised mesoscale wind circulations then the correlation will be reduced and the data from the reference site may not be suitable for long-term adjustment.

8.2 Measure–Correlate–Predict: An Overview

A frequently used method to adjust short-term data to correct for long-term variation is that known as *measure–correlate–predict* (MCP). A schematic of how the method is applied is shown in Figure 8.3.

In the first step, the correlation is determined between ten-minute or hourly wind speed measurements from the target site and concurrent measurements from an appropriate reference site. The reference site could be a physical meteorological station, or it could be data from a geographically nearby grid point from a reanalysis dataset. There are several ways that the correlation is determined, but in its simplest form a linear regression relationship is formulated between reference and target site, usually broken down by several direction sectors. In the next step, these regression relationships are used to transform the long-term data (usually averaged over 20–30 years) to a prediction of what the long-term wind speed would be at the target site had measurements been made over the historic long-term period. As the method is normally applied when a site is to be developed as a wind farm, it is used as a prediction of what the long-term wind speed would be for the future 20- to 30-year period. This presupposes that the climatology at both reference and target site is stationary and will not change in the future. The impact of global heating on wind speeds is very uncertain, indeed more uncertain than any prediction of future global temperatures. Although some studies have suggested declining wind speeds in some areas of the world, climate signals appear to be small compared with interannual variability (Watson, 2014).

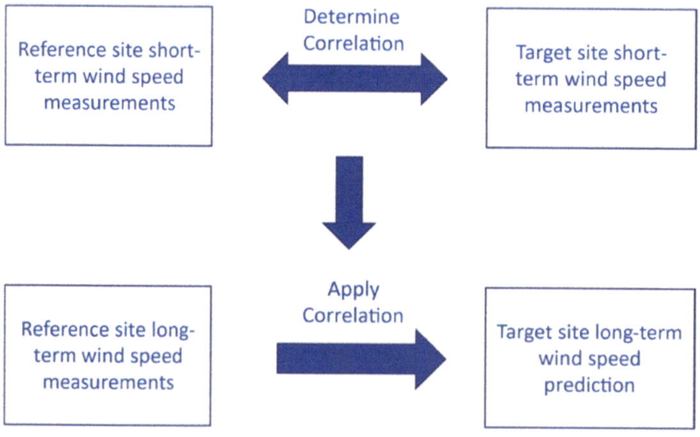

Figure 8.3 Schematic representation of the measure–correlate–predict (MCP) method.

8.2.1 Linear Regression

Though there are many variants of the MCP process (Carta et al., 2013), the commonest approach is to assume a linear regression equation by reference site wind direction sector i which is used to estimate the long-term wind speed at the target site

$$\left(u_{i\tau}\right)_T^{LT} = m_i\left(u_{i\tau}\right)_R^{LT} + c_i + \varepsilon_{i\tau} \tag{8.2}$$

where $\left(u_{i\tau}\right)_T^{LT}$ is the estimate of the long-term wind speed at the target site at time τ, $\left(u_{i\tau}\right)_R^{LT}$ is the measured long-term wind speed at time τ at the reference site, and $\varepsilon_{i\tau}$ is a Gaussian white noise term. Normally, a value of the long-term target site wind speed at a given time is not required, rather an estimate of the long-term *mean* is required, in which case Eq. (8.2) becomes

$$\left(\bar{u}_i\right)_T^{LT} = m_i\left(\bar{u}_i\right)_R^{LT} + c_i \tag{8.3}$$

where $(\bar{u}_i)_T^{LT}$ is the estimate of the long-term mean wind speed at the target site and $(\bar{u}_i)_R^{LT}$ is the measured long-term mean wind speed at the reference site. The Gaussian white noise term averages out to zero in this case. The measured long-term mean wind speed at the reference site is typically based on measurements made over a period of 20–30 years, though opinions differ as to what is a suitable long-term period to represent a stationary mean. As an alternative to values measured at a meteorological station, values from a nearby grid point (or interpolated from the grid) of a reanalysis dataset, such as ERA5, may be used. Caution should be exercised for sites which are coastal in this case. The grid point chosen should be representative of the target site, so if the site is offshore, an offshore grid point should be chosen that may not be the closest grid point to the site. Interpolation from the grid may also not be appropriate in this case. Reanalysis data have been shown to give an acceptable level of correlation with data measured at a meteorological mast (Jourdier, 2020; Santos et al., 2020). This is looked at in more detail in Section 9.4. Whether using data from a meteorological station or a reanalysis, care should be taken for any discontinuities in the data over a long term-period. For example, when using data from a meteorological station, instruments can be changed or moved, which can slightly change the average wind speed value. In the case of a reanalysis, changes to the model over time can have a similar effect. The long-term period should thus be chosen such that it does not straddle such a discontinuity or at least any discontinuity should be carefully corrected for.

The regression parameters m_i and c_i are determined from the concurrent short-term data at target and reference sites based most commonly on ordinary linear regression (OLR) (Carta et al., 2013)

$$m_i = \frac{\sum_{\tau=1}^{N} (u_{i\tau})_R^{ST} (u_{i\tau})_T^{ST} - (1/N)\sum_{\tau=1}^{N} (u_{i\tau})_R^{ST} (u_{i\tau})_T^{ST}}{\sum_{\tau=1}^{N} \left(u_{i\tau}^2\right)_R^{ST} - (1/N)\left[\sum_{\tau=1}^{N} (u_{i\tau})_R^{ST}\right]^2} = \frac{\left(s_{(u_i)_R (u_i)_T}\right)^{ST}}{\left(s^2_{(u_i)_R}\right)^{ST}} \tag{8.4}$$

$$c_i = 1/N\left[\sum_{\tau=1}^{N} (u_{i\tau})_T^{ST} - m_i \sum_{\tau=1}^{N} (u)_R^{ST}\right] \tag{8.5}$$

where $\left(s_{(u_i)_R (u_i)_T}\right)^{ST}$ is the covariance between $\left(u_i(n)\right)_R^{ST}$ and $\left(u_i(n)\right)_T^{ST}$, $\left(s^2_{(u_i)_R}\right)^{ST}$ is the variance of $\left(u_{i\tau}\right)_R^{ST}$, and $\left(u_{i\tau}\right)_R^{ST}$ and $\left(u_{i\tau}\right)_T^{ST}$ are concurrent measurements at reference and target sites, respectively at a time τ during a short-term measurement campaign at the target site where N valid measurements are collected, typically over a period of 12–18 months. It could be argued that the offset value c_i should be set to zero as it would be expected that if there are calm conditions at one site this should also be the case at the other site. However, setting the offset value to zero generally does not provide a good representation of the higher wind speeds which are more important from a wind power generation point of view (Pramod Jain, 2016), thus it is generally recommended to fit a nonzero value of c_i. OLR assumes that all error in the data is on the dependent variable. What counts as the 'correct' value is open to interpretation but may be considered as that which results only from a global pressure gradient which simultaneously drives the wind speed at reference and target sites. Any error, aside from physical measurement uncertainty, will then be mostly down to local turbulent variation. An alternative regression fit which considers the variance of the errors on both the dependent and independent variable is known as *general orthogonal regression* (GOR) (Fuller, 1987) where

$$m_i = \frac{\left(s^2_{(u_i)_T}\right)^{ST} - \left(s^2_{(u_i)_R}\right)^{ST} + \sqrt{\left[\left(s^2_{(u_i)_T}\right)^{ST} - \lambda_i\left(s^2_{(u_i)_R}\right)^{ST}\right]^2 + 4\lambda_i\left(\left(s^2_{(u_i)_R (u_i)_T}\right)^{ST}\right)^{ST}}}{2\left(s_{(u_i)_r (u_i)_t}\right)^{ST}} \tag{8.6}$$

$$c_i = 1/N\left[\sum_{\tau=1}^{N} (u_{i\tau})_T^{ST} - m_i \sum_{\tau=1}^{N} (u_{i\tau})_R^{ST}\right] \tag{8.7}$$

$$\lambda_i = \frac{s^2_{(\varepsilon u_i)_T}}{s^2_{(\varepsilon u_i)_R}} \tag{8.8}$$

where $\left(s^2_{(u_i)_T}\right)^{ST}$ is the variance of $\left(u_{i_T}\right)^{ST}_T$, $(\varepsilon u_i)_T$ is the error term on the target site wind speed, $(\varepsilon u_i)_R$ is the error term on the reference site wind speed, and $s^2_{(\varepsilon u_i)_T}$ and $s^2_{(\varepsilon u_i)_R}$ are the variances of these error terms, respectively, based on the short-term data. Determining these variance terms may not be straightforward, but a reasonable estimate could be provided by using the mean turbulence values from each site (based on ten-minute values, for example) or, failing that, the variance of the short-term wind speed values at each site.

An alternative approach to OLR or GOR is the *variance ratio* method (Rogers et al., 2005). This gets around the problem that the predicted wind speed distribution will tend to have a smaller variance than would be observed (by a factor of R^2 of the regression fit)

$$\left(u_{i_T}\right)^{LT}_T = \left[\left(\bar{u}_i\right)^{ST}_T - \left(\frac{s^{ST}_{(u_i)T}}{s^{ST}_{(u_i)R}}\right)\left(u_i\right)^{ST}_R\right] + \left(\frac{s^{ST}_{(u_i)T}}{s^{ST}_{(u_i)R}}\right)\left(u_{i_T}\right)^{LT}_R \tag{8.9}$$

where $s^{ST}_{(u_i)T}$ and $s^{ST}_{(u_i)R}$ are the standard deviations of the short-term wind speed values at the target and reference sites, respectively.

The accuracy of the regression parameters depends on the level of correlation between the short-term measurements. An acceptable level of correlation is still a matter of debate, but a value of the Pearson correlation coefficient of 0.7 or more is generally considered acceptable (Bass, 2009). An example of a regression fit with a good correlation (R^2=0.78) is shown in Figure 8.4. Note that the values at the reference site are recorded in integer knot values which is why the data points are discretely clustered.

The most common number of direction sectors chosen is twelve 30° sectors, though if certain sectors are very sparsely populated then an alternative is to determine sector sizes based on equal numbers of observations by sector (King and Hurley, 2005; Riedel and Strack, 2001).

It is normally assumed that the long-term wind rose at the target site is identical to the short-term wind rose, though variants exist which attempt to correct for the long term, based on the long-term directional statistics at the reference site. One such variant is known as *matrix MCP* (Woods and Watson, 1997). This method also provides a way to estimate long-term target site wind speeds in situations where there is not a one-to-one

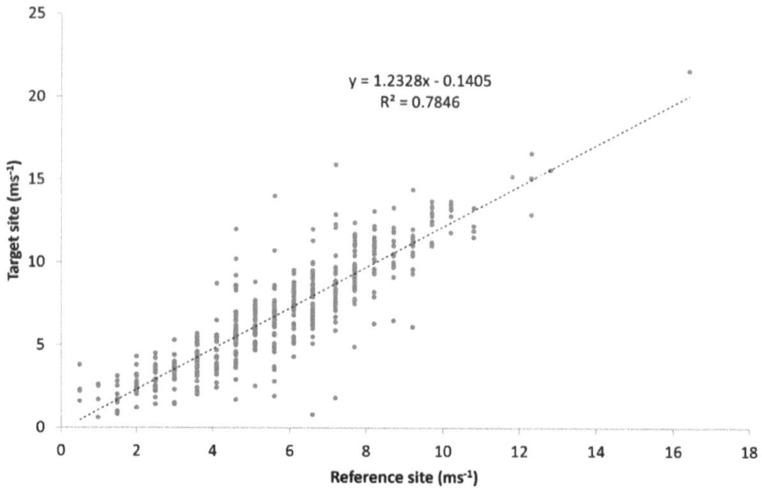

Figure 8.4 Example of a regression fit to concurrent short-term wind speed data measured at a target and reference site for a particular direction sector at the reference site. An OLR line fit to the data is shown with the equation of the line and the R^2 coefficient of determination.

correspondence between the direction at the target site and the reference site. This may arise in complex terrain where channelling is present in one site but not the other. This method uses a percentage weighting matrix to map the wind direction at the target site onto that at the reference site. The matrix can be formulated in two ways: (i) percentage weightings summing to 100% based on the reference site data, W_{ij}; or (ii) percentage weightings summing to 100% based on the target site data, Z_{ij}, where i is the reference site wind direction sector (as before) and j is the target site wind direction sector. The first step is to bin the short-term data by direction at the target and reference sites, an example of which is shown in Table 8.1. Any bins that contribute less than 5% of the total in a target site direction sector are then discarded. This cut-off is slightly arbitrary, and in some cases it may make sense to lower this threshold or base any discarding on the level of correlation between the measurements in a bin (Carta et al., 2013). The values are then converted to percentages using the two methods described above. The matrices obtained from the values in Table 8.1, W_{ij} and Z_{ij}, are shown in Tables 8.2 and 8.3, respectively.

An estimate of the long-term wind direction probability distribution (wind rose) $\left(p_j\right)_T^{LT}$ at the target site can be determined based on W_{ij} and the long-term wind rose at the reference site $\left(p_i\right)_R^{LT}$ assuming the relationship

Table 8.1 Example of wind speed data counts binned by sector at a reference and target site.

Target site sector, *j*	Reference site sector, *i*											
	1	2	3	4	5	6	7	8	9	10	11	12
1	480	118	25	3	3	3	6	—	1	9	17	95
2	42	175	115	37	12	8	2	—	—	1	10	9
3	14	47	130	96	12	19	2	1	—	—	3	3
4	14	—	79	125	73	17	6	—	2	—	4	1
5	8	20	57	143	349	265	17	—	1	4	2	7
6	—	4	8	7	39	763	70	—	—	1	3	2
7	3	6	7	2	3	191	932	62	5	6	10	1
8	—	—	2	1	1	4	178	744	165	18	7	1
9	—	—	—	1	—	2	38	70	567	148	9	3
10	1	1	—	1	—	3	14	10	70	550	76	6
11	2	—	—	1	—	4	10	2	3	87	340	61
12	119	4	5	3	2	1	8	3	1	17	108	443

Table 8.2 Percentage weighting matrix W_{ij}, based on the values in Table 8.1, after discarding bins with values of less than 5% of the total in the target site wind direction sector. All values sum to 100% for a given reference site wind direction sector.

Target site sector, *j*	Reference site sector, *i*											
	1	2	3	4	5	6	7	8	9	10	11	12
1	74.9	34.7	—	—	—	—	—	—	—	—	—	—
2	6.6	51.5	30.2	9.2	—	—	—	—	—	—	—	—
3	—	13.8	34.1	23.9	—	—	—	—	—	—	—	—
4	—	—	20.7	31.2	17.3	—	—	—	—	—	—	—
5	—	—	15.0	35.7	82.7	21.7	—	—	—	—	—	—
6	—	—	—	—	—	62.6	5.9	—	—	—	—	—
7	—	—	—	—	—	15.7	79.0	—	—	—	—	—
8	—	—	—	—	—	—	15.1	91.4	20.6	—	—	—
9	—	—	—	—	—	—	—	8.6	70.7	18.9	—	—

(Continued)

Table 8.2 (Continued)

Target site sector, *j*	Reference site sector, *i*											
	1	**2**	**3**	**4**	**5**	**6**	**7**	**8**	**9**	**10**	**11**	**12**
10	—	—	—	—	—	—	—	—	8.7	70.1	14.5	—
11	—	—	—	—	—	—	—	—	—	11.1	64.9	10.2
12	18.6	—	—	—	—	—	—	—	—	—	20.6	74.0

Table 8.3 Percentage weighting matrix Z_{ij}, based on the values in Table 8.1, after discarding bins with values of less than 5% of the total in the target site wind direction sector. All values sum to 100% for a given target site wind direction sector.

Target site sector, *j*	Reference site sector, *i*											
	1	**2**	**3**	**4**	**5**	**6**	**7**	**8**	**9**	**10**	**11**	**12**
1	69.3	17.0	—	—	—	—	—	—	—	—	—	—
2	11.4	47.4	31.2	10.0	—	—	—	—	—	—	—	—
3	—	17.2	47.6	35.2	—	—	—	—	—	—	—	—
4	—	—	28.5	45.1	26.4	—	—	—	—	—	—	—
5	—	—	7.0	17.6	42.9	32.6	—	—	—	—	—	—
6	—	—	—	—	—	91.6	8.4	—	—	—	—	—
7	—	—	—	—	—	17.0	83.0	5.2	—	—	—	—
8	—	—	—	—	—	—	16.4	68.4	15.2	—	—	—
9	—	—	—	—	—	—	—	8.9	72.2	18.9	—	—
10	—	—	—	—	—	—	—	—	10.1	79.0	10.9	—
11	—	—	—	—	—	—	—	—	—	17.8	69.7	12.5
12	17.8	—	—	—	—	—	—	—	—	—	16.1	66.1

between the wind rose at the reference and target sites does not change over time

$$\left(p_j\right)_T^{LT} = \frac{\sum_{i=1}^{12} \left(p_i\right)_R^{LT} W_{ij}}{100} \tag{8.10}$$

The long-term mean wind speed at the target site can be estimated in two ways

$$\left(\bar{u}_j\right)_T^{LT} = \frac{\sum_{i=1}^{12} Z_{ij}\left(m_i\left(\bar{u}_i\right)_R^{LT} + c_i\right)}{100} \tag{8.11}$$

$$\left(\bar{u}_j\right)_T^{LT} = m_j \frac{\sum_{i=1}^{12} Z_{ij}\left(\bar{u}_i\right)_R^{LT}}{100} + c_j \tag{8.12}$$

In the first case, the regression parameters are determined by binning the concurrent short-term data according to the reference site wind direction and in the second case, binning by the target site wind direction. The choice of equation is best determined based on the correlations obtained between the short-term data values. If the correlation is good, there is little to choose between the formulae in terms of accuracy, but where the correlation is less good, the second formula tends to give better results.

8.2.2 Nonlinear Regression

Alternatives to a linear regression fit by direction sector have been proposed for those relatively rare cases where a linear fit is not appropriate on inspection of the data. One such method suggests the use of a power law (Derrick, 1992)

$$\left(u_{i\tau}\right)_T^{ST} = a_i\left[\left(u_{i\tau}\right)_R^{ST}\right]^{b_i} \tag{8.13}$$

where a_i and b_i are parameters which are fitted using the short-term data by direction sector i, $\left(u_{i\tau}\right)_T^{ST}$ is the short-term measured wind speed at the target site, and $\left(u_{i\tau}\right)_R^{ST}$ is the short-term measured wind speed at the reference site at time τ. The most straightforward way to determine a_i and b_i is to take the logarithm of each side of the equation to linearise it before making the fit using OLR or similar.

Where measurements are made at different heights at the two sites, it has been suggested that the effect of atmospheric stability be taken into account (Joensen et al., 1999). If temperature data are available at two heights at the reference site, the following equation is suggested

$$\left(u_{i\tau}\right)_T^{LT} = m_{1i}\left(u_{i\tau}\right)_R^{LT} + m_{2i}\left(u_{i\tau}\right)_R^{LT}\left(\Delta T_{i\tau}^{LT}\right)_R + c_i + \varepsilon_{i\tau} \tag{8.14}$$

where m_{1i} and m_{2i} are coefficients fitted using the short-term data and $\left(\Delta T_{i\tau}^{LT}\right)_r$ is the temperature difference between two heights at the reference site. This formula is based on the assumption that the diabatic wind profile depends on an interaction between wind speed and vertical temperature difference. If no temperature difference data are available, an alternative is suggested

$$\left(u_{i_\tau}\right)_T^{LT} = m_{3i}\left(u_{i_\tau}\right)_R^{LT} + m_{4i}\left[\left(u_{i_\tau}\right)_R^{LT}\right]^2 + c_i + \varepsilon_{i_\tau} \tag{8.15}$$

where again m_{3i} and m_{4i} are coefficients fitted using the short-term data. The rationale behind this alternative formulation is based on the fact that, at high wind speeds, the surface layer tends to neutral conditions, whereas at low wind speeds the stability is dominated by surface heat flux, which, when compared with the stability dependent wind profile, leads to a parabolic relationship with wind speed.

It should be noted that there has been little validation of these nonlinear regression relationships. Indeed, most researchers suggest that the simple linear relationship performs best in most circumstances.

8.3 Analogue Ensembles

An alternative to MCP is to use a method based on analogue ensembles (Vanvyve et al., 2015). This is illustrated schematically in Figure 8.5. Like MCP, the method is based on the use of an observed short-term wind speed and direction dataset at the target site of interest and a long-term meteorological reference dataset of values both concurrent with the short-term dataset and for a significant period in the past. This could be data from a climatologically similar (e.g. nearby) meteorological station or from a long-term reanalysis. The long-term dataset should contain wind speed and direction values but may also contain other relevant meteorological variables that may affect the correlation between the target and reference wind speed values. The method aims to reconstruct every time step in the long-term period (*period to reconstruct*) based on analogues in the short-term period (*training period*).

The method is carried out in several steps:

- For each (normally hourly) time-step t in the period to reconstruct (marked as a red star in Figure 8.5), the wind speed, wind direction, and other predictor meteorological variables which may affect the corelation between target and reference are extracted for a small window of times (*analogue trend*) around t from the historical dataset.
- Analogous meteorological values from the historical dataset in the training period are selected for the same time of day as the target

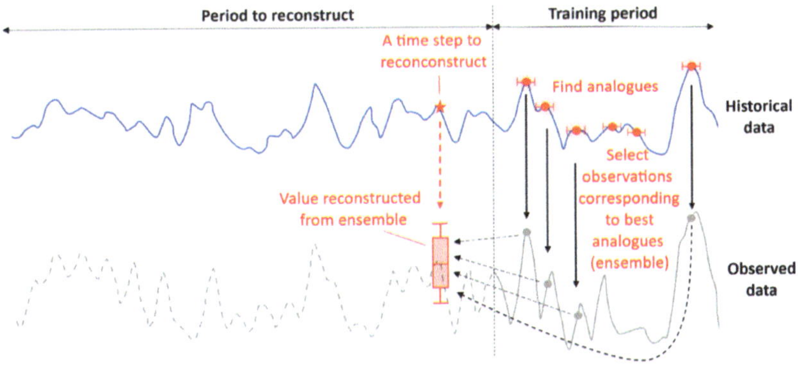

Figure 8.5 The method of analogue ensembles to reconstruct a long-term time series at a site of interest based on a short-term series of observed values and a long-term reference dataset. *Source:* adapted from Vanvyve et al. (2015).

reconstruction time and for a window of times (*search window*) around this (time τ). In principle, analogues can come from any time in the training period, shown as red dots in the upper time series in Figure 8.5.

- The closeness of fit of each analogue is determined using Eq. (8.16) and the K closest analogues are determined.
- From these K analogues, the corresponding values (in time) from the concurrent observed values are selected to form an ensemble (grey dots in the lower time series in Figure 8.5). This ensemble is used to reconstruct the predicted value at the target site at time t as well as its uncertainty.

The metric $m_{t\tau}$ used to rank the analogues is given by

$$m_{t\tau} = \sum_{p=1}^{P} \frac{w_p}{\sigma_p} \sqrt{\sum_{i=-\delta}^{\delta} \left[H_{p,t+i} - H_{p,\tau+i}\right]^2} \tag{8.16}$$

where $H_{p,t+i}$ is the pth predictor meteorological value from the historical dataset at time t in the period to be reconstructed, $H_{p,\tau+i}$ is the corresponding candidate analogue value at time τ in the training period, P is the total number of predictor meteorological variables, w_p is the weighting of each p predictor variable, σ_p is the standard deviation of each p predictor variable over the period of reconstruction (used to scale each variable), and δ is half the number of values in the analogue trend window.

For each of the analogues chosen based on the K smallest values of $m_{t\tau}$, the predicted wind speed value at time t at the target site A_t' can be calculated based on the ensemble of observed wind speed values concurrent with the times of the K analogues

$$A_t' = \sum_{k=1}^{K} \gamma_k O_k \tag{8.17}$$

where γ_k is the weighting of the kth ensemble analogue and O_k is the observed wind speed corresponding to the time τ of the k th ensemble analogue. Each γ_k weighting value depends inversely on the value of the ranking metric m_{t,τ_k} of each k th ensemble analogue

$$\gamma_k = \frac{m_{t,\tau_k}^{-1}}{\sum_{i=1}^{K} m_{t,\tau_k}^{-1}} \tag{8.18}$$

with

$$\sum_{k=1}^{K} \gamma_k = 1 \tag{8.19}$$

The corresponding value of the standard deviation σ_t' of the predicted wind speed based on the ensemble is given by

$$\sigma_t' = \left[\sum_{k=1}^{K} \gamma_k \left(O_k - A_t' \right)^2 \right]^{\frac{1}{2}} \tag{8.20}$$

In terms of the application of this method in practice according to Vanvyve et al. (2015):

- Meteorological variables other than wind speed and direction have little effect in terms of improving the accuracy of prediction.
- Direction should be converted to a circular variable or variables (e.g. by taking sine and cosine values).
- Using a search window around τ makes little difference to prediction accuracy and can be set to zero (so τ is matched exactly to the same time of day as t).
- An ensemble of $K = 25$ values gives satisfactory performance.
- In the absence of any other information w_p can be set to 1 for all meteorological variables (equal weightings).
- The length of the analogue trend window required depends on the degree of correlation between the raw historical and target site observed data: where the correlation is good, a window of ±2 h gives satisfactory performance; if the correlation is poor, this may need to be increased to at least ±12 h.

The method can give significantly more accurate predictions compared with MCP, particularly when the correlation between raw historical and target site observed data is poor. It should, however, be noted that to give satisfactory results there needs to be a large enough range of analogues in the training period to be representative of the values in the reconstruction period. For example, extremely high wind speed values are likely to be poorly reconstructed if such extreme values are not seen in the training period.

8.4 Wind Indices

Annual mean wind speeds can vary significantly from one year to another. This presents a challenge to predict the long-term wind speed at a potential wind farm site for which the MCP process was developed, but it also makes it challenging to determine whether a wind farm is performing as expected in a given year. To address this problem, a benchmark windiness index (or *wind index*) can be used. A wind index can also be used to assess whether there are climatic changes that may give rise to trends in the long-term annual mean wind speed, though care is needed to check that such trends are not the result of changes in the exposure of wind speed measuring equipment or the position of the equipment. Indeed, one of the principal challenges of producing a robust and reliable wind index is to ensure continuity of the measurements on which it is based. For this reason, wind indices are often based on parameters that are less sensitive to local conditions, such as air pressure, or use values from a reanalysis dataset.

A wind index I_j is normally defined as a regional average of wind speed measurements from one or more sites or grid points (if a reanalysis dataset is used) over a given time interval (e.g. yearly or monthly) normalised to a long-term average, typically over a period of several decades (Watson et al., 2015)

$$I_j = \frac{\frac{1}{n}\sum_{k=1}^{n}\left(\frac{1}{m_j}\sum_{i=1}^{m_j} U_{ijk}\right)}{\frac{1}{l}\sum_{j=1}^{l}\left[\sum_{k=1}^{n}\left(\frac{1}{m_j}\sum_{i=1}^{m_j} U_{ijk}\right)\right]} \tag{8.21}$$

where U_{ijk} is the wind speed at site or grid point k, at a time i within the desired averaging period for the index j; m_j is the total number of values in the averaging period; n is the total number of regional sites or grid points contributing to the wind index; and l is the total number of long-term averaging periods. Typically, U_{ijk} will be a ten-minute or one-hourly value depending on the type of data used to generate the index.

Figure 8.6 shows an example of a wind index for the UK based on seven onshore meteorological stations and the seven nearest grid points from the

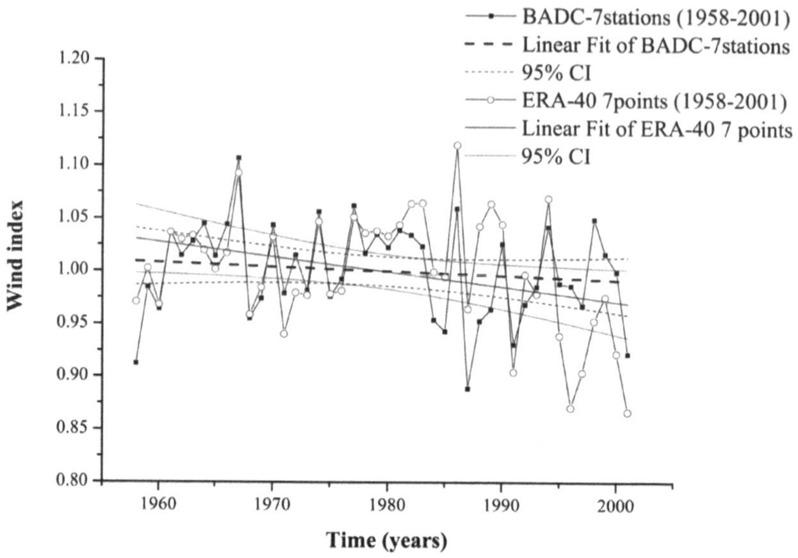

Figure 8.6 Wind index based on seven UK meteorological stations and the nearest ERA-40 grid points. *Source:* Watson et al., (2015) with permission of John Wiley & Sons.

ERA-40 reanalysis dataset (Watson et al., 2015). Although the general pattern of variation is similar between the two data sources, there are differences in the trend, particularly in the latter half of the index.

In the case of offshore sites, satellite data provide an alternative source of information to generate an index. A comparison was made between wind speed measurements from buoy and mast data and those from the QuikSCAT scatterometer for several sites in the Mediterranean Sea (Furevik et al., 2011). The indices derived are shown in Figure 8.7. In this case, although the patterns are similar, the in situ data show a larger degree of variation. This might be expected as the satellite measurements are smoothed areal averages of the wind speeds at the sites, whereas the buoy and mast data measurements are effectively point values.

8.5 The Robust Coefficient of Variation

In an assessment of the uncertainty or variation in the expected output of a wind farm, a measure is required which can accurately provide this which is both robust (insensitive to the distribution of values, e.g. does not necessarily assume a Gaussian distribution) and resistant (insensitive to outliers).

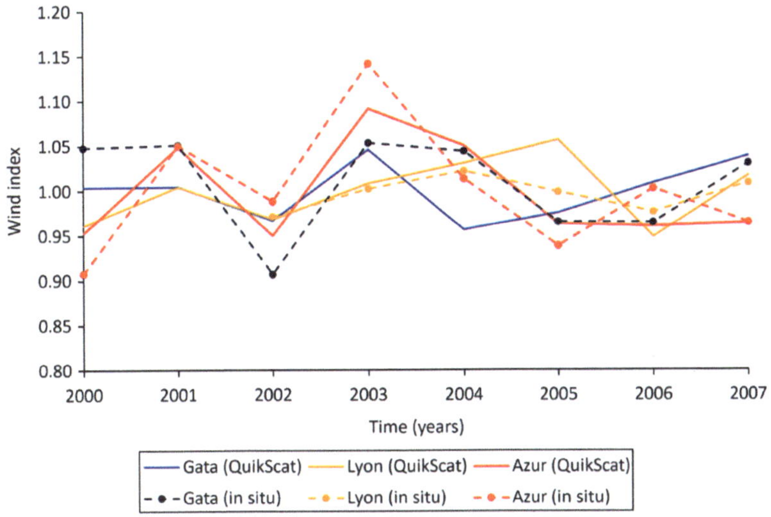

Figure 8.7 Wind index based on three sites in the Mediterranean Sea using satellite (QuikSCAT) and mast/buoy (in situ) data. *Source:* adapted from Furevik et al. (2011).

A metric which fulfils these criteria and which is often used in parameterising variability in wind speed and potential wind power is the robust coefficient of variation (RCoV) (Gunturu and Schlosser, 2012; Lee et al., 2018). The RCoV is defined as the *median absolute deviation* (MAD) from the median value divided by the median of a distribution. The advantage of using this value to quantify wind speed variability is, because of its robustness and resistance, that it gives a good correlation with wind power at a given site. It converges relatively quickly in terms of the amount of data required to give a reliable estimate of the long-term variability. It is recommended in Lee et al. (2018) that to provide a reliable estimate of the variability of the wind resource at a site that it be based on data with a temporal resolution which is at least monthly.

Figure 8.8 shows an example taken from an analysis in Lee et al. (2018) of how the RCoV can be used, together with a mean wind speed atlas, to assess the suitability of sites for wind power generation across a wide area. The analysis is based on mean monthly wind speed for the US over a period of 37 years. Figure 8.8(a) shows how many years of mean monthly wind speed values are required to provide a reliable RCoV. Figure 8.8(b) and (c) show the long-term RCoV and mean wind speed values at 80 m, respectively, and Figure 8.8(d) classifies the suitability of sites in the US for wind power generation. The darker red areas represent those sites which have the highest

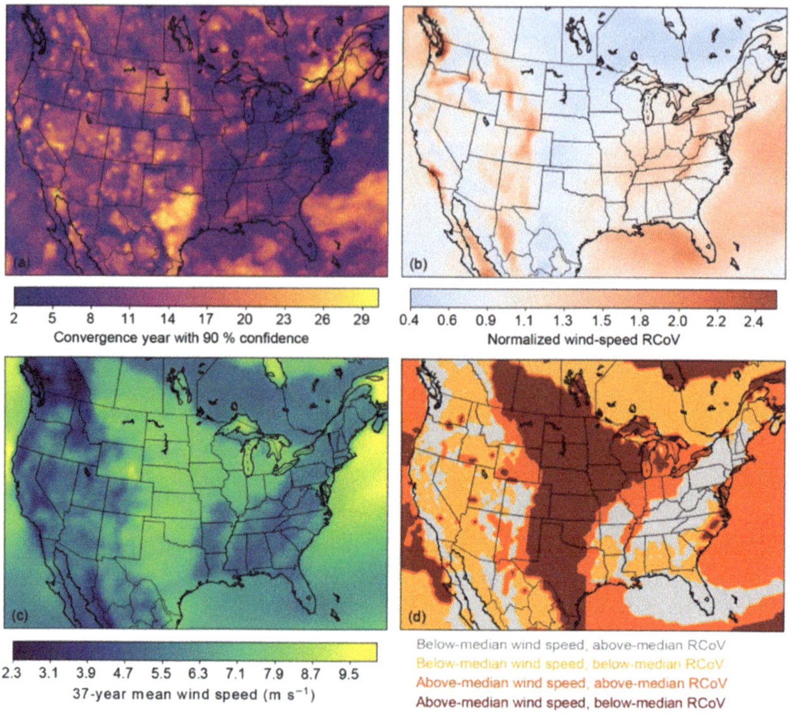

Figure 8.8 Variability of the wind resource in the US: (a) the number of years of mean monthly wind speed values to provide an estimate of the RCoV at each grid point of the map which differs by no more than 10% from the long-term value with a 90% confidence; (b) RCoV normalised by the mean RCoV for the contiguous US ; (c) the mean long-term wind speed at 80 m for the US based on monthly values; (d) classification of the wind resource over the US based on (c) and (d). *Source:* Lee et al. (2018) / Copernicus Publications / CC BY 4.0.

wind speeds and the least variability (more reliable generation) and would be most suitable for wind power generation.

8.6 Spatial Interpolation

Numerical models provide a physically based way of interpolating the variation of wind speed in the presence of changing terrain conditions. A simpler, though less accurate, method involves the use of more basic functions to interpolate wind speeds based on geographically spaced measurements based on simple relationships, such as location, height, or distance from a measurement point. In this section, some of these methods are reviewed.

8.6.1 Trend Surface Analysis

In trend surface analysis (TSA), the estimated wind speed $\hat{u}$ is assumed to be approximated by a two-dimensional (2D) polynomial function (effectively a surface) of one or higher powers (Agterberg, 1984)

$$\hat{u}(x,y) = a_{00} + a_{10}x + a_{01}y + a_{20}x^2 + a_{11}xy + \ldots + a_{pq}x^p y^q \tag{8.22}$$

where x and y are Cartesian or longitudinal/latitudinal coordinates and a_{pq} are constants which are fitted using observations from spatially separated locations over the area of interest. The surface fit can be made to either a set of local points (local polynomial) or the whole area of interest (global polynomial). If there is small-scale variation in the wind speed over a short distance, a local fit within a moving window around each point of interest may be more appropriate. The orders of the polynomial equation p and q are largely a matter of choice, though too high an order will create an over-fitted surface and will degrade accuracy. It was found that a fit of order three was optimal ($p = q = 3$) when applying TSA both locally and globally for England and Wales (Luo et al., 2008).

8.6.2 Inverse Distance Weighting

Inverse distance weighting (IDW) involves estimating the target wind speed $\hat{u}$ at a given point (x_t, y_t) based on weightings of N observations u at locations (x_i, y_i) close to the point where the weightings w reduce with distance d_i (Luo et al., 2008)

$$\hat{u}(x_t, y_t) = \frac{\sum_{i=1}^{N} w(d_i) u(x_i, y_i)}{\sum_{i=1}^{N} w(d_i)} \tag{8.23}$$

where

$$d_i = \sqrt{(x_t - x_i)^2 + (y_t - y_i)^2} \tag{8.24}$$

There are different choices for the weighting function w but commonly an inverse power law is used (i.e. $w = d_i^{-p}$, where p is the power chosen as appropriate depending on the desired distance of influence of the observations).

8.6.3 Splines

Spines, sometimes referred to as *radial basis functions*, are 2D piecewise polynomials which are joined together at specific points, or *knots*, to represent the surface of interest (e.g. wind speed). At the join, the polynomials of order n are continuous for derivatives up to an order of $n-2$, where n is frequently 3. A 2D spline can be thought of as an elastic sheet that provides the best fit to the points of interest.

The general spline function takes the form

$$\hat{u}(x_t, y_t) = p(x_i, y_i) + \sum_{i=1}^{N} w_i \phi(d_i) \tag{8.25}$$

where $p(x_i, y_i)$ is a polynomial up to order k fitted using least squares to the N observations, w_i is a weighting function for each observation i and $\phi(d_i)$ is a basis function. The second term on the right hand side of Eq. (8.25) represents the proximity of the polynomial to the data points.

The basis function can take various forms. A type of spline which has been used to interpolate meteorological variables (Luo et al., 2008) is a thin plate spline (TPS) which has the basis function

$$\phi(d_i) = (\sigma_t d_i)^2 \ln(\sigma_t d_i) \tag{8.26}$$

where σ_t is a tension parameter.

Alternative spline functions include:

The multiquadric function

$$\phi(d_i) = \left(d_i^2 + \sigma_t^2\right)^{\frac{1}{2}} \tag{8.27}$$

The completely regularised spline (CRS)

$$\phi(d_i) = -\sum_{n=1}^{\infty} \frac{(-1)^n (\sigma_t d_i)^{2n}}{n!n} = \ln\left(\frac{\sigma_t d_i}{2}\right)^2 + E_1\left(\frac{\sigma_t d_i}{2}\right)^2 + \gamma_e \tag{8.28}$$

where E_1 is the exponential integral function and γ_e is Euler's constant.

The spline with tension

$$\phi(d_i) = \ln\left(\frac{\sigma_t d_i}{2}\right) + K_0(\sigma_t d_i) + \gamma_e \tag{8.29}$$

where K_0 is the modified Bessel function.

8.6.4 Kriging

Kriging (Krige, 1966) is a variation on IDW where the spatial correlation between variables at different locations are accounted for. This method therefore provides a more unbiased spatial estimator than some other spatial interpolation methods. Ordinary kriging applied to estimation of wind speed at a target site assumes a weighted combination of wind speed measurements at N specified locations (Um and Kim, 2017)

$$\hat{u}(x_t, y_t) = \sum_{i=1}^{N} w_i u(x_i, y_i) \tag{8.30}$$

subject to the conditions that

$$\sum_{i=1}^{N} w_i \operatorname{cov}(u(x_i, y_i), u(x_j, y_j)) + \lambda = \operatorname{cov}(u(x_t, y_t), u(x_i, y_i)) \tag{8.31}$$

$$\sum_{i=1}^{N} w_i = 1 \tag{8.32}$$

where the weightings w_i are scalar values, λ is the Lagrange multiplier, and $\operatorname{cov}(a, b)$ is the covariance between a and b.

Other versions of kriging are used, such as universal kriging, where the predicted variable is assumed to have a trend in the form of a deterministic function, or simple kriging, where it is assumed that the predicted variable has a known mean value and kriging is performed on the residual.

8.6.5 Co-kriging

Co-kriging is a variation on ordinary kriging that considers the predictive ability of a secondary variable to provide an improved prediction. In the case of wind speed estimation, height elevation (height above sea level) at the location of the wind speed measurements used in the kriging estimation can show a strong correlation with the wind speed, though not in all cases. This is so as sloping terrain tends to cause wind speed-up, as discussed in Chapter 5. In this case, the weighted estimate of the wind speed becomes (Um and Kim, 2017)

$$\hat{u}(x_t, y_t) = \sum_{i=1}^{N} w_i u(x_i, y_i) + \sum_{j=1}^{M} z_j t(x_j, y_j) \tag{8.33}$$

where z_j is the weighting value for the secondary variable t (e.g. terrain elevation) for those M locations where this variable shows a strong correlation with u. In this case, the estimated value of the wind speed at the target site is subject to the conditions

$$\sum_{i=1}^{N} w_i \operatorname{cov}(u(x_i,y_i),u(x_i,y_i))+\sum_{j=1}^{M} x_j \operatorname{cov}(u(x_i,y_i),t(x_j,y_j))-\lambda_1$$
$$=\operatorname{cov}(u(x_t,y_t),u(x_l,y_l)) \tag{8.34}$$
$$l=1,2,\ldots,N$$

$$\sum_{i=1}^{N} w_i \operatorname{cov}(t(x_i,y_i),u(x_i,y_i))+\sum_{j=1}^{M} x_j \operatorname{cov}(t(x_i,y_i),t(x_j,y_j))-\lambda_2$$
$$=\operatorname{cov}(u(x_t,y_t),t(x_l,y_l)) \tag{8.35}$$
$$l=1,2,\ldots,M$$

$$\sum_{i=1}^{N} w_i = 1 \tag{8.36}$$

$$\sum_{j=1}^{M} x_j = 1 \tag{8.37}$$

where λ_1 and λ_2 are Lagrange multipliers.

8.6.6 A Comparison of Spatial Interpolation Techniques

To give an idea of the relative accuracy of the different interpolation techniques, a comparison is presented between several interpolation methods to estimate 10 m wind speeds throughout England and Wales (Luo et al., 2008) and the Flanders area of Belgium (Van Ackere et al., 2015). For England and Wales, averaged wind speed data from 189 meteorological stations for one day were used for spatial interpolation onto a regular 5 × 5 km grid. In the case of Flanders, annual averaged wind speed data for 43 stations were interpolated into a 250 × 250 m grid. Mean and root-mean-squared errors were calculated by removing the data from one of the stations, calculating the residual error and repeating for all other stations, a procedure sometimes referred to as *leave one out cross-validation*. The results are shown in Table 8.4 along with minimum and maximum interpolated wind speeds for the England and Wales case. The best results are given by the various kriging techniques, with co-kriging standing out particularly in the case of the results for England and Wales. The global polynomial interpolation gives the worst results as well as significantly underpredicting the wind speed range for England and Wales. The use of local polynomial interpolation gives a significant improvement in accuracy for both regions. In the case of

Table 8.4 A comparison of mean and root-mean-squared errors of the 10 m wind speed predicted for England and Wales (for one day) and Flanders (annual mean) using various spatial interpolation techniques. Minimum and maximum interpolated wind speeds are also shown for England and Wales. Data taken from Luo et al., (2008) for England and Wales (EW) and (Van Ackere et al., 2015) for Flanders (FL).

Method	Min. wind speed ($m \cdot s^{-1}$)	Max. wind speed ($m \cdot s^{-1}$)	Mean error ($m \cdot s^{-1}$)		Root-mean-squared error ($m \cdot s^{-1}$)	
	EW	EW	EW	FL	EW	FL
Global polynomial	4.6	9.2	−0.02	−0.26	1.93	0.69
Local polynomial	0.1	11.7	0.01	−0.26	1.69	0.50
IDW (p optimised)	3.1	12.1	−0.09	—	1.74	—
IDW ($p = 1$)	—	—	—	−0.14	—	0.58
IDW ($p = 2$)	—	—	—	−0.13	—	0.52
IDW ($p = 3$)	—	—	—	−0.12	—	0.51
IDW ($p = 4$)	—	—	—	−0.13	—	0.52
IDW ($p = 5$)	—	—	—	−0.13	—	0.54
TPS	1.5	13.8	−0.05	—	1.89	—
CRS	—	—	—	−0.13	—	0.48
Standard kriging	—	—	—	−0.03	—	0.48
Ordinary kriging	2.6	13.4	−0.01	−0.13	1.61	0.49
Universal kriging	3.2	9.8	0.01	−0.14	1.71	0.48
Co-kriging	2.6	13.6	−0.01		1.47	

IDW, the choice of an inverse power exponent makes little difference but a value of $p = 3$ seems to perform the best, at least in the case of Flanders. The results for spline interpolation contrast between the two regions: for England and Wales this method was one of the worst, but for Flanders it performs favourably compared with the kriging methods. It should be noted that in the Flanders case the completely regularised spline function was used in contrast to England and Wales where the TPS was used.

In the case of Flanders, the spatially interpolated values are shown in Figure 8.9. Most notably it can be seen that IDW with an inverse power

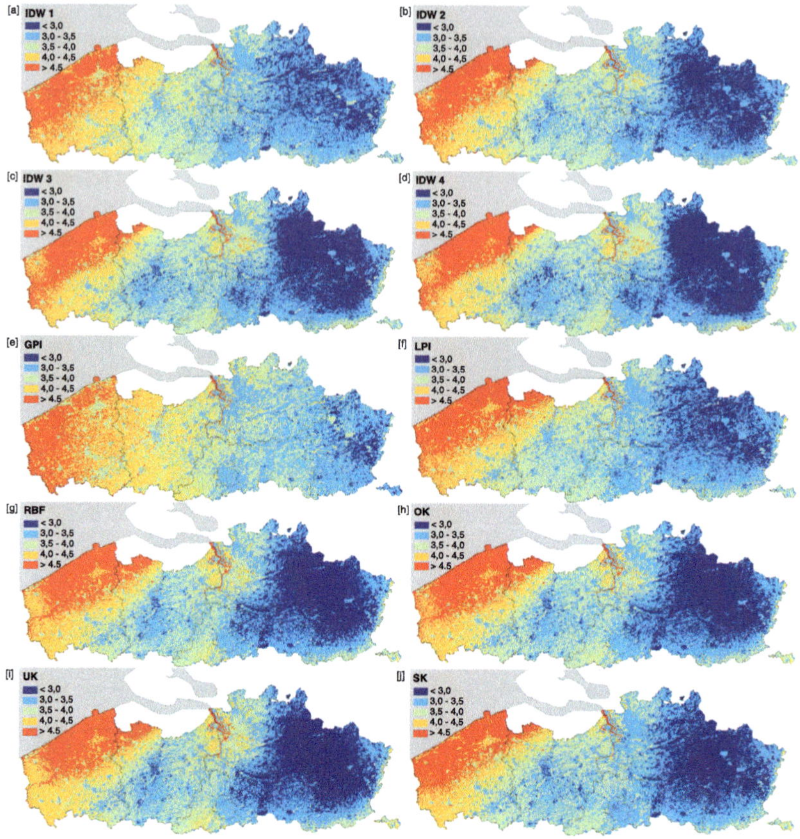

Figure 8.9 Spatially interpolated mean annual 10 m wind speeds (in ms^{-1}) for Flanders using different methods: (a) IDW (p = 1), (b) IDW (p = 2), (c) IDW (p = 3), (d) IDW (p = 4), (e) global polynomial, (f) local polynomial, (g) completely regularised spline, (h) ordinary kriging, (i) universal kriging, and (j) standard kriging. *Source:* Van Ackere et al. 2015 / MDPI / CC BY 4.0.

coefficient of 1 and the global polynomial interpolation tend to smooth out the data most and give the worst result in terms of root-mean-squared error.

8.6.7 Machine Learning

Machine learning (ML) is a technique which is being used increasingly in many disciplines where the relationship between several input variables and one or more output variables is unknown and a simple linear relationship may not be adequate to predict the output. The increasing availability of computing resources makes it feasible for ML to be used to train a nonlinear algorithm based on a large dataset which can then be used for prediction. This training process involves tuning several fundamental parameters on which the algorithm is based known as *hyperparameters*. ML is starting to be used in wind resource assessment to statistically downscale meteorological variables where low-resolution data at a sparse number of points need to be converted to a higher resolution, based on additional information such as terrain height, terrain type, distance from the coast, etc. Many ML techniques are available and a full description of these is beyond the scope of this book. However, many publicly accessible educational training resources are available. One such technique which can be applied to downscale low-resolution data, uses a series of decision trees. Figure 8.10 shows an example decision tree based on high-resolution terrain features and wind speed values from a lower spatial resolution dataset. Depending on the path through the tree, different wind speed scaling factors in the final leaves are chosen to correct the lower resolution wind speed values for the location of interest. The number of branches, the degree of branching, the decision features, etc., are some of the hyperparameters that need to be chosen and tuned. A set of trees is produced iteratively based on a set of training data and a weighted average of the tree-based correction factors used to predict the final wind speed value at a particular location. By way of example, in Oh et al. (2022), high-resolution WRF-generated wind speed values are used to train a decision-tree-based model, and inputs to the model are low-resolution global model values from the ERA5 reanalysis (described in Section 9.2.10), terrain height, terrain coverage data, and derived variables.

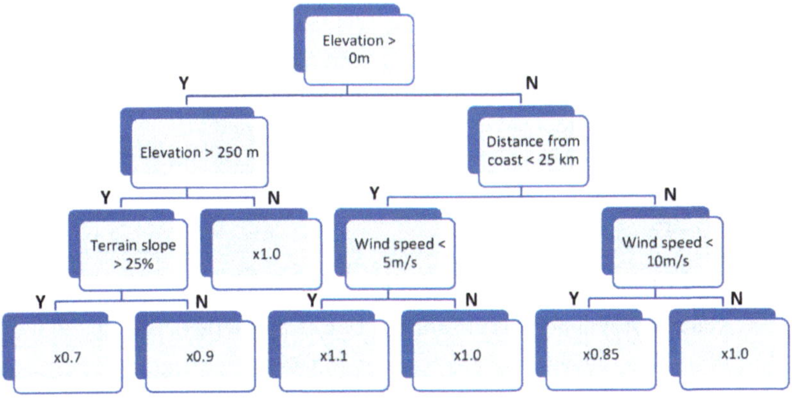

Figure 8.10 An example decision tree where low-resolution wind speed values are to be downscaled to a higher-resolution map based on terrain features. *Source:* adapted from Oh et al. (2022).

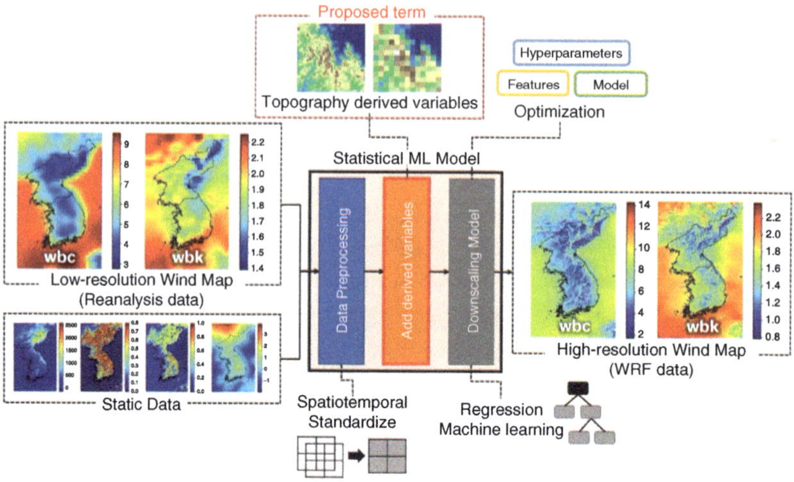

Figure 8.11 Overview of a procedure to downscale low-resolution numerical model wind speed data to produce high-resolution maps based on an ML approach. *Source:* Oh et al. (2022) with permission of John Wiley & Sons.

Figure 8.11 shows an overview of the process. First of all, the ERA5 reanalysis, terrain height, and terrain coverage data are standardised to be of the same resolution. Additional variables are derived from these data, such as distance from the coast, terrain type, etc. The ML model is trained using a test dataset taken from a limited sample of the high-resolution WRF data. The trained model is then used to predict wind speed values at a higher resolution over a wider area than the test dataset. Although this example uses numerical model data, equally, a sparse network of data from measurement sites could be used along with additional information, such as satellite data. The use of ML in resource assessment is still in its infancy but its application is likely to expand in the coming years.

References

Agterberg, F.P., 1984. Trend surface analysis, in: *Spatial Statistics and Models.* G.L., Gaile, C.J. Willmott, (eds) Springer, Dordrecht, Netherlands, pp. 147–171. https://doi.org/10.1007/978-94-017-3048-8_8.

Bass, J., 2009. MCP: pitfalls and common mistakes, in: *AWEA Wind Resource & Project Assessment Workshop, 30th September–1st October 2009.* Minneapolis, MN.

Carta, J.A., Velázquez, S., Cabrera, P., 2013. A review of measure-correlate-predict (MCP) methods used to estimate long-term wind characteristics at a target site. *Renew. Sustain. Energy Rev.* https://doi.org/10.1016/j.rser.2013.07.004.

Derrick, A., 1992. Development of the measure-correlate-predict strategy for site assessment, in: B. Clayton (Ed.), *Fourteenth British Wind Energy Association Conference 25th–27th March 1992*, Nottingham, UK. Mechanical Engineering Publications, London,, pp. 259–265.

Fuller, W., 1987. *Measurement Error Models.* Wiley. https://doi.org/10.1002/9780470316665.

Furevik, B.R., Sempreviva, A.M., Cavaleri, L., Lefèvre, J.-M., Transerici, C., 2011. Eight years of wind measurements from scatterometer for wind resource mapping in the Mediterranean Sea. *Wind Energy* 14, 355–372. https://doi.org/10.1002/we.425.

Gunturu, U.B., Schlosser, C.A., 2012. Characterization of wind power resource in the United States. *Atmos. Chem. Phys.* 12, 9687–9702. https://doi.org/10.5194/acp-12-9687-2012.

Hill, D.C., McMillan, D., Bell, K.R.W., Infield, D., 2012. Application of auto-regressive models to UK wind speed data for power system impact studies. *IEEE Trans. Sustain. Energy* 3, 134–141. https://doi.org/10.1109/TSTE.2011.2163324.

Joensen, A., Landberg, L., Madsen, H., 1999. A new measure-correlate-predict approach for resource assessment, in: *European Wind Energy Conference. Nice, France* (eds. G. C. Larsen, K. Westermann and P. Ndrgard). Riso National Laboratory Roskilde, Denmark, pp. 165–168.

Jourdier, B., 2020. Evaluation of ERA5, MERRA-2, COSMO-REA6, NEWA and AROME to simulate wind power production over France. *Adv. Sci. Res.* 17, 63–77. https://doi.org/10.5194/asr-17-63-2020.

King, C., Hurley, B., 2005. The SpeedSort, DynaSort and scatter wind correlation methods. *Wind Eng.* 29, 217–241. https://doi.org/10.1260/030952405774354868.

Krige, D.G., 1966. *Two-dimensional Weighted Moving Average Trend Surfaces for Ore Evaluation.* South African Institute of Mining and Metallurgy.

Lee, J.C.Y., Jason Fields, M., Lundquist, J.K., 2018. Assessing variability of wind speed: comparison and validation of 27 methodologies. *Wind Energy Sci.* 3, 845–868. https://doi.org/10.5194/wes-3-845-2018.

Luo, W., Taylor, M.C., Parker, S.R., 2008. A comparison of spatial interpolation methods to estimate continuous wind speed surfaces using irregularly distributed data from England and Wales. *Int. J. Climatol.* 28, 947–959. https://doi.org/10.1002/joc.1583.

Oh, M., Lee, J., Kim, J.Y., Kim, H.G., 2022. Machine learning-based statistical downscaling of wind resource maps using multi-resolution topographical data. *Wind Energy* 25, 1121–1141. https://doi.org/10.1002/WE.2718.

Pramod Jain, P.D., 2016. *Wind Energy Engineering*, 2nd edition. McGraw-Hill Education, New York.

Riedel, V., Strack, M., 2001. Robust approximation of functional relationships between meteorological data: alternative measure-correlate-predict algorithms, in: *European Wind Energy Conference*. Wind Energy for the New Millennium. pp. 806–809.

Rogers, A.L., Rogers, J.W., Manwell, J.F., 2005. Comparison of the performance of four measure-correlate-predict algorithms. *J. Wind Eng. Ind. Aerodyn.* 93, 243–264. https://doi.org/10.1016/j.jweia.2004.12.002.

Santos, J.S., Sakagami, Y., Haas, R., Passos, J.C., Machuca, M.N., Radünz, W.C., Dias, E., Lima, M.M., 2020. Wind speed evaluation of MERRA-2, ERA-interim and ERA-5 reanalysis data at a wind farm located in Brazil, in: *Proceedings of the ISES Solar World Congress 2019 and IEA SHC International Conference on Solar Heating and Cooling for Buildings and Industry 2019*. International Solar Energy Society, pp. 2265–2274. https://doi.org/10.18086/swc.2019.45.10.

Schlez, W., Infield, D., 1998. Horizontal, two point coherence for separations greater than the measurement height. *Boundary-Layer Meteorol* 87, 459–480. https://link.springer.com/article/10.1023/A:1000997610233.

St. Martin, C.M., Lundquist, J.K., Handschy, M.A., 2015. Variability of interconnected wind plants: correlation length and its dependence on variability time scale. *Environ. Res. Lett.* 10. https://doi.org/10.1088/1748-9326/10/4/044004.

Um, M.-J., Kim, Y., 2017. Estimating potential wind energy from sparsely located stations in a mountainous coastal region. *Meteorol. Appl.* 24, 279–289. https://doi.org/10.1002/met.1629.

Van Ackere, S., Van Eetvelde, G., Schillebeeckx, D., Papa, E., Van Wyngene, K., Vandevelde, L., 2015. Wind resource mapping using landscape roughness and spatial interpolation methods. *Energies* 8, 8682–8703. https://doi.org/10.3390/en8088682.

Vanvyve, E., Delle Monache, L., Monaghan, A.J., Pinto, J.O., 2015. Wind resource estimates with an analog ensemble approach. *Renew. Energy* 74, 761–773. https://doi.org/10.1016/J.RENENE.2014.08.060.

Watson, S., 2014. Quantifying the variability of wind energy. *Wiley Interdiscip. Rev. Environ.* 3, 330–342. https://doi.org/10.1002/wene.95.

Watson, S.J., Kritharas, P., Hodgson, G.J., 2015. Wind speed variability across the UK between 1957 and 2011. *Wind Energy* 18, 21–42. https://onlinelibrary.wiley.com/doi/abs/10.1002/we.1679.

Woods, J.C., Watson, S.J., 1997. A new matrix method of predicting long-term wind roses with MCP. *J. Wind Eng. Ind. Aerodyn.* 66, 85–94. https://doi.org/10.1016/S0167-6105(97)00009-3.

9

Atmospheric Reanalyses and Wind Atlases

For over 30 years, reanalyses on a global scale of a range of climate variables have been available on a regular grid and at a regular temporal resolution. The initial reanalysis datasets available had a relatively coarse spatial and temporal resolution, making them limited for the purposes of a detailed wind resource assessment, though they were useful for producing regional resource maps. With the advent of more powerful computing resources, subsequent global reanalyses have increased in their resolution, making the data more accessible and more appropriate for carrying out detailed site assessments, particularly where no on-site measurements are available. Reanalyses have the feature that values of variables such as wind speed are not influenced by very localised effects, for example the location and mounting of instruments on a mast or nearby trees, making the data appropriate for a regional assessment. On the other hand, the influence of small-scale terrain features may not be captured. Nonetheless, this makes the data suitable for downscaling using a mesoscale or microscale model to produce a wind atlas (N.B. model chains are covered in Section 5.5). This chapter explains what is meant by reanalysis, describes the best-known reanalysis datasets, and briefly reviews two well-known wind atlases which have been numerically downscaled from reanalysis data.

9.1 What Is a Reanalysis?

To understand what a *reanalysis* is, first the concept of an *analysis* should be understood. Figure 9.1 shows the steps in an atmospheric analysis. First of all, a short-term (up to a few hours ahead) forecast of the state of the atmosphere is made using a global forecasting model. This is commonly known as the *background*. The background forecast is then statistically combined with concurrent observations, using a process known as

Handbook of Wind Resource Assessment, First Edition. Simon Watson.

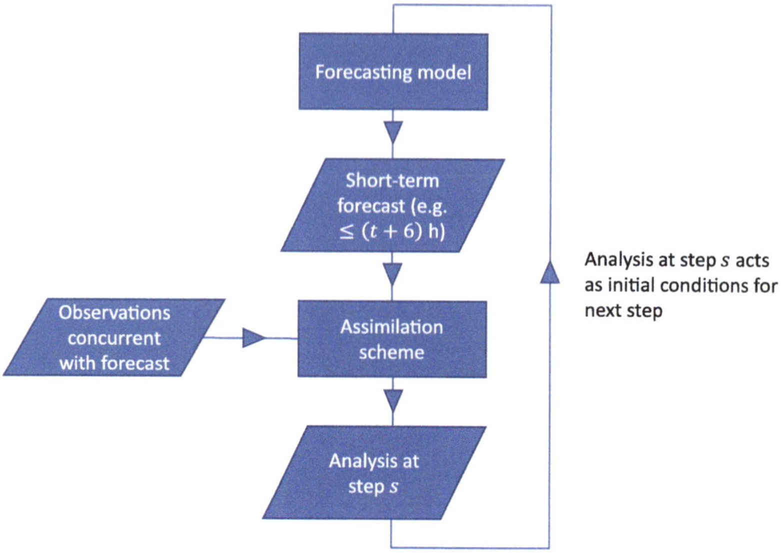

Figure 9.1 Steps in an atmospheric analysis.

assimilation, based on an appropriate error minimisation metric which considers the variation of both observations and forecasts, to provide the best estimate of the state of the atmosphere. Common assimilation schemes are: 3DVAR which is a three-dimensional (3D) variational data assimilation scheme where observations close to the forecast analysis time are assumed to occur, and thus are assimilated, *at* that model time; and 4DVAR, which is a 4D scheme (three spatial dimensions plus time) where the observations close to the forecast analysis time are assimilated at the time they are actually measured. The latter scheme is potentially more accurate but more computationally intensive. The output of the process (the *analysis*) is available on a regular spatial grid and at specified time intervals depending on the forecast model. This process has the advantage of providing estimates of meteorological variables that had not been or could not be directly measured. The analysis at each step acts as the initial conditions for the next analysis step so that the process can be repeated several times to provide a time series of analyses. This process is frequently used in real-time operational forecasting to ensure that each forecast produced is initialised with the best information available regarding the state of the atmosphere based on model output and observations. This process can also be used retrospectively using historic data and model *hindcasts* (forecasts made after the event). The term *reanalysis* refers to a special case of analysis where a long series of analysis steps, typically over several decades in the past, are calculated based on a fixed version of an assimilation scheme and forecast model. The reanalysis so produced is then consistent and homogeneous over a long period making it suitable for long-term climate study or, in the case of wind resource assessment, to provide a coherent long-term wind speed reference for procedures such as measure–correlate–predict (MCP). The most-well-known reanalyses are reviewed in the following sections in approximate chronological order. Note that reanalyses are sometimes referred to as first, second, third generation, etc. The distinction is somewhat arbitrary but generally reflects an evolution where more data (especially satellite), improved models, and better assimilation schemes have become available over time.

9.2 Global Reanalyses

Global scale reanalyses, as the name suggests, can be used to provide reference data for anywhere in the world, making them universally applicable for wind resource assessment. Because of their global coverage, their resolution is generally restricted to tens of kilometres. The model output is frequently produced on a Gaussian grid, where the grid lines are equally spaced in latitude. In most cases a *regular* Gaussian grid is used where the grid lines in longitude are approximately (but not exactly) equally spaced where the grid point spacing is determined using Gaussian quadrature and the number of grid points along the longitudes is approximately double those along the latitudes. The equal grid line spacing means that the distance between grid points along a line of latitude is constant but reduces along a line of longitude moving from the equator to the poles. In a reduced Gaussian grid, which is sometimes used, the number of grid points is reduced between the equator and the poles to give approximately equally spaced grid points. There are no grid points at the poles. The models used to produce global reanalyses generally solve some of the governing equations in the frequency domain as at global scale this has some advantages in terms of speed and accuracy over spatial grid point models. Such models are called *spectral models*. This means that the spectral solutions need to be transformed to a grid to produce usable data. The resolution of the grid is dictated by the number of waves of different wavelengths (spherical harmonics) which the model uses in its spectral solutions (the *model resolution*). Above this number, the solution is said to be *truncated*. A popular type of truncation is known as *triangular*, or *T*, *truncation*. The resolution of the Gaussian grid is therefore often referred to using the type and level of spectral truncation using a format, such as T*xx*, where *xx* represents the

Figure 9.2 A T62 Gaussian grid. *Source:* M. W. Toews / Wikimedia Commons / CC BY 4.0.

maximum wavenumber above which the solution is truncated (this also represents the size of the smallest phenomena that can be resolved by the model). Figure 9.2 shows an example of a regular Gaussian grid (in this case T62). In the case of a reduced Gaussian, or linear, grid where the grid point spacing is approximately constant in both latitude and longitude, it may be denoted TL*xx*.

The following sections describe some of the main features of the most-well-known global reanalyses.

9.2.1 NCEP/NCAR

The National Centers for Environmental Prediction (NCEP) and the National Center for Atmospheric Research (NCAR) in the US collaborated on the first widely available reanalysis of the global climate covering the 40-year period 1957–1996 (Kalnay et al., 1996). This was subsequently known as *Reanalysis (R-1)* and is often referred to as a *first-generation reanalysis*. The project was inspired by the need to produce a long-term reliable time series of climate variables where climate-related changes could be determined without being obscured by model changes or changes to an assimilation scheme. This can be a problem, if historical analyses from past operational weather forecasts are used for climate study. Forecasting models are modified and improved over time, as are data assimilation schemes. Unfortunately, this can give rise to discontinuities (e.g. jumps) in meteorological variables from analyses before and after such changes. Clearly, these changes could be confused with climate-related artefacts which are not actually present. The NCEP/NCAR project was conceived to eliminate as far as possible such discontinuities by using a single frozen model and fixed data assimilation scheme for the entire 40-year time series of analyses. Each reanalysis produced by the project also had the advantage of being able to use a much more comprehensive set of observational data for model initialisation than would have been available for an operational forecast analysis made at the time.

The forecasting model used was the NCEP global operational forecast model frozen at its stage of development on 10 January 1995 (Kanamitsu, 1989; Kanamitsu et al., 1991). The operational model had a horizontal resolution of approximately 105 km at the equator, though the reanalysis used a coarser resolution (210 km). This uses a T62 Gaussian grid with 192 lines of longitude and 94 lines of latitude. The vertical grid had 28 σ (sigma) levels with five levels in the boundary layer. Data used as input to the model included:

- Global rawinsonde data.
- Surface marine data including those from ships, buoys and ocean stations.
- Aircraft data.
- Land surface observations from meteorological stations.
- Satellite sounder data.

A plan to use satellite microwave observations to infer ocean wind speeds was abandoned due to problems with the data.

Reanalyses were output as instantaneous values for four times per day (00, 06, 12, and 18 UTC, or Universal Time Coordinated, i.e. the time on the 0° line of longitude without adjustment for summertime). Several datasets are available as detailed in Kalnay et al. (1996) including data at 17 pressure levels on 144 lines of longitude by 73 lines of latitude (2.5 × 2.5°) and data on the original T62 grid including surface flux values and values on the 28 sigma (σ) levels. Zonal and meridional wind speed values are available at 10 m height on the T62 grid. Monthly averages of the data are also available.

Subsequently, Reanalysis 2 (R-2) improved upon and extended the period of the original R-1 project (Kanamitsu et al., 2002) covering the years 1979 to the present (2022 at the time of writing). Several errors and discrepancies came to light that were corrected in this second reanalysis. Updated versions of both the forecasting model and assimilation scheme were used but the data were produced at the same temporal and spatial resolutions as R-1.

9.2.2 ERA-15

Although the first reanalysis project of the European Centre for Medium-Range Weather Forecasts (ECMWF) was developed much earlier than the NCEP/NCAR reanalysis based on the 1980 forecasting model of the time using data from the First Global Atmospheric Research Program (GARP) Global Experiment (FGGE), this only covered the year 1979 and had a relatively coarse spatial resolution (Bengtsson et al., 1982).

The first multiyear reanalysis produced by the ECMWF covered the period 1979–1993 and used the Integrated Forecast System (IFS) version of the ECMWF forecasting model. This was known as ERA-15 (Gibson et al., 1999), where ERA stands for ECMWF Reanalysis. The model employed a

T106 grid with 320 lines of longitude and 160 lines of latitude giving a resolution of 125 km at the equator and 31 vertical hybrid levels. In this sense, *hybrid* means that the vertical levels follow the terrain in the lower part of the atmosphere and follow the pressure levels further aloft with a smooth transition between the two.

Data input to the model included:

- Satellite radiance data.
- Cloud track winds.
- Surface meteorological station observations.
- Radiosonde data.
- Ship and buoy data.
- Aircraft data.
- Sea surface temperature and sea ice coverage from satellite observations.

Data are available at six-hourly and monthly averages on the model (T106) grid as well as a coarser $2.5 \times 2.5°$ grid at the 31 model levels, 17 pressure levels, and surface and sigma levels. Wind speed components are available at 10 m.

The ERA-15 dataset is not recommended for use in resource assessment as it was superseded by further reanalyses from the ECMWF which are described in Sections 9.2.3, 9.2.5 and 9.2.10.

9.2.3 ERA-40

ERA-40 (Uppala et al., 2005) was a second-generation reanalysis from ECMWF using IFS that benefited from the significant increase in availability of satellite data, aircraft, and ocean observations. It was the first reanalysis to directly assimilate satellite radiance data leading to less measurement bias error in contrast to ERA-15, where the data were preprocessed. It was also the first reanalysis to couple an ocean wave model with an atmospheric model.

The resolution of the model output was greater than previous analyses and benefited from improved computing resources. The reanalysis was made for the period 1957 to 2002. Table 9.1 shows the availability of various datasets assimilated by the ERA-40 reanalysis for different time periods.

A steady increase in surface observations is seen, but the rapid increase in satellite data and aircraft measurements over the period of the reanalysis is quite apparent. Also noteworthy is the peak in the availability of observations from weather/pilot balloons and radiosondes during the middle of the analysis but a decline in later years which has more than been made up for by other data sources.

Table 9.1 The availability over time of different data sources assimilated in the ERA-40 reanalysis. Values are observations per day.

Obs. Type	1958–1966	1967–1972	1973–1978	1979–1990	1991–2001
Synop./ship	15 313	26 615	28 187	33 902	37 049
Radiosondes	1821	2605	3341	2274	1456
Pilot balloons	679	164	1721	606	676
Aircraft	58	79	1544	4085	26 341
Buoys	0	1	69	1462	3991
Satellite radiance	0	6	35 069	131 209	181 214
Satellite winds	0	0	61	6598	45 671
Scatterometer	0	0	0	0	7575
PAOBs	0	14	1031	297	277

Synop.: synoptic surface weather stations. PAOB: synthetic surface pressure observations from satellite images.
Source: Uppala et al. (2005) / Royal Meteorological Society.

The ERA-40 reanalysis used a 3DVAR assimilation scheme and a T159 grid with 480 lines of longitude and 240 lines of latitude giving a resolution of 83 km at the equator and 60 vertical hybrid levels. The reanalysis data are available at six-hourly intervals.

9.2.4 Climate Forecast System Reanalysis

Following on from the NCEP/NCAR R-1 and R-2 reanalyses, NCEP used their Climate Forecast System (CFS) model to produce the Climate Forecast System Reanalysis (CFSR) (Saha et al., 2010). This incorporated many of the advances of other third-generation reanalyses including comprehensive satellite datasets, atmosphere-ocean model coupling, and an interactive sea ice model. Data were output hourly on a T382 grid with 1152 lines of longitude and 576 lines of latitude giving a resolution of 38 km at the equator. The model uses 64 vertical hybrid levels. Data are available for the period 1979 to 2011.

9.2.5 ERA-Interim

ERA-Interim (Berrisford et al., 2011; Dee et al., 2011) was devised as a bridge between the ERA-40 reanalysis and the next generation which later became ERA5. It is sometimes referred to as a *third-generation reanalysis*. The IFS forecasting model used had improved physics compared with that used for

ERA-40, certain biases were corrected (e.g. in the satellite data), and the data assimilation scheme was improved, in this case using 4DVAR. The reanalysis was made for the period 1979 to 2019 with data available on a T255 grid with 512 lines of longitude and 256 lines of latitude giving a resolution of 60 km at the equator and 60 hybrid levels. Reanalysis fields are available six-hourly, though some forecast surface parameters are available three-hourly. A database of monthly means of several parameters is also available.

9.2.6 Modern-Era Retrospective Analysis for Research and Application

The Modern-Era Retrospective Analysis for Research and Application (MERRA) was a third-generation reanalysis carried out by the US National Aeronautics and Space Administration (NASA) using version 5.2.0 of the Goddard Earth Observing System (GEOS) (Rienecker et al., 2008). The intention of this analysis was to put the observations from NASA's Earth Observing System (EOS) satellites into a climate context and to improve on the implementation of the hydrological cycle in the forecast model which created known deficiencies in previous generations of reanalysis (Rienecker et al., 2011). The aim was to produce climate data directly of relevance to the wider science community. In contrast to previous reanalyses, error statistics are available based on the difference between the background forecast and the observations and the difference between the background forecast and the analysis. This provides useful information to assess uncertainties. Output data are available hourly for the period 1979 to 2016 on a grid with 540 lines of longitude and 360 of latitude giving a horizontal resolution at the equator of around 60 km with 72 hybrid levels.

9.2.7 JRA-55

The Japanese Meteorological Agency (JMA) conducted their first reanalysis based on a 25-year period known as the Japanese ReAnalysis (JRA-25) covering the period 1979–2004 (Onogi et al., 2007). This was considered rather short for climatological studies and thus led to the follow-up reanalysis project JRA-55 initially covering a 55-year period from 1958 based on the first availability of radiosonde data, though analyses are ongoing. The reanalysis is based on JRA's operational forecasting system using the Global Spectral Model (GSM) as of December 2009. Data are available every six hours on a TL319 grid with 640 lines of longitude and 320 lines of latitude giving an approximate horizontal resolution at the equator of 55 km, with 60 hybrid levels.

9.2.8 CFSR Version 2

CFSR was superseded by CFSR Version 2, which used an improved version of the CFS forecast model as well as an improved assimilation scheme (Saha et al., 2014). Spatial and temporal resolution remained the same. Data are available from 2011 to present as of time of writing.

9.2.9 MERRA-2

To address shortcomings in MERRA, such as discontinuities in data output due to changes in measurement biases and reliance on old legacy satellite observations, NASA produced its next generation of reanalysis, MERRA-2, which used an improved version of the GEOS operational model and a better data assimilation scheme (Gelaro et al., 2017). MERRA-2 benefitted from a significantly enhanced range of satellite measurements compared with its predecessor. Figure 9.3 illustrates the difference in the number of observations used in the MERRA-2 reanalysis compared to MERRA. It also serves to illustrate the growth in observations in general available for forecasting and analysis, driven mainly by the increase in satellite measurements. The increased number of satellite measurements available after 2008 is quite apparent. Data are available from 1979 to present day as of time of writing. The model output is hourly on a grid with 576 lines of longitude and 360 of latitude giving a horizontal resolution at the equator of around 60 km with 72 hybrid levels.

9.2.10 ERA5

The most up-to-date reanalysis produced by the ECMWF at the time of writing is that known as ERA5 (Hersbach et al., 2020). Although this might be considered a fourth-generation reanalysis, somewhat confusingly it is referred to by ECMWF as *fifth generation*. This reflects the fact that it is ECMWF's fifth-generation reanalysis, if the first reanalysis based on FGGE is included. Needless to say, ERA5 benefits from improved spatial resolution, forecasting, and assimilation schemes but in addition has several innovations compared to ERA-Interim, which it replaced, including:

- 100 m wind speeds.
- Output data at an hourly resolution.
- Uncertainty estimates at a three-hourly resolution based on an ensemble average of ten model forecasts.[1]
- Availability of data within five days of real time (fully quality checked after two months).

---

1 An ensemble is produced by running a forecast model multiple times, each time perturbing the input data to reflect their uncertainty. The ensemble of forecasts thus produced can be used to calculate an estimate of the forecast uncertainty for the meteorological variables of interest.

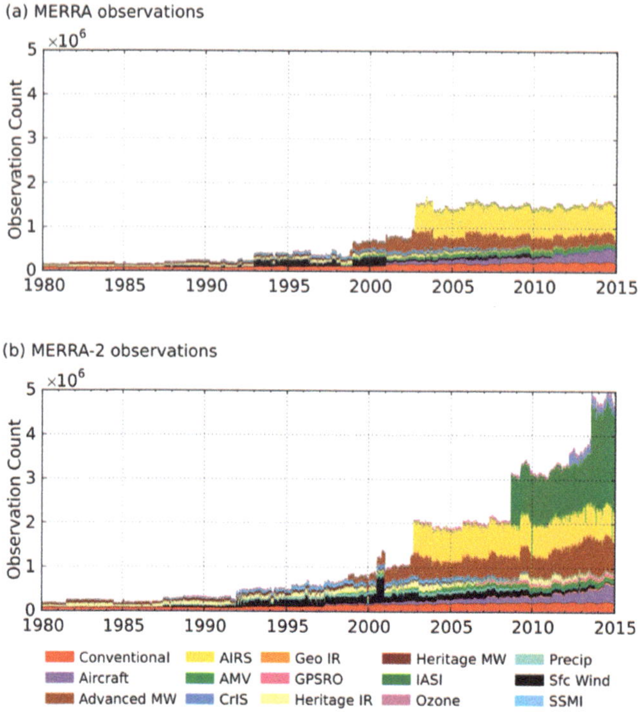

Figure 9.3 The number and type of observations assimilated by (a) MERRA and (b) MERRA-2. Key: MW: microwave; AIRS: Advanced Infrared Sounder; AMV: Atmospheric Motion Vector; CrIS: Cross-Track Infrared Sounder; IR: infrared; GPSRO: Global Positioning System Radio Occultation; IASI: Infrared Atmospheric Sounding Interferometer; Precip: precipitation; Sfc Wind: surface wind speeds; SSMI: Special Sensor Microwave Imager. *Source:* Gelaro et al. (2017) / with permission of American Meteorological Society.

The hourly data are output on a T639 grid with 1280 lines of longitude and 640 lines of latitude giving a resolution of 31 km at the equator and 137 hybrid vertical levels. The period of coverage will eventually range from 1950 to present day, though the earliest available data are from 1959 at the time of writing.

As this dataset is now one of the most widely used of the reanalysis datasets, it is worth pointing out a couple of known problems in relation to wind speed (Hersbach et al., 2020), namely:

- The analysis window used is 12 hours and there are known jumps in boundary layer wind speed (e.g. at 10 m) at the transition between windows. Though the jumps are small (of the order of 0.25 ms^{-1}), they are noticeable in long-term averaging, for example at certain hours of the day and at certain locations.

- Up to once or twice a year, there are unrealistic spikes (up to 300 ms^{-1}) in surface wind speed lasting two to three hours at certain global locations. Quality control is ongoing to remove these anomalies, but users should be aware.

9.2.11 Twentieth-century Reanalyses

The various generations of reanalyses described in this chapter have endeavoured to assimilate ever-greater volumes of data to provide the best estimate of the state of the atmosphere at any given time. There is another type of global reanalysis which deliberately restricts the type of model input variables. This type of reanalysis has the aim of producing a very long data record (>100 years) for long-term climatological studies. This is perhaps of less relevance for wind resource assessment but may be of use to study potential long-term climatological trends leading to systematic increases or decreases in wind speed or changes in extremes. Because such reanalyses stretch back typically to the early twentieth or late nineteenth centuries, when the number and type of meteorological measurements was much restricted compared to present day, it is necessary to restrict model input variables to ensure a homogeneous long-term record without biases introduced as new data sources (e.g. satellite) become available even if the output data are less accurate than the latest generation reanalyses, such as CFSR and ERA5. Such biases can be significant when trying to reconstruct climatic variation over such a long period.

The most comprehensive attempt to date to provide a twentieth-century reanalysis was an international project known as 20CR (Compo et al., 2011) led by the Physical Sciences Laboratory (PSL) of the National Oceanic and Atmospheric Administration (NOAA) and the Cooperative Institute for Research in Environmental Sciences (CIRES) at the University of Colorado. The first reanalysis made use of the April 2008 experimental version of the NCEP Global Forecast System (GFS) and an Ensemble Kalman filter assimilation scheme, which are more suitable than schemes such as 3DVAR and 4DVAR, when there are greater uncertainties in model output and assimilation variables. The project only used sea level pressure measurements in the data assimilation as this is the longest and most widely spread meteorological variable recorded with extensive records stretching back to at least the late nineteenth century. Observed sea surface temperature and sea ice temperature were also used but only as boundary conditions for the model; they were not used in the assimilation.

Three versions of 20CR have been produced. The latest, 20CRv3, covers the period 1836 to 2015 (with an experimental extension back to 1806) and uses the 2017 version of the GFS forecast system. Eighty ensemble forecasts are used for each analysis to provide uncertainty estimates. The model is run using a T254 grid with 64 vertical levels. Analysis data are available as three-hourly values on a grid of 360 lines of longitude by 180 lines of latitude with a resolution of around 110 km at the equator. Data are available at 28 vertical pressure levels. Wind speed components are also available at multiple surface levels between 12 m and 500 m. As well as three-hourly values, daily and monthly values are available for download.

Despite the very limited data used for the boundary conditions and assimilation, the 20CRv3 reanalysis compares favourably to reanalyses such as ERA5 that use a full suite of available data (Slivinski et al., 2021). Figure 9.4

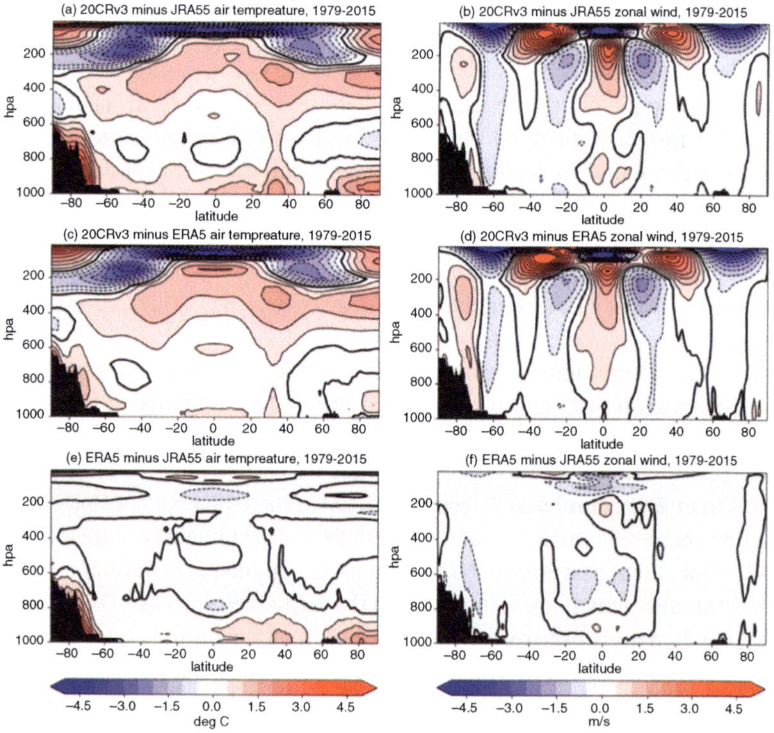

Figure 9.4 Comparison between 20CRv3, ERA5, and JRA55 reanalyses of (left) mean temperature and (right) mean zonal wind speed over the period 1979–2015. *Source:* Slivinski et al. (2021) / American Meteorological Society / CC BY 4.0.

shows a comparison between the 20CRv3, ERA5, and JRA55 reanalyses of mean temperature and mean zonal wind speed averaged over the period 1979–2015 for different heights (pressure levels) and latitudes. In the case of the zonal (or westerly) wind speed component, 20CRv3 compares well with ERA5 close to the ground with differences within $\pm 0.5\,ms^{-1}$. Higher up, there are larger discrepancies due to the model being driven by surface pressure, but for the purposes of long-term wind resource analysis, where surface conditions are most important, the accuracy would seem to be acceptable.

9.3 Regional Reanalyses

The best global reanalyses have spatial resolutions of a few tens of kilometres. This is still a relatively coarse resolution for the purposes of wind resource assessment. Regional reanalyses using mesoscale models driven by the output of global reanalyses as well as additional observations can help, at least in part, to address this lack of resolution. Several regional reanalyses have been performed in different parts of the world and these are described in this section.

9.3.1 NARR

The NCEP North American Regional Reanalysis (NARR) was a downscaling of the NCEP/NCAR R-2 global reanalysis (GR2) using the NCEP Eta model (Mesinger et al., 2006). The reanalysis assimilated several of the observational datasets used by GR2 but also incorporated a number of additional datasets. The data are available at three-hourly intervals on a grid 349×277 covering the US (see Figure 9.5) and 29 vertical pressure levels giving a horizontal resolution of 32 km at the lowest latitude. Both daily and monthly means are also available. Several surface variables are available, including wind speed at 10 m. The reanalysis runs from 1979 to present day at the time of writing and updates are provided monthly.

9.3.2 ASR

The NCAR-led Arctic System Reanalysis (ASR) was a reanalysis of the north polar region using the WRF-ARW mesoscale model forced at the lateral boundaries using ERA-Interim (Bromwich et al., 2016). The WRF model was run using two nested grids. The first version of this reanalysis (ASRv1) employed an inner domain of 361×361 grid points giving a resolution of 30 km and data every three hours at 71 hybrid levels, with monthly averages also available spanning the period 2001–2012. A later reanalysis (ASRv2)

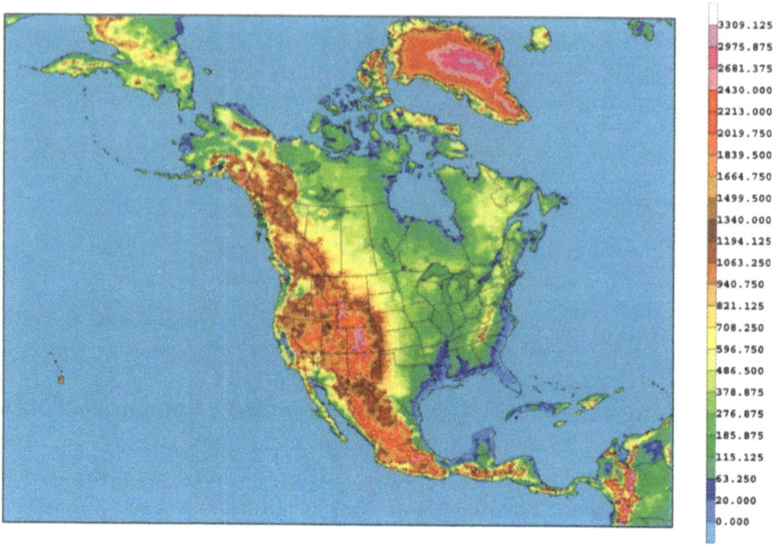

Figure 9.5 The NARR domain and orography (at 32 km resolution). The height contours shown in the legend are in metres. *Source:* Mesinger et al. (2006). with permission of American Meteorological Society.

was produced at a higher horizontal resolution inner domain of 721 × 721 giving a resolution of 15 km (Bromwich et al., 2018). The lowest output level of data is at 4 m. Although the reanalyses were primarily to study the particular climate of the Arctic, the extent of the inner domain (see Figure 9.6) covered a wide area of North America, Europe, and Asia and thus has relevance for wind resource assessment in these regions. The range of data assimilated in ASRv1 and ASRv2 is broadly the same as those used by ERA-Interim. As well as having a higher resolution, ASRv2 benefitted from improved model physics, had a better assimilation scheme, and the period was extended to 2016.

9.3.3 UERRA

Uncertainties in Ensembles of Regional ReAnalyses (UERRA) was a project funded under the European Union's Framework Programme 7 (FP7) where comparisons were made between several European regional reanalyses and model uncertainties were determined by the generation of ensembles (Kaiser-Weiss et al., 2019). This project built on several regional reanalyses developed in an earlier FP7 project, EUropean Reanalysis and Observations for Monitoring

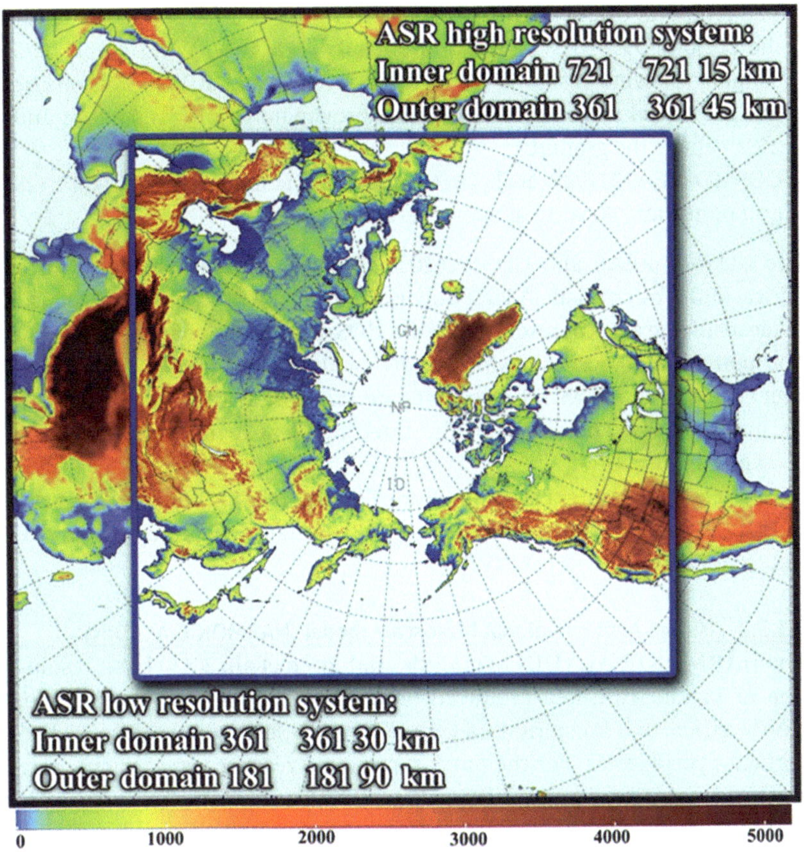

Figure 9.6 The inner and outer domains used by WRF for the ASR. The low-resolution grid corresponds to ASRv1 and the high-resolution to ASRv2. The height contours shown in the legend are in metres. *Source:* Bromwich et al. (2016). With permission of the Royal Meteorological Society.

(EURO4M). All the models used ERA-Interim to provide boundary conditions and additional satellite, upper air, and surface measurements were assimilated. Analyses were output at six-hourly time-steps initially for the period 2006–2010. The models used for the comparisons were:

9.3.3.1 Consortium for Small-scale Modelling (COSMO)

The COSMO regional reanalysis was initially run at two resolutions:

- COSMO-REA6 (Bollmeyer et al., 2015) at 6 km spacing on a grid of 880 × 856 grid points (longitude by latitude) and 40 vertical hybrid levels.

The temporal extent of the reanalysis has since been expanded to cover the period 1995–2019 and data are available hourly. Wind speed data suitable for wind resource assessment are available at 10 m and on the first six model levels corresponding approximately to between 20 m and 250 m above the ground.

- COSMO-REA12 (Bach et al., 2016) at 12 km spacing on a grid of 440 × 428 grid points and 40 vertical hybrid levels.

The COSMO-REA6 and COSMO-REA12 domains cover Europe, North Africa, and Western Asia. A later downscaled version of COSMO-REA6 was made available at 2 km resolution (COSMOS-REA2) with data at 724 × 780 grid points on 50 vertical hybrid levels for the period 2007–2013 centred over Central Europe. Wind speed data are available at 10 m and on the first six model levels. The extent of the three regional reanalysis domains is shown in Figure 9.7.

The COSMO model is run by the Hans-Ertel-Centre for Weather Research (HErZ) at the University of Bonn and the German meteorological service Deutscher Wetterdienst (DWD).

9.3.3.2 Hirlam Aladin Regional Mesoscale model (HARMONIE-ALADIN)

The HARMONIE-ALADIN reanalysis (Dahlgren et al., 2016) has a resolution of 11 km with data produced on a grid of 565 × 565 over Europe, North Africa, and Western Asia (see Figure 9.8, left) with surface and 65 vertical hybrid levels. For the purposes of wind resource assessment, data are also interpolated onto 11 levels between 15 and 500 m. The reanalysis

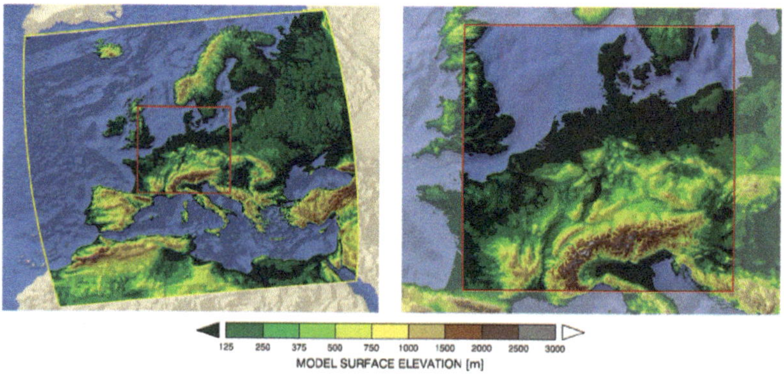

Figure 9.7 Yellow box: COSMO-REA6 and COSMO-REA12 domain. Red box: COSMO-REA2 domain. *Source:* adapted from Wahl et al. (2017).

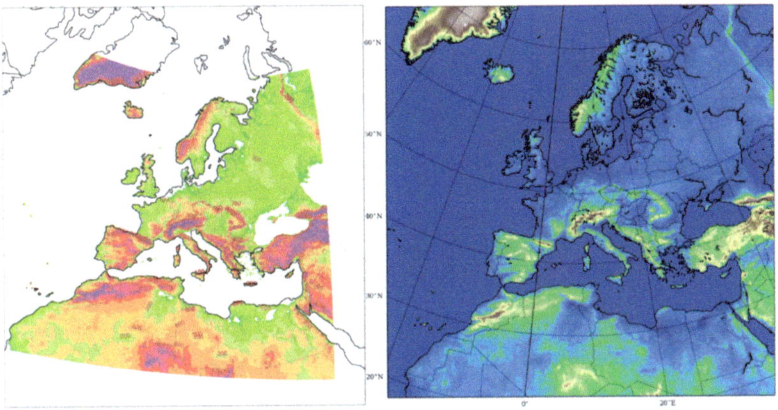

Figure 9.8 Left: HARMONIE-ALADIN reanalysis domain; Right: MESCAN reanalysis domain. *Source:* produced with permission from Gé Verver, KNMI from Verver (2017).

was extended to cover the period 1961–2019. The model is run by the Swedish Meteorological and Hydrological Institute (SMHI).

9.3.3.3 MESCAN (Mesoscale Analysis)

The MESCAN reanalysis (Soci et al., 2016) provided a further downscaling of the HARMONIE-ALADIN reanalysis increasing the resolution to 5.5 km on a grid of 1069 × 1069. The model domain is shown in Figure 9.8 (right). This model is run by the French meteorological service Météo-France. The reanalysis was extended to cover the period 1979–2019.

9.3.3.4 Unified Model (UM)

The UK Meteorological Office Unified Model (UM) (Jermey and Renshaw, 2016) was run at two resolutions: 12 km to provide a deterministic reanalysis and at 36 km to provide 20 ensemble members. The spatial extent of the reanalysis is shown in Figure 9.9. The reanalysis was extended to cover the period 1979–2016. Wind speed data are also available at 11 levels between 15 and 500 m.

9.3.3.5 UERRA Ensemble

In addition to those described in this section, a multimodel ensemble using COSMO-REA12, HARMONIE-ALADIN, MESCAN, and UM was produced by the UERRA project.

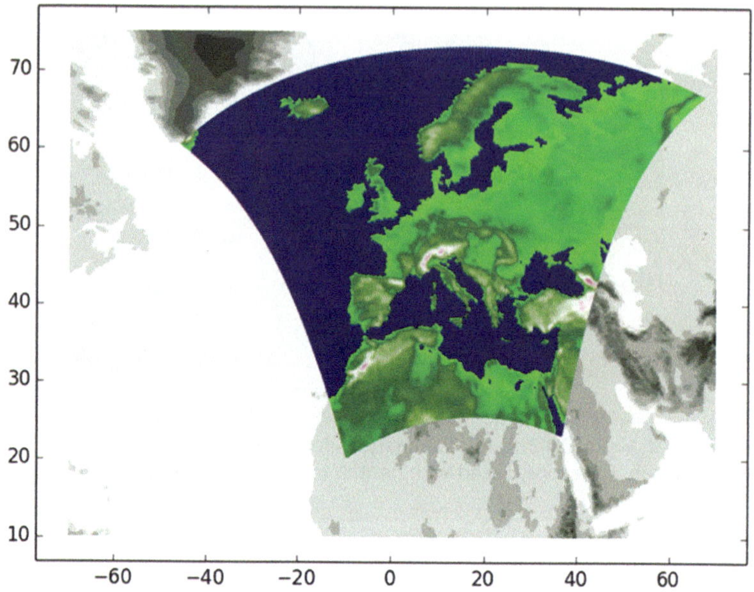

Figure 9.9 The UM reanalysis domain. Latitude and longitude values are in degrees. Produced with permission from Gé Verver, KNMI from (Verver, 2017).

9.3.4 CERRA

The Copernicus Regional Reanalysis for Europe (CERRA) (Unden et al., 2019) is the most up-to-date regional reanalysis (RRA) covering Europe (see Figure 9.10). The RRA is driven using ERA5 to provide boundary conditions and uses HARMONIE-ALADIN to output data hourly on a horizontal grid of 1069 × 1069 grid points and 106 hybrid levels giving a horizontal resolution of approximately 5.5 km. Ten model ensemble runs are also provided at 11 km resolution to estimate uncertainty bounds for the output variables. A wide range of observational data are used in the assimilation. At the time of writing, the reanalysis is still being produced and will run from the early 1980s to close to real time.

9.4 Comparing Reanalyses

Studies have been made to assess the accuracy of reanalyses in predicting site-specific observations including wind speed. Comparisons have been made between absolute wind speeds, wind speed distributions, and the correlation between reanalysis and measured data. A further example of where

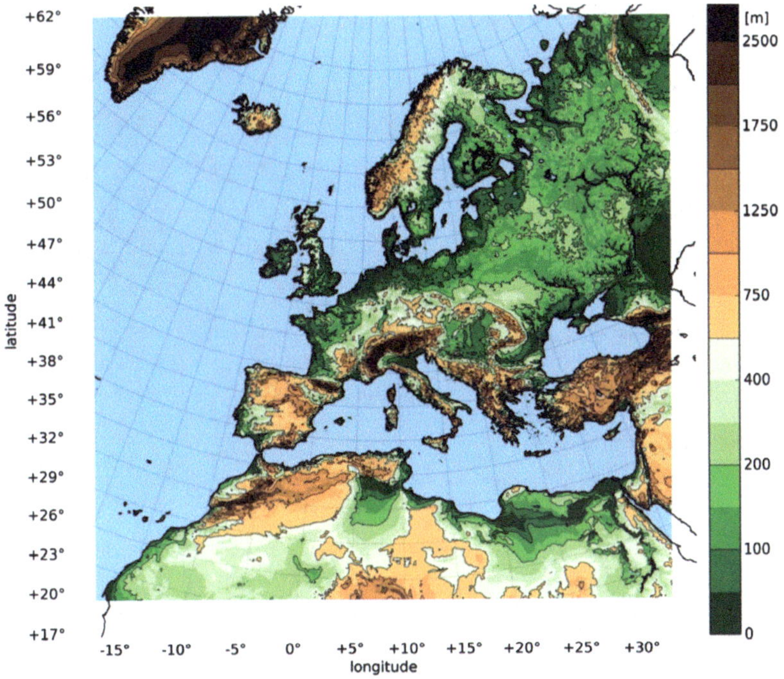

Figure 9.10 The CERRA domain and orography map. Latitude and longitude values are in degrees. *Source:* Unden et al. (2019) / Copernicus Publications / CC BY 4.0.

reanalyses have been compared is in their ability to model storm tracks. Although these are not a direct measure of wind speed, they are the driving mechanism for weather systems and thus surface wind speed, particularly in midlatitudes. Figure 9.11 shows frequency plots of the correlation between reanalysis estimates and satellite observations of outgoing longwave radiation over storm track regions (Harada et al., 2016). A higher correlation suggests a more accurate representation of storm tracks which might be expected to lead to a better representation of surface parameters such as wind speed.

The increase in average correlation is quite clear across the generations of reanalysis with a significant improvement between NCEP/NCAR and the more recent CFSR and ERA-Interim reanalyses.

Due to the limited spatial resolution of even the newest reanalyses, a close match between absolute site-specific wind speeds is not to be expected. However, as mentioned in Section 8.2.1, there is increasing interest in using reanalysis wind speeds as a reference for MCP procedures where a regional

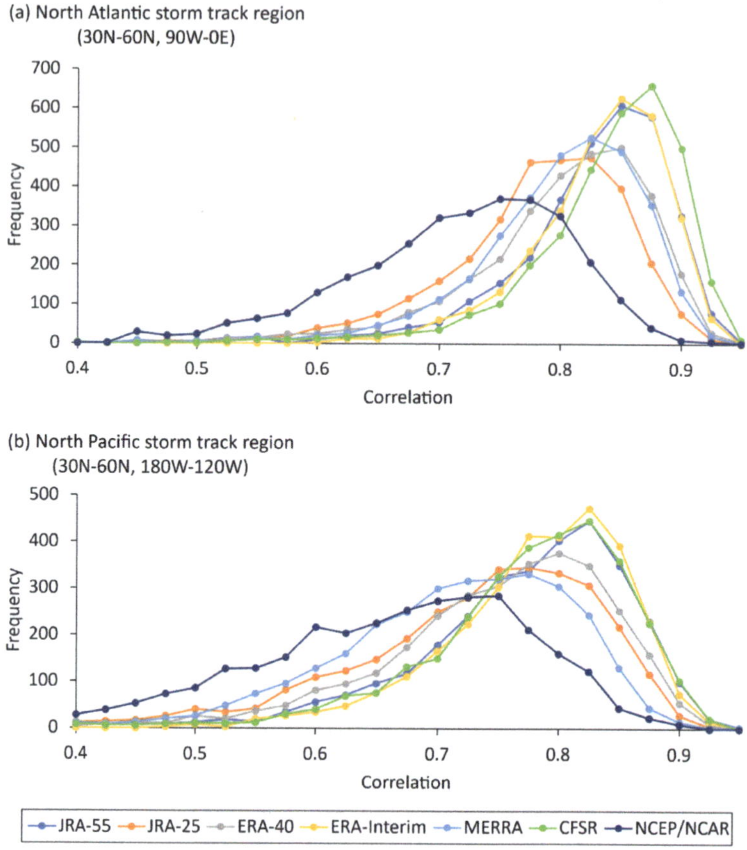

Figure 9.11 Frequency plots of the spatial correlation between reanalysis modelled values of outgoing longwave radiation and satellite measured values over storm tracks during the northern hemisphere cold season (November–March) for the period 1980/1981–2000/2001 in (a) the North Atlantic and (b) the North Pacific. *Source:* data taken from Harada et al. (2016).

long-term climatic trend is required to adjust short-term on-site measurements. In this case, the correlation is a suitable metric. This can be in the form of the Pearson correlation coefficient, ρ

$$\rho = \frac{\sum_{i=1}^{N}\left[\left(u_i - \overline{u}\right)\left(v_i - \overline{v}\right)\right]}{\sqrt{\sum_{i=1}^{N}\left(u_i - \overline{u}\right)^2 \sum_{i=1}^{N}\left(v_i - \overline{v}\right)^2}} \tag{9.1}$$

where u_i and v_i are the ith of N concurrent measured and modelled wind speed values, respectively, and $\overline{u}$ and $\overline{v}$ are the means of the N values

$$\bar{u} = \sum_{i=1}^{N} u_i \tag{9.2}$$

$$\bar{v} = \sum_{i=1}^{N} v_i \tag{9.3}$$

A value of 1 represents perfect correlation, −1 represents perfect anticorrelation, and a value of 0 represents no correlation. An alternative metric is R^2, the coefficient of determination, which can be considered as how well a linear regression fits the dataset of u_i and v_i values

$$R^2 = 1 - \frac{\sum_{i=1}^{N}\left(u_i - \widehat{u_i}\right)^2}{\sum_{i=1}^{N}\left(u_i - \bar{u}\right)^2} \tag{9.4}$$

where $\widehat{u_i}$ is an estimate of u_i based on a linear regression fit using v_i as the independent variable

$$\widehat{u_i} = mv_i + c \tag{9.5}$$

and (similar to Eqs. 8.4 and 8.5)

$$m = \frac{\left(\sum_{i=1}^{N} u_i\right)\left(\sum_{i=1}^{N} v_i^2\right) - \left(\sum_{i=1}^{N} v_i\right)\left(\sum_{i=1}^{N} u_i v_i\right)}{\left(N\sum_{i=1}^{N} v_i^2\right) - \left(\sum_{i=1}^{N} v_i\right)^2} \tag{9.6}$$

$$c = \frac{N\left(\sum_{i=1}^{N} u_i v_i\right) - \left(\sum_{i=1}^{N} v_i\right)\left(\sum_{i=1}^{N} u_i\right)}{N\left(\sum_{i=1}^{N} v_i^2\right) - \left(\sum_{i=1}^{N} v_i\right)^2} \tag{9.7}$$

A value of $R = 1$ would represent all N u_i and v_i values lying on a straight line and 0 would suggest that the straight line is no better at predicting u_i than assuming $\widehat{u_i} = \bar{v}$. It could be argued that, if the intention is to use reanalysis grid point data to provide a reference for MCP, the R^2 metric is the preferred metric.

There have been many studies to determine ρ and R^2 for numerous sites around the world when comparing reanalysis grid point values with measurements made both onshore and offshore. The resulting ρ and R^2 values for several of these studies and a selection of the most popular reanalysis datasets are presented in Figures 9.12 and 9.13, respectively. The first thing to note is that the greater number have used ρ as the preferred metric. The

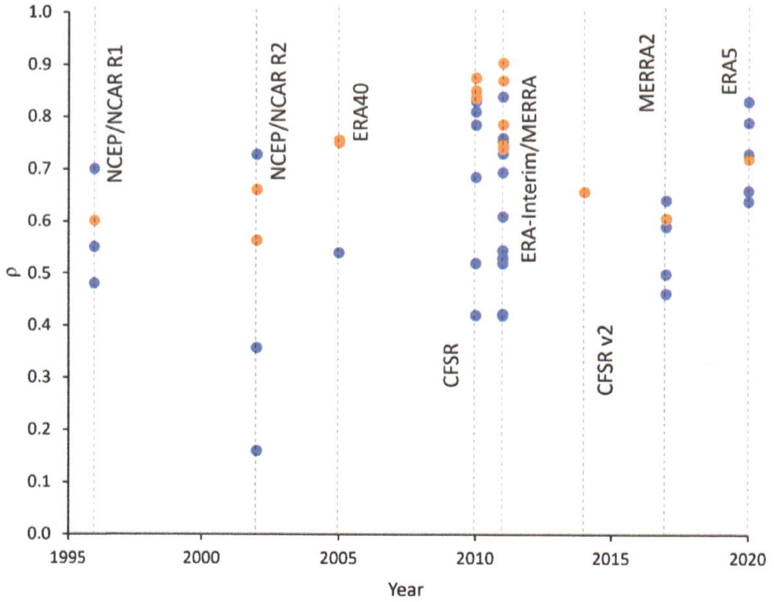

Figure 9.12 Pearson correlation coefficient between measured site-specific wind speeds and those estimated using various reanalysis datasets based on Alvarez et al. (2014), Cannon et al. (2015), de Assis Tavares et al. (2020), Dong et al. (2020), Foussekis and Gkarakis (2014), Gkarakis (2014), Kaspar et al. (2020), Korres et al. (2011), Li et al. (2013), Liléo and Petrik (2011), Minola et al. (2020), Molina et al. (2021), Morales-Ruvalcaba et al. (2020), Rabbani and Zeeshan (2020), Rehman Tahir et al. (2018), Santos et al. (2020), Sharp et al. (2015), Stopa and Cheung (2014), Yu et al. (2019). Onshore sites are marked in blue and offshore in orange. Results are ordered by the year in which the first publication describing each reanalysis was published. These are marked by the vertical dotted lines.

results are presented along a timeline of when the first paper describing a particular reanalysis was published. The fewer studies shown in Figure 9.13 which used R^2 as the correlation metric do not necessarily cover all the reanalyses shown in Figure 9.12. There is a large amount of scatter in the values but the correlation with offshore sites (which are predominantly buoys) is higher than for onshore sites. This might be expected given the limited resolution of the models used which cannot resolve small-scale terrain features.

There is some evidence for an improvement over the different generations though the scatter makes this hard to confirm. Comparisons are difficult given the different temporal resolutions of the different generations of reanalyses. Nevertheless, a significant number of the sites show a value of ρ

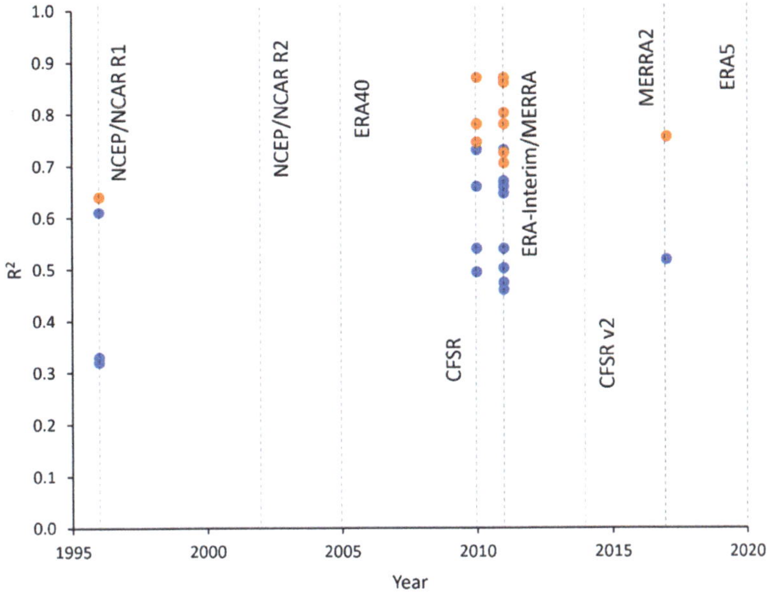

Figure 9.13 As Figure 9.12, but using R^2 as the correlation metric based on Bethke et al. (2014), Brower et al. (2013), Carvalho et al. (2014), Chawla et al. (2013), Eichelberger et al. (2013), Kim et al. (2018), Pinto et al. (2006), Rose and Apt (2016), The Crown Estate (2014).

greater than 0.7, which suggests that reanalysis grid point data could be used as a proxy for long-term data instead of surface meteorological stations for some sites, particularly offshore. This is especially beneficial for regions where meteorological station data are sparse.

Regional reanalyses have the potential to predict site-specific wind speeds more accurately due to their greater spatial resolution. Figure 9.14 shows a comparison from the UERRA project of the Pearson correlation coefficient calculated between six-hourly values from five reanalysis datasets (four of the regional reanalyses used in UERRA and ERA-Interim as boundary conditions for these as described in Section 9.3.3) and hourly data from six masts (Kaiser-Weiss et al., 2015). For the offshore masts, all the regional reanalyses (COSMOE-REA6, HARMONIE, UM, and COSMOE-REA12) perform better than ERA-Interim. For the onshore sites, not all perform better but some do, particularly UM. Overall, the results show the value of the regional downscaling to provide more accurate estimates of site-specific wind speeds.

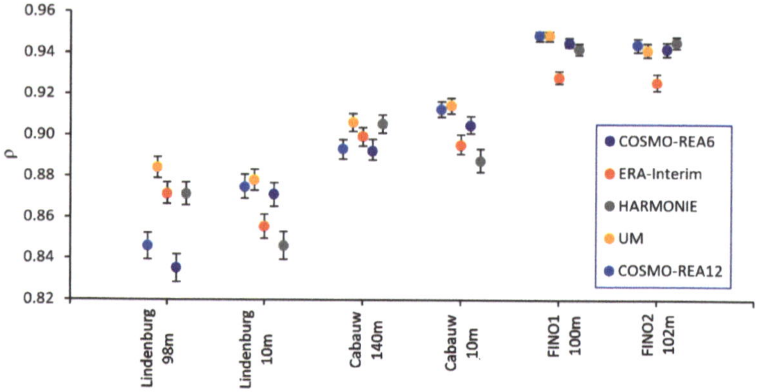

Figure 9.14 Correlation (plotted with 95% confidence limits) between six-hourly reanalysis data and hourly mast data from two onshore masts (Lindenburg and Cabauw) and two offshore masts (FINO1 and FINO2). *Source:* data taken from Kaiser-Weiss et al. (2015).

9.5 Wind Atlases

Reanalysis datasets can be used in a model chain to produce detailed maps of regional or global wind resource. In Section 5.5, the use of model chains to produce wind atlases is briefly described, taking the European Wind Atlas as an example. In this section, some of the more well-known wind atlases are described and compared in terms of coverage, resolution, and accuracy.

9.5.1 The New European Wind Atlas

A project was carried out using a model chain combination of WRF and WAsP (WRF-WAsP) to create both a mesoscale and higher-resolution microscale map of wind speeds covering western, central, and a large part of eastern Europe (Dörenkämper et al., 2020; Hahmann et al., 2020) known as the New European Wind Atlas (NEWA). The WRF model was driven using boundary conditions provided by the ERA5 reanalysis for the 30-year period 1989–2018. Sea surface temperature and sea ice conditions (which are important driving parameters for a mesoscale model) were provided from the Operational Sea Surface Temperature and Sea Ice Analysis (OSTIA) system (Donlon et al., 2012). Orographic height information was provided using Global Multi-resolution Terrain Elevation Data 2010 with a 30 arcsecond resolution (Danielson and Gesch, 2011) and surface roughness length information inferred from the CORINE (coordination of information on the environment) Land Cover (CLC) database (Copernicus, 2018). The WRF model was run in a three-level nested grid configuration with the smallest inner grid having a resolution of 3 km. Ten different overlapping 3 km resolution domains were used to cover the area of interest, as shown in Figure 9.15(a). The data from these domains were merged to provide the final dataset, as shown in Figure 9.15(b).

Each WRF model run was for a seven-day period with an additional 24-hour *spin-up period* so that for each simulation period there was a one-day overlap between each run. A spin-up period is commonly used when running a numerical weather prediction as it takes some time for the model to reach a proper equilibrium with the prevailing synoptic conditions from a cold start. The spin-up period is discarded, and it is assumed the period following the spin-up is an accurate representation of the atmospheric conditions. Atmospheric variables from WRF were output every 30 minutes.

The downscaling to the microscale was carried out following the methodology described in Section 5.5, whereby the WAsP linearised model was used to effectively remove the effect of orography and roughness at the 3 km resolution used by the WRF model to give an estimate of the generalised climatology following which, WAsP was used to calculate the wind speeds at a 50 m resolution using higher spatial resolution orography, i.e. the Shuttle Radar Topography Mission Coverage Version 3 (SRTM v3) dataset (Farr et al., 2007) for locations south of 60°N and the ViewFinder Digital Elevation Model (DEM) dataset (de Ferranti, 2014) for locations north of 60°N both of which have a three-arcsecond resolution (approximately 90 m). Again, the CORINE land use dataset was used to determine the surface roughness length but in this case at its highest spatial resolution of 100 m.

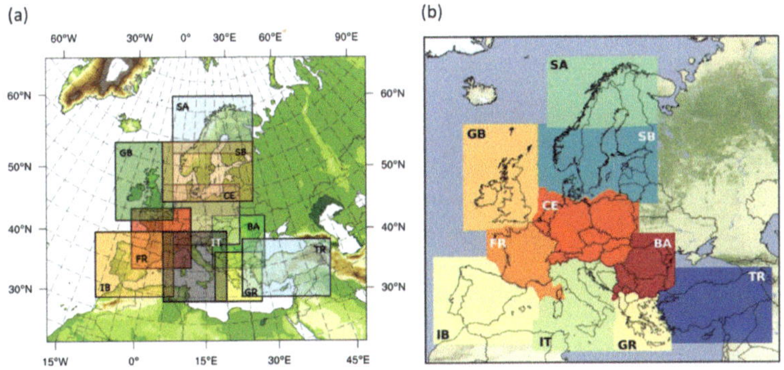

Figure 9.15 NEWA WRF high-resolution domains: (a) computational domains; (b) final merged dataset. *Source:* Dörenkämper et al. (2020) / Copernicus Publications / CC BY 4.0.

A validation of NEWA was carried out using mast data between 40 and 150 m above ground level for a range of a sites across Europe in terrain of varying complexity based on the RIX number described in Section 5.2.3, classified as either low (RIX = 0%), medium (0% < RIX < 2%), or high (RIX ≥ 2%). Figure 9.16 shows the resulting wind speed bias distributions for the raw ERA5 data extrapolated to the masts, the WRF mesoscale predictions, and the WRF-WAsP microscale predictions. Both the mesoscale and mesoscale/microscale downscaling show a significant improvement on the ERA5 data. In terrain with a low RIX value, the microscale downscaling shows an improvement on the mesoscale-only downscaling both in terms of mean and standard deviation bias. For the medium RIX sites, the magnitude of the mean bias is less for the mesoscale-only downscaling, though the standard deviation is reduced with the additional microscale downscaling. In the case of the high RIX values, the failure to capture flow separation results in overprediction using WAsP, and both the magnitude of the mean and the standard deviation bias are larger than the mesoscale-only downscaled data.

9.5.2 The Global Wind Atlas

The latest version of the Global Wind Atlas (GWA3.1) (https://globalwindatlas.info) provides, as the name suggests, wind resource estimates for the entire globe, following a very similar process as that used to produce the NEWA. The same orographic data were used as for NEWA (i.e. SRTM/DEM, projected onto a 150 m grid) and surface roughness was based on the European Space Agency's Climate Change Initiative Land Cover (CCI-LC) dataset v2.0.7 with a 10 arcsecond (~300 m) resolution. In regions where no ESA CCI-LC data

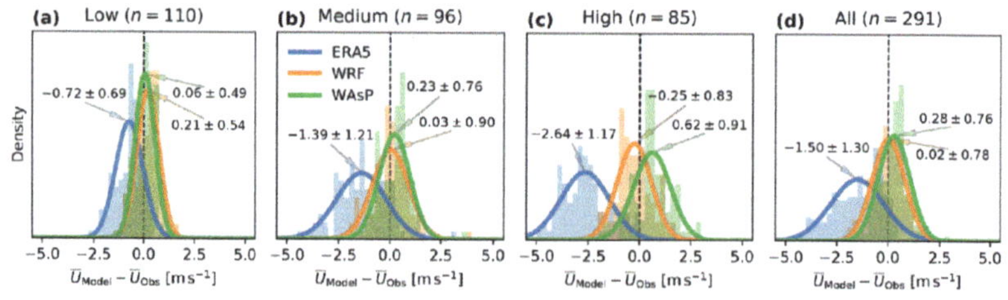

Figure 9.16 NEWA wind speed biases compared for ERA5, WRF, and WRF-WAsP (labelled *WAsP*) with dependency on RIX value. For the three distributions, a Gaussian fit is shown labelled with the mean bias ± one standard deviation in ms^{-1}. *Source:* Dörenkämper et al. (2020) / Copernicus Publications / CC BY 4.0.

were available, slightly lower resolution MODIS (Moderate Resolution Imaging Spectroradiometer)-based Global Land Cover Climatology was used. ERA5 data for the ten-year period 2008–2017 were used to drive WRF, using a nested grid approach to produce wind speed estimates on a 3 km grid which were further downscaled using WAsP to a 250 m resolution at five heights: 10, 50, 100, 150, and 200 m. An example of the wind resource resulting from the GWA analysis is shown in Figure 9.17.

Validation of the GWA downscaled wind speed predictions using the WRF/WAsP model chain is relatively limited having been carried out at only 35 masts in Bangladesh, Maldives, Pakistan, Papua New Guinea, Vietnam, and Zambia. Using the same RIX classification as in Section 9.5.1, Figure 9.18 shows the distribution of wind speed biases based on these masts. Note that the validation dataset contains no sites with a high RIX value. The bias distribution based on this limited dataset is broadly similar to NEWA, which is not surprising given the almost identical methodology and input data.

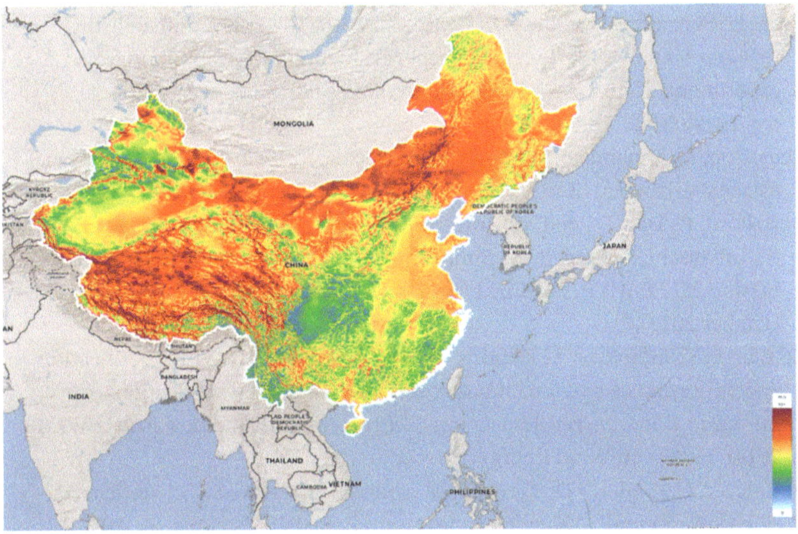

Figure 9.17 Wind resource at 100 m for China taken from the Global Wind Atlas 3.1, a free, web-based application developed, owned and operated by the Technical University of Denmark (DTU). The Global Wind Atlas 3.1 is released in partnership with the World Bank Group, utilising data provided by Vortex, using funding provided by the Energy Sector Management Assistance Program (ESMAP). For additional information: https://globalwindatlas.info *Source:* Global Wind Atlas 3.0 / CC BY 4.0.

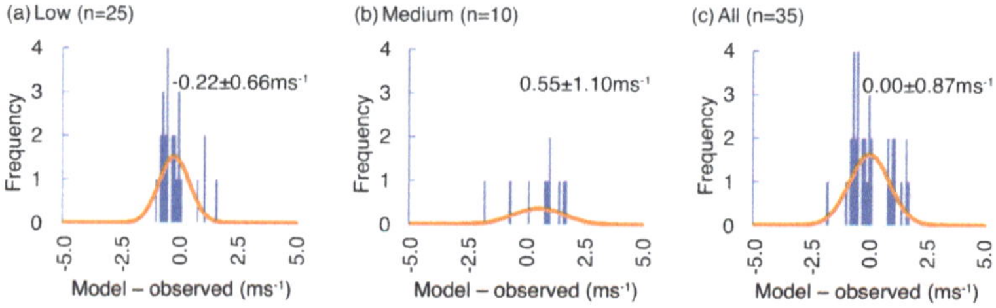

Figure 9.18 GWA3.1 wind speed biases (in blue) with dependency on RIX value. The Gaussian fit (in orange) is shown labelled with the mean bias ± one standard deviation.

References

Alvarez, I., Gomez-Gesteira, M., DeCastro, M., Carvalho, D., 2014. Comparison of different wind products and buoy wind data with seasonality and interannual climate variability in the southern Bay of Biscay (2000–2009). *Deep. Res. Part II Top. Stud. Oceanogr.* 106, 38–48. https://doi.org/10.1016/j.dsr2.2013.09.028.

Bach, L., Schraff, C., Keller, J.D., Hense, A., 2016. Towards a probabilistic regional reanalysis system for Europe: evaluation of precipitation from experiments. *Tellus, Ser. A Dyn. Meteorol. Oceanogr.* 68. https://doi.org/10.3402/tellusa.v68.32209.

Bengtsson, L., Kanamitsu, M., Kallberg, P., Uppala, S., 1982. FGGE 4-dimensional data assimilation at ECMWF. *Bull. Am. Meteorol. Soc.* 63, 29–43.

Berrisford, P., Dee, D., Poli, P., Brugge, R., Fielding, K., Fuentes, M., Kallberg, P., Kobayashi, S., Uppala, S., Simmons, A., 2011. *The ERA-Interim archive: Version 2.0 ECMWF*. Reading https://www.ecmwf.int/node/8174 (accessed 4 October 2022).

Bethke, J., Kampmeyer, J., Kilburg, M., Mengelkamp, H.T., 2014. Verification and performance of wind climatologies in wind energy applications, in: *European Wind Energy Association Conference and Exhibition 2014 (EWEA 2014)*. European Wind Energy Association.

Bollmeyer, C., Keller, J.D., Ohlwein, C., Wahl, S., Crewell, S., Friederichs, P., Hense, A., Keune, J., Kneifel, S., Pscheidt, I., Redl, S., Steinke, S., 2015. Towards a high-resolution regional reanalysis for the European CORDEX domain. *Q. J. R. Meteorol. Soc.* 141, 1–15. https://doi.org/10.1002/qj.2486.

Bromwich, D.H., Wilson, A.B., Bai, L., Liu, Z., Barlage, M., Shih, C.F., Maldonado, S., Hines, K.M., Wang, S.H., Woollen, J., Kuo, B., Lin, H.C., Wee, T.K., Serreze, M.C., Walsh, J.E., 2018. The arctic system reanalysis, version 2. *Bull. Am. Meteorol. Soc.* 99, 805–828. https://doi.org/10.1175/BAMS-D-16-0215.1.

Bromwich, D.H., Wilson, A.B., Bai, L.S., Moore, G.W.K., Bauer, P., 2016. A comparison of the regional Arctic System Reanalysis and the global ERA-Interim Reanalysis for the Arctic. *Q. J. R. Meteorol. Soc.* 142, 644–658. https://doi.org/10.1002/qj.2527.

Brower, M.C., Lledó, L., Dubois, J., 2013. A study of wind speed variability using global reanalysis data. *AWS Truepower Report*, 1–12.

Cannon, D.J., Brayshaw, D.J., Methven, J., Coker, P.J., Lenaghan, D., 2015. Using reanalysis data to quantify extreme wind power generation statistics: a 33 year case study in Great Britain. *Renew. Energy* 75, 767–778. https://doi.org/10.1016/j.renene.2014.10.024.

Carvalho, D., Rocha, A., Gómez-Gesteira, M., Silva Santos, C., 2014. Comparison of reanalyzed, analyzed, satellite-retrieved and NWP modelled winds with buoy data along the Iberian Peninsula coast. *Remote Sens. Environ.* 152, 480–492. https://doi.org/10.1016/j.rse.2014.07.017.

Chawla, A., Spindler, D.M., Tolman, H.L., 2013. Validation of a thirty year wave hindcast using the Climate Forecast System Reanalysis winds. *Ocean Model* 70, 189–206. https://doi.org/10.1016/j.ocemod.2012.07.005.

Compo, G.P., Whitaker, J.S., Sardeshmukh, P.D., Matsui, N., Allan, R.J., Yin, X., Gleason, B.E., Vose, R.S., Rutledge, G., Bessemoulin, P., Bronnimann, S., Brunet, M., Crouthamel, R.I., Grant, A.N., Groisman, P.Y., Jones, P.D., Kruk, M.C., Kruger, A.C., Marshall, G.J., Maugeri, M., Mok, H.Y., Nordli, O., Ross, T.F., Trigo, R.M., Wang, X.L., Woodruff, S.D., Worley, S.J., 2011. The Twentieth Century Reanalysis Project. *Q. J. R. Meteorol. Soc.* 137, 1–28. https://doi.org/10.1002/qj.776.

Copernicus, 2018. CORINE Land Cover. *Copernicus Land Monitoring Service.L. Monit. Serv.*https://land.copernicus.eu/pan-european/corine-land-cover (accessed 2 March 2022).

The Crown Estate, 2014. *UK MERRA validation with offshore meteorological data*. https://www.thecrownestate.co.uk/media/1802/ei-uk-merra-validation-with-offshore-meteorological-data.pdf (accessed 4 October 2022).

Dahlgren, P., Landelius, T., Kållberg, P., Gollvik, S., 2016. A high-resolution regional reanalysis for Europe: part 1: three-dimensional reanalysis with the regional HIgh-Resolution Limited-Area Model (HIRLAM). *Q. J. R. Meteorol. Soc.* 142, 2119–2131. https://doi.org/10.1002/qj.2807.

Danielson, J.J., Gesch, D.B., 2011. Global multi-resolution terrain elevation data 2010 (GMTED2010). *U.S. Geological Survey: Open-File Report 2011-1073.* https://pubs.usgs.gov/of/2011/1073/pdf/of2011-1073.pdf (accessed 4 October 2022).

de Assis Tavares, L.F., Shadman, M., de Freitas Assad, L.P., Silva, C., Landau, L., Estefen, S.F., 2020. Assessment of the offshore wind technical potential for the Brazilian Southeast and South regions. *Energy* 196, 117097. https://doi.org/10.1016/j.energy.2020.117097.

de Ferranti, J., 2014. Digital elevation data. *Viewfind Panoramas*. http://viewfinderpanoramas.org/dem3.html (accessed 4 March 2022).

Dee, D.P., Uppala, S.M., Simmons, A.J., Berrisford, P., Poli, P., Kobayashi, S., Andrae, U., Balmaseda, M.A., Balsamo, G., Bauer, P., Bechtold, P., Beljaars, A.C.M., van de Berg, L., Bidlot, J., Bormann, N., Delsol, C., Dragani, R., Fuentes, M., Geer, A.J., Haimberger, L., Healy, S.B., Hersbach, H., Hólm, E.V., Isaksen, L., Kållberg, P., Köhler, M., Matricardi, M., Mcnally, A.P., Monge-Sanz, B.M., Morcrette, J.J., Park, B.K., Peubey, C., de Rosnay, P., Tavolato, C., Thépaut, J.N., Vitart, F., 2011. The ERA-Interim reanalysis: configuration and performance of the data assimilation system. *Q. J. R. Meteorol. Soc.* 137, 553–597. https://doi.org/10.1002/qj.828.

Dong, X., Wang, Y., Hou, S., Ding, M., Yin, B., Zhang, Y., 2020. Robustness of the recent global atmospheric reanalyses for Antarctic near-surface wind speed climatology. *J. Clim.* 33, 4027–4043. https://doi.org/10.1175/JCLI-D-19-0648.1.

Donlon, C.J., Martin, M., Stark, J., Roberts-Jones, J., Fiedler, E., Wimmer, W., 2012. The Operational Sea Surface Temperature and Sea Ice Analysis (OSTIA) system. *Remote Sens. Environ.* 116, 140–158. https://doi.org/10.1016/J.RSE.2010.10.017.

Dörenkämper, M., Olsen, B.T., Witha, B., Hahmann, A.N., Davis, N.N., Barcons, J., Ezber, Y., Garciá-Bustamante, E., Fidel González-Rouco, J., Navarro, J., Sastre-Marugán, M., Sile, T., Trei, W., Žagar, M., Badger, J., Gottschall, J., Rodrigo, J.S., Mann, J., 2020. The Making of the New European Wind Atlas: part 2: production and evaluation. *Geosci. Model Dev.* 13, 5079–5102. https://doi.org/10.5194/GMD-13-5079-2020.

Eichelberger, S., Stoelinga, M., McCaa, J., 2013. Performance of new reanalysis data sets for estimating the temporal and spatial variability of wind resource, in: EWEA (Ed.), *Proceedings of the EWEA Annual Event 2013 4th–7th February 2013*. EWEA, Vienna.

Farr, T.G., Rosen, P.A., Caro, E., Crippen, R., Duren, R., Hensley, S., Kobrick, M., Paller, M., Rodriguez, E., Roth, L., Seal, D., Shaffer, S., Shimada, J., Umland, J., Werner, M., Oskin, M., Burbank, D., Alsdorf, D.E., 2007. The Shuttle Radar Topography Mission. *Rev. Geophys.* 45, 2004. https://doi.org/10.1029/2005RG000183.

Foussekis, D., Gkarakis, K., 2014. Comparison of reanalysis data with long-term measurements in complex and coastal terrains, in: *European Wind Energy Association Conference and Exhibition 2014*, EWEA 2014.

Gelaro, R., McCarty, W., Suárez, M.J., Todling, R., Molod, A., Takacs, L., Randles, C.A., Darmenov, A., Bosilovich, M.G., Reichle, R., Wargan, K., Coy, L., Cullather, R., Draper, C., Akella, S., Buchard, V., Conaty, A., da Silva, A.M., Gu, W., Kim, G.K., Koster, R., Lucchesi, R., Merkova, D., Nielsen, J.E.,

Partyka, G., Pawson, S., Putman, W., Rienecker, M., Schubert, S.D., Sienkiewicz, M., Zhao, B., 2017. The Modern-Era Retrospective Analysis for Research and Applications, Version 2 (MERRA-2). *J. Clim.* 30, 5419–5454. https://doi.org/10.1175/JCLI-D-16-0758.1.

Gibson, J.K., Kållberg, P., Uppala, S., Hernandez, A., Nomura, A., Serrano, E., 1999. ECMWF re-analysis project report series 1: ERA-15 description (version 2: January 1999). *European Centre for Medium-Range Weather Forecasts*. https://www.ecmwf.int/sites/default/files/elibrary/1997/9584-era-description.pdf (accessed 4 October 2022).

Gkarakis, K.C., 2014. ConWx mesoscale data and MERRA reanalysis data in comparison to thirty ground based meteorological masts in Greece, in: *European Wind Energy Association Conference and Exhibition 2014*, EWEA 2014.

Hahmann, A.N., Sile, T., Witha, B., Davis, N.N., Dörenkämper, M., Ezber, Y., Garciá-Bustamante, E., Fidel González-Rouco, J., Navarro, J., Olsen, B.T., Söderberg, S., 2020. The making of the New European Wind Atlas: part 1: model sensitivity. *Geosci. Model Dev.* 13, 5053–5078. https://doi.org/10.5194/GMD-13-5053-2020.

Harada, Y., Kamahori, H., Kobayashi, C., Endo, H., Kobayashi, S., Ota, Y., Onoda, H., Onogi, K., Miyaoka, K., Takahashi, K., 2016. The JRA-55 Reanalysis: Representation of Atmospheric Circulation and Climate Variability. *J. Meteorol. Soc. Japan. Ser. II* 94, 269–302. https://doi.org/10.2151/jmsj.2016-015.

Hersbach, H., Bell, B., Berrisford, P., Hirahara, S., Horányi, A., Muñoz-Sabater, J., Nicolas, J., Peubey, C., Radu, R., Schepers, D., Simmons, A., Soci, C., Abdalla, S., Abellan, X., Balsamo, G., Bechtold, P., Biavati, G., Bidlot, J., Bonavita, M., De Chiara, G., Dahlgren, P., Dee, D., Diamantakis, M., Dragani, R., Flemming, J., Forbes, R., Fuentes, M., Geer, A., Haimberger, L., Healy, S., Hogan, R.J., Hólm, E., Janisková, M., Keeley, S., Laloyaux, P., Lopez, P., Lupu, C., Radnoti, G., de Rosnay, P., Rozum, I., Vamborg, F., Villaume, S., Thépaut, J.N., 2020. The ERA5 global reanalysis. *Q. J. R. Meteorol. Soc.* 146, 1999–2049. https://doi.org/10.1002/qj.3803.

Jermey, P.M., Renshaw, R.J., 2016. Precipitation representation over a two-year period in regional reanalysis. *Q. J. R. Meteorol. Soc.* 142, 1300–1310. https://doi.org/10.1002/qj.2733.

Kaiser-Weiss, A.K., Borsche, M., Niermann, D., Kaspar, F., Lussana, C., Isotta, F.A., van den Besselaar, E., van der Schrier, G., Undén, P., 2019. Added value of regional reanalyses for climatological applications. *Environ. Res. Commun.* 1. https://doi.org/10.1088/2515-7620/ab2ec3.

Kaiser-Weiss, A.K., Kaspar, F., Heene, V., Borsche, M., Tan, D.G.H., Poli, P., Obregon, A., Gregow, H., 2015. Comparison of regional and global reanalysis near-surface winds with station observations over Germany. *Adv. Sci. Res.* 12, 187–198. https://doi.org/10.5194/asr-12-187-2015.

Kalnay, E., Kanamitsu, M., Kistler, R., Collins, W., Deaven, D., Gandin, L., Iredell, M., Saha, S., White, G., Woollen, J., Zhu, Y., Chelliah, M., Ebisuzaki, W., Higgins, W., Janowiak, J., Mo, K.C., Ropelewski, C., Wang, J., Leetmaa, A., Reynolds, R., Jenne, R., Joseph, D., 1996. The NCEP/NCAR 40-year reanalysis project. *Bull. Am. Meteorol. Soc.* 77(3), 437–472.

Kanamitsu, M., 1989. Description of the NMC Global Data Assimilation and Forecast System. *Weather Forecast* 4, 335–342. https://doi.org/10.1175/1520-0434(1989)004<0335:dotngd>2.0.co;2.

Kanamitsu, M., Alpert, J.C., Campana, K.A., Caplan, P.M., Deaven, D.G., Iredell, M., Katz, B., Pan, H.-L., Sela, J., White, G.H., 1991. Recent changes implemented into the Global Forecast System at NMC. *Weather Forecast* 6, 425–435. https://doi.org/10.1175/1520-0434(1991)006<0425:rciitg>2.0.co;2.

Kanamitsu, M., Ebisuzaki, W., Woollen, J., Yang, S.K., Hnilo, J.J., Fiorino, M., Potter, G.L., 2002. NCEP-DOE AMIP-II reanalysis (R-2). *Bull. Am. Meteorol. Soc.* 83, 1631–1644. https://doi.org/10.1175/bams-83-11-1631.

Kaspar, F., Niermann, D., Borsche, M., Fiedler, S., Keller, J., Potthast, R., Rösch, T., Spangehl, T., Tinz, B., 2020. Regional atmospheric reanalysis activities at Deutscher Wetterdienst: review of evaluation results and application examples with a focus on renewable energy. *Adv. Sci. Res.* 17, 115–128. https://doi.org/10.5194/asr-17-115-2020.

Kim, H.G., Kim, J.Y., Kang, Y.H., 2018. Comparative evaluation of the third-generation reanalysis data for wind resource assessment of the southwestern offshore in South Korea. *Atmosphere (Basel)* 9, 73. https://doi.org/10.3390/atmos9020073.

Korres, G., Papadopoulos, A., Katsafados, P., Ballas, D., Perivoliotis, L., Nittis, K., 2011. A 2-year intercomparison of the WAM-Cycle4 and the WAVEWATCH-III wave models implemented within the Mediterranean Sea. *Mediterr. Mar. Sci.* 12, 129–152. https://doi.org/10.12681/mms.57.

Li, M., Liu, J., Wang, Z., Wang, H., Zhang, Z., Zhang, L., Yang, Q., 2013. Assessment of sea surface wind from NWP reanalyses and satellites in the Southern Ocean. *J. Atmos. Ocean. Technol.* 30, 1842–1853. https://doi.org/10.1175/JTECH-D-12-00240.1.

Liléo, S., Petrik, O., 2011. Investigation on the use of NCEP/NCAR, MERRA and NCEP/CFSR reanalysis data in wind resource analysis, in: *European Wind Energy Conference and Exhibition 2011, EWEC 2011*. European Wind Energy Association, pp. 181–185.

Mesinger, F., DiMego, G., Kalnay, E., Mitchell, K., Shafran, P.C., Ebisuzaki, W., Jović, D., Woollen, J., Rogers, E., Berbery, E.H., Ek, M.B., Fan, Y., Grumbine, R., Higgins, W., Li, H., Lin, Y., Manikin, G., Parrish, D., Shi, W., 2006. North

American regional reanalysis. *Bull. Am. Meteorol. Soc.* 87, 343–360. https://doi.org/10.1175/BAMS-87-3-343.

Minola, L., Zhang, F., Azorin-Molina, C., Safaei Pirooz, A.A., Flay, R.G.J., Hersbach, H., Chen, D., 2020. Near-surface mean and gust wind speeds in ERA5 across Sweden: towards an improved gust parametrization. *Clim. Dyn.* 55, 887–907. https://doi.org/10.1007/s00382-020-05302-6.

Molina, M.O., Gutiérrez, C., Sánchez, E., 2021. Comparison of ERA5 surface wind speed climatologies over Europe with observations from the HadISD dataset. *Int. J. Climatol.* https://doi.org/10.1002/joc.7103.

Morales-Ruvalcaba, C.F., Rodríguez-Hernández, O., Martínez-Alvarado, O., Drew, D.R., Ramos, E., 2020. Estimating wind speed and capacity factors in Mexico using reanalysis data. *Energy Sustain. Dev.* 58, 158–166. https://doi.org/10.1016/j.esd.2020.08.006.

Onogi, K., Tsutsui, J., Koide, H., Sakamoto, M., Kobayashi, S., Hatsushika, H., Matsumoto, T., Yamazaki, N., KAMAHORI, H., TAKAHASHI, K., Kadokura, S., Wada, K., Kato, K., Oyama, R., OSE, T., Mannoji, N., Taira, R., 2007. The JRA-25 Reanalysis. *J. Meteorol. Soc. Japan. Ser. II* 85, 369–432. https://doi.org/10.2151/jmsj.85.369.

Pinto, C., Guedes, R., Pinto, P., Ferreira, M., 2006. NCEP/NCAR reanalysis data for the Portuguese mainland, in: *European Wind Energy Conference and Exhibition 2006, EWEC 2006*. pp. 2035–2039.

Rabbani, R., Zeeshan, M., 2020. Exploring the suitability of MERRA-2 reanalysis data for wind energy estimation, analysis of wind characteristics and energy potential assessment for selected sites in Pakistan. *Renew. Energy* 154, 1240–1251. https://doi.org/10.1016/j.renene.2020.03.100.

Rehman Tahir, Z.U., Asim, M., Jamil, S., Shad, R., Hayat, N., Moaz, A., Akram, M.T., Safyan, M., 2018. Comparison of reanalysis, analysis and forecast datasets with measured wind data for a wind power project in Jhimpir, Pakistan. *J. Phys. Conf. Ser.* 1102, 012004. https://doi.org/10.1088/1742-6596/1102/1/012004.

Rienecker, M.M., Suarez, M.J., Gelaro, R., Todling, R., Bacmeister, J., Liu, E., Bosilovich, M.G., Schubert, S.D., Takacs, L., Kim, G.K., Bloom, S., Chen, J., Collins, D., Conaty, A., Da Silva, A., Gu, W., Joiner, J., Koster, R.D., Lucchesi, R., Molod, A., Owens, T., Pawson, S., Pegion, P., Redder, C.R., Reichle, R., Robertson, F.R., Ruddick, A.G., Sienkiewicz, M., Woollen, J., 2011. MERRA: NASA's modern-era retrospective analysis for research and applications. *J. Clim.* 24, 3624–3648. https://doi.org/10.1175/JCLI-D-11-00015.1.

Rienecker, M.M., Suarez, M.J., Todling, R., Bacmeister, J., Takacs, L., Liu, H.-C., Gu, W., Sienkiewicz, M., Koster, R.D., Gelaro, R., Stajner, I., Nielsen, J.E., 2008. The GEOS-5 Data Assimilation System: Documentation of Versions 5.0.1, 5.1.0, and 5.2.0. *NASA Technical Report NASA/*

TM-2008-104606, Vol. 27. https://gmao.gsfc.nasa.gov/pubs/docs/Rienecker369.pdf (accessed 4 October 2022).

Rose, S., Apt, J., 2016. Quantifying sources of uncertainty in reanalysis derived wind speed. *Renew. Energy* 94, 157–165. https://doi.org/10.1016/j.renene.2016.03.028.

Saha, S., Moorthi, S., Pan, H.L., Wu, X., Wang, J., Nadiga, S., Tripp, P., Kistler, R., Woollen, J., Behringer, D., Liu, H., Stokes, D., Grumbine, R., Gayno, G., Wang, J., Hou, Y.T., Chuang, H.Y., Juang, H.M.H., Sela, J., Iredell, M., Treadon, R., Kleist, D., Van Delst, P., Keyser, D., Derber, J., Ek, M., Meng, J., Wei, H., Yang, R., Lord, S., Van Den Dool, H., Kumar, A., Wang, W., Long, C., Chelliah, M., Xue, Y., Huang, B., Schemm, J.K., Ebisuzaki, W., Lin, R., Xie, P., Chen, M., Zhou, S., Higgins, W., Zou, C.Z., Liu, Q., Chen, Y., Han, Y., Cucurull, L., Reynolds, R.W., Rutledge, G., Goldberg, M., 2010. The NCEP climate forecast system reanalysis. *Bull. Am. Meteorol. Soc.* 91, 1015–1057. https://doi.org/10.1175/2010BAMS3001.1.

Saha, S., Moorthi, S., Wu, X., Wang, J., Nadiga, S., Tripp, P., Behringer, D., Hou, Y.T., Chuang, H.Y., Iredell, M., Ek, M., Meng, J., Yang, R., Mendez, M.P., Van Den Dool, H., Zhang, Q., Wang, W., Chen, M., Becker, E., 2014. The NCEP climate forecast system version 2. *J. Clim.* 27, 2185–2208. https://doi.org/10.1175/JCLI-D-12-00823.1.

Santos, J.S., Sakagami, Y., Haas, R., Passos, J.C., Machuca, M.N., Radünz, W.C., Dias, E., Lima, M.M., 2020. Wind speed evaluation of MERRA-2, ERA-interim and ERA-5 reanalysis data at a wind farm located in Brazil, in: *Proceedings of the ISES Solar World Congress 2019 and IEA SHC International Conference on Solar Heating and Cooling for Buildings and Industry 2019*. International Solar Energy Society, pp. 2265–2274. https://doi.org/10.18086/swc.2019.45.10

Sharp, E., Dodds, P., Barrett, M., Spataru, C., 2015. Evaluating the accuracy of CFSR reanalysis hourly wind speed forecasts for the UK, using in situ measurements and geographical information. *Renew. Energy* 77, 527–538. https://doi.org/10.1016/j.renene.2014.12.025.

Slivinski, L.C., Compo, G.P., Sardeshmukh, P.D., Whitaker, J.S., McColl, C., Allan, R.J., Brohan, P., Yin, X., Smith, C.A., Spencer, L.J., Vose, R.S., Rohrer, M., Conroy, R.P., Schuster, D.C., Kennedy, J.J., Ashcroft, L., Brönnimann, S., Brunet, M., Camuffo, D., Cornes, R., Cram, T.A., Domínguez-Castro, F., Freeman, J.E., Gergis, J., Hawkins, E., Jones, P.D., Kubota, H., Lee, T.C., Lorrey, A.M., Luterbacher, J., Mock, C.J., Przybylak, R.K., Pudmenzky, C., Slonosky, V.C., Tinz, B., Trewin, B., Wang, X.L., Wilkinson, C., Wood, K., Wyszynski, P., 2021. An evaluation of the performance of the Twentieth Century Reanalysis version 3. *J. Clim.* 34, 1417–1438. https://doi.org/10.1175/JCLI-D-20-0505.1.

Soci, C., Bazile, E., Besson, F.O., Landelius, T., 2016. High-resolution precipitation re-analysis system for climatological purposes. *Tellus, Ser. A Dyn. Meteorol. Oceanogr.* 68. https://doi.org/10.3402/tellusa.v68.29879.

Stopa, J.E., Cheung, K.F., 2014. Intercomparison of wind and wave data from the ECMWF Reanalysis Interim and the NCEP Climate Forecast System Reanalysis. *Ocean Model* 75, 65–83. https://doi.org/10.1016/j.ocemod.2013.12.006.

Unden, P., Schimanke, S., Schyberg, H., 2019. Copernicus Climate Change Service Regional Reanalysis C3S_322 Lot 1: Europe Lot 2: Arctic, in: *C3S General Assembly 2019*. European Commission. https://climate.copernicus.eu/sites/default/files/2019-11/12%20Unden_C3S_RRA_Oct19.pdf (accessed 4 October 2022).

Uppala, S.M., Kållberg, P.W., Simmons, A.J., Andrae, U., da Costa Bechtold, V., Fiorino, M., Gibson, J.K., Haseler, J., Hernandez, A., Kelly, G.A., Li, X., Onogi, K., Saarinen, S., Sokka, N., Allan, R.P., Andersson, E., Arpe, K., Balmaseda, M.A., Beljaars, A.C.M., van de Berg, L., Bidlot, J., Bormann, N., Caires, S., Chevallier, F., Dethof, A., Dragosavac, M., Fisher, M., Fuentes, M., Hagemann, S., Hólm, E., Hoskins, B.J., Isaksen, L., Janssen, P.A.E.M., Jenne, R., McNally, A.P., Mahfouf, J.F., Morcrette, J.J., Rayner, N.A., Saunders, R.W., Simon, P., Sterl, A., Trenberth, K.E., Untch, A., Vasiljevic, D., Viterbo, P., Woollen, J., 2005. The ERA-40 re-analysis. *Q. J. R. Meteorol. Soc.* https://doi.org/10.1256/qj.04.176.

Verver, G., 2017. *FP7 Project 607193 UERRA: Uncertainties in Ensembles of Regional Re-Analyses Deliverable D8.4: User Guidance.*

Wahl, S., Bollmeyer, C., Crewell, S., Figura, C., Friederichs, P., Hense, A., Keller, J.D., Ohlwein, C., 2017. A novel convective-scale regional reanalysis COSMO-REA2: improving the representation of precipitation. *Meteorol. Zeitschrift* 26, 345–361. https://doi.org/10.1127/metz/2017/0824.

Yu, J., Zhou, T., Jiang, Z., Zou, L., 2019. Evaluation of near-surfacewind speed changes during 1979 to 2011 over China based on five reanalysis datasets. *Atmosphere (Basel)* 10, 804. https://doi.org/10.3390/ATMOS10120804.

10

Mesoscale Phenomena

The process of wind resource assessment is generally concerned with long-term estimates of wind speed and their variability over multiple years. The analysis presented in the foregoing chapters of this book has primarily focussed on this. However, there are some transient phenomena that can add uncertainty on the short term and affect the spatial variability of the resource over a wide area, especially if such phenomena occur frequently at a given location. This may be important when assessing the variability of the output from a large wind farm, for example. This chapter gives an overview of the more common transient mesoscale phenomena and their potential impact on wind farm performance.

10.1 Thermally Driven Winds

Large-scale synoptic motions are driven primarily by the balance between pressure and Coriolis forces. However, mesoscale winds can be driven by thermal gradients, especially when the background synoptic flow is relatively weak (<5–$6\,\mathrm{ms}^{-1}$). Usually, thermally driven winds are relatively small in terms of wind resource (with some exceptions) but, nonetheless, may affect the overall wind conditions at a site particularly in certain orographic situations which are described in this section. To fully understand the characteristics of these phenomena, a mesoscale model is required.

10.1.1 Sea Breezes

It is described in Section 2.2 that the land and sea have very different thermal properties whereby the land heats up rapidly during the day, particularly on a cloudless summer day, whereas the sea shows little variation in diurnal temperature. This has implications for surface layer stability in

Handbook of Wind Resource Assessment, First Edition. Simon Watson.

near-coastal offshore regions but also can cause a thermally driven circulation to occur. During the day, as the land heats up, this creates a local area of low pressure at the ground as buoyancy drives convection and the air moves aloft, as shown in Figure 10.1. At the same time, this will draw in air from the cooler sea. The vertical convection current above the land will eventually cool and descend again over the sea, creating a local area of high pressure at the sea surface. This circulation pattern is known as a *sea breeze*. A true sea breeze blows directly from sea to land in opposition to any weak background synoptic flow. However, if there are synoptic components along the coast, this can cause the sea breeze circulation to develop a helical motion (Miller et al., 2003). The circulation will increase in magnitude during the day and tend to peak during the afternoon. Sea breezes normally reach a magnitude of 6–10 ms^{-1}, although in some locations they can significantly exceed this value where temperature gradients are high and there is some form of orographic channelling. The vertical extent of a sea breeze circulation can reach 1 km with a maximum horizontal extent of up to 100 km once the circulation has progressed fully inland. As the afternoon progresses and the sea breeze develops, the Coriolis force may cause a deflection of the flow direction. During the night-time period, when the land cools, a reverse circulation may develop (a *land breeze*), though this is generally of a lower magnitude and extent to a sea breeze as the temperature gradient and turbulent mixing are lower.

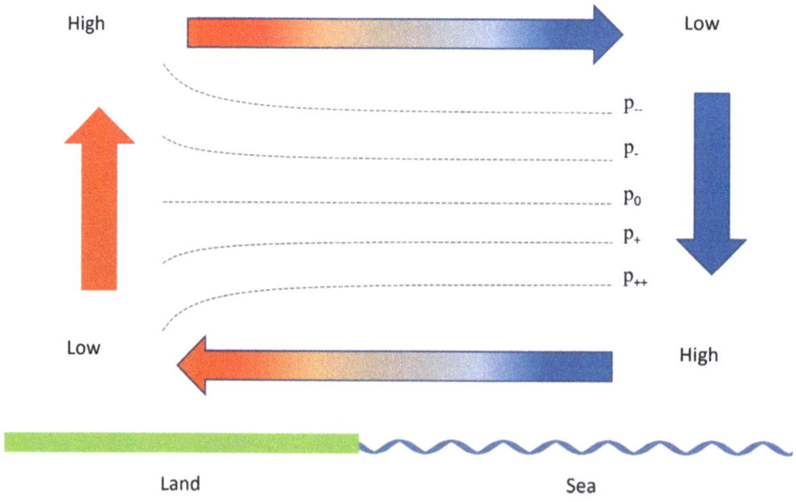

Figure 10.1 A sea breeze circulation developing during the day showing areas of locally high and low pressure and the local pressure isobars.

10.1.2 Anabatic and Katabatic Winds

During the day, a terrain slope will tend to heat up, heating the layer of air immediately above. The air expands and its density decreases, causing a buoyancy force that, if the degree of heating is sufficient, will overcome gravity, causing the air to rise up the slope pulling in cooler air from the region below. This creates an *upslope* or *anabatic wind*, as shown in Figure 10.2(a). Aloft there is a reverse circulation where the air cools and descends again into the lower region. The term *anabatic* generally refers to any upslope wind. If the flow is associated with a mountain slope, this type of anabatic wind is also known as a *mountain wind* or *mountain breeze*. At the top of the mountain, the updraft may create clouds as the water vapour in the air condenses, as also shown schematically in Figure 10.2(a). During the night, the slope will tend to cool radiatively and cooled air immediately above will increase in density. This denser air will tend to sink, aided by gravity, in this case giving rise to a *downslope* or *katabatic wind*, as shown in Figure 10.2(b). In the case of a mountain slope, this is also known as a *valley wind* or *valley breeze*.

Like sea breezes, anabatic and katabatic winds normally only arise when the background synoptic wind speed is low. They are shallow flows reaching a depth of a few tens of metres above a slope. Anabatic winds can reach speeds of a few metres per second (normally $< 5\,\text{ms}^{-1}$), and katabatic winds are normally weaker as the turbulent surface heating process during the day is generally more energetic than the radiative cooling process. The one exception to this is the katabatic winds that develop over Antarctica. The extreme cooling of the central uplands of the continent can cause very strong katabatic winds. The strength of the winds is also helped by the fact that the surface is relatively smooth. In this case, the winds can reach

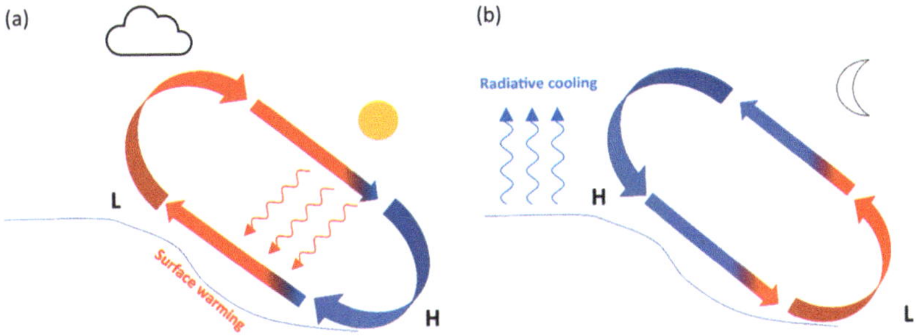

Figure 10.2 (a) An anabatic wind flowing up a slope during the day; (b) a katabatic wind flowing down a slope during the night. Areas of low (L) and (H) pressure are indicated.

$25\,ms^{-1}$ or more with a depth of 300 m. Needless to say, however, there is little demand for wind energy generation in Antarctica.

10.2 Internal Gravity Waves

The concept of lee waves was introduced briefly in Section 5.6. It will be recalled that, in a stable atmosphere, displacement of a parcel of air vertically will result in the air parcel having a lower potential temperature (or higher density) than its surroundings and so the parcel will return to its original position. Potentially, this restoring movement can result in an overshoot from the equilibrium position resulting in an oscillatory motion, rather like a mass on a spring. An air mass can be perturbed vertically by a large orographic feature such as a hill or mountain. When the air flows over the obstacle, it is pushed upwards and in a stable atmosphere will tend to oscillate downwind of the obstacle creating a mountain wave, which is a type of atmospheric or *internal gravity wave* (IGW). The air mass undergoes *simple harmonic motion* so that if it is displaced vertically a distance δz it will obey the equation of motion (Nappo, 2013)

$$\frac{d^2(\delta z)}{dt^2} = -\frac{g}{\theta}\frac{\partial \theta}{\partial z}\delta z \tag{10.1}$$

This gives the solution

$$\delta z(t) = A_0 e^{iNt} + B_0 e^{-iNt} \tag{10.2}$$

where N is the Brunt–Väisälä frequency introduced in Eq. (2.44) and A_0 and B_0 are constants that define the amplitude of the wave. From the definition of N, if $\frac{\partial \theta}{\partial z} < 0$, N is imaginary and the second term in Eq. (10.2) increases indefinitely, resulting in a convective instability meaning that IGWs can only exist in a stable atmosphere. More generally, the Taylor–Goldstein equation derived from linear wave theory is used to describe IGW motion (Goldstein, 1931; Nappo, 2013; Taylor and Research, 1931). If the wind velocity perturbations from their mean values at height z in the horizontal (u_1) and vertical (w_1) are assumed to have periodic solutions

$$u_1(x,z,t) = \tilde{u}(z)e^{i(kx-\omega t)} \tag{10.3}$$

$$w_1(x,z,t) = \tilde{w}(z)e^{i(kx-\omega t)} \tag{10.4}$$

then $\tilde{w}(z)$ is given by the solution of

$$\frac{d^2\widehat{w}}{dt^2}+\left[\frac{N^2}{\left(c-u_0\right)^2}+\frac{1}{\left(c-u_0\right)}\frac{d^2u_0}{dt^2}-\frac{g}{R_dT\left(c-u_0\right)}\frac{du_0}{dt}-\left(\frac{g}{2R_dT}\right)^2-k^2\right]\widehat{w}=0 \tag{10.5}$$

where

$$\tilde{w}=e^{gz/2R_dT}\,\widehat{w}. \tag{10.6}$$

and u_0 is the background wind speed, c is the wave phase speed and k is the horizontal wave number ($=2\pi/\lambda_x$, where λ_x is the horizontal wavelength, assuming propagation aligned to the x-axis). The angular frequency ω is given by

$$\omega=ck \tag{10.7}$$

Replacing the terms in square brackets by m^2, where m is the vertical wave number ($=2\pi/\lambda_z$, where λ_z is the vertical wavelength), Eq. (10.5) simplifies to

$$\frac{d^2\widehat{w}}{dt^2}+m^2\widehat{w}=0 \tag{10.8}$$

which, if m is a constant, leads to

$$\widehat{w}=A_0e^{imz}+B_0e^{-imz} \tag{10.9}$$

The Brunt–Väisälä frequency N provides an upper limit to the frequency of IGWs for given wind speed and temperature profile.

The real value of $\tilde{u}$ is given by

$$\mathrm{Re}(\tilde{u})=-A_0\frac{m}{k}\cos mz \tag{10.10}$$

A mountain wave will propagate perpendicular to the streamline of the air parcel as it flows over the obstacle (Nappo, 2013). So, for a three-dimensional (3D) obstacle, waves will propagate in multiple directions, horizontally and vertically. If the temperature stratification increases sufficiently with height and the wind speed is moderate then any vertically propagating wave is reflected downwards, and the mountain wave can become trapped and only propagate horizontally staying relatively close to the ground. This leads to the lee waves. In the case of lee waves, to the observer, the waves appear to be stationary. This means that c = 0, although in the frame of reference of the moving wind the intrinsic speed of the lee wave is equal and opposite to the background wind speed (i.e. $-u_0$).

In Nappo (2013), an analysis is made of mountain waves propagating downwind of a Gaussian shaped ridge with height h_G as a function of horizontal distance x of the form

$$h_G(x) = He^{-x^2/b} \tag{10.11}$$

where H is the maximum height of the ridge and b is its half-height. Theoretically, a range of waves of different values of k and w will be excited by the ridge, with the most dominant horizontal wavelength excited being when $\lambda_x \approx 4.4b$. This corresponds to around 98% of the width of the ridge and thus the most excited wave scales with the width of the ridge. The propagation of waves downwind of the ridge depends on b and the value of N / u_0 which is known as the *Scorer parameter*, L_{Sc}. A typical situation where mountain waves may propagate downwind of a ridge is shown in Figure 10.3, where $L_{Sc}b \approx 1$ ($N = 0.01s^{-1}$, $u_0 = 10$ ms^{-1}, and $b = 1$ km). In this case, the wave amplitude decreases relatively rapidly above and downstream of the

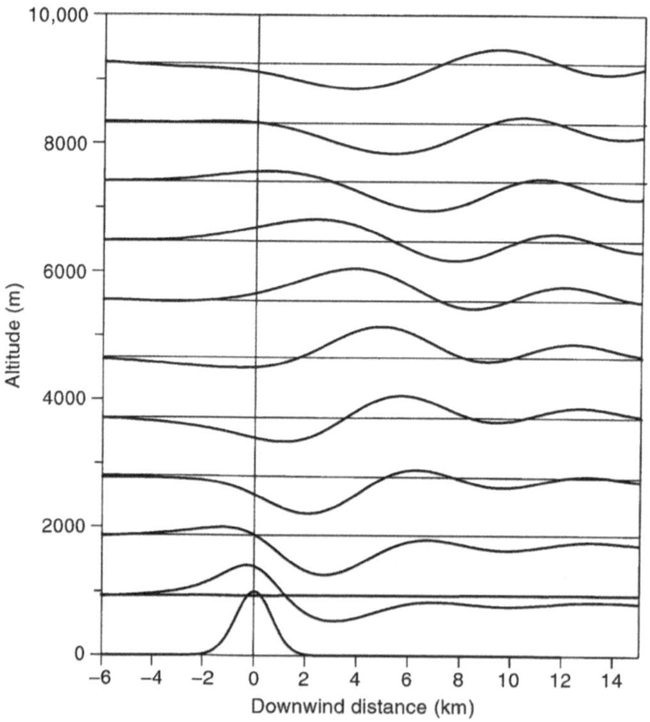

Figure 10.3 Mountain waves downwind of a ridge where $L_{Sc}b \approx 1$ (B_0, $u_0 = 10$ ms^{-1} and $b = 1$ km). *Source:* Nappo (2013) with permission of Elsevier.

ridge, but depending on the wind speed, the scale of the ridge, and the rate of increase of temperature stratification aloft, the waves may be increased in magnitude and duration.

Although IGWs are most frequently caused by orographic features creating mountain or lee waves, they can also be triggered by islands, convection, frontal systems, and jets (Plougonven and Zhang, 2014). From the point of view of wind resource assessment, although they have little impact on the mean wind speed at a site, they can cause sporadic periodic modulations to the near-surface wind speed over horizontal distances of tens or hundreds of kilometres. IGWs are difficult to visualise directly but, as mentioned in Section 5.6, mountain waves can sometimes be made visible through the condensation of water vapour at their peaks causing rows of clouds. These rows of clouds, in contrast to clouds that may result from roll vortices (see Section 10.5), are aligned perpendicular to the wind direction. Over the sea, IGWs can be detected by a satellite instrument such as a synthetic aperture radar (SAR), if the waves remain close to the surface to create a *footprint*. There are numerous examples of this in the literature in various areas of the world (e.g. Alpers et al., 2008; Cheng and Alpers, 2010; da Silva and Magalhães, 2009; Evdoshenko, 2020; Li et al., 2011; Thomson et al., 1992; Vachon et al., 1994; Zheng et al., 1998). An example of IGWs over the North Sea taken from a SAR satellite image is shown in Figure 10.4.

The image shows that the IGWs modulate wind speed from a maximum of around 12 ms^{-1} to a minimum of less than 5 ms^{-1} at 10 m above the sea surface over a distance (half a wavelength) of around 7 km. The IGWs persist at least 60 km from the coast. Given that a modern large-scale offshore wind farm may have dimensions that span more than 20 km, this can have a significant impact on the variation in output of the wind turbines across the wind farm when IGWs are present, as shown in Figure 5.16.

10.3 Low-level Jets

There is no precise definition of a low-level jet (LLJ), but it is generally characterised by a peak in wind speed within the atmospheric boundary layer (ABL) with a reduction in wind speed aloft thus creating a wind profile that can deviate significantly from a log or power law profile in the surface layer if the LLJ is sufficiently close to the ground. LLJs are caused by a transition whereby pressure gradient, Coriolis, and surface frictional forces transition from a balanced to an unbalanced state where there is an acceleration of the flow relative to the geostrophic balance, such as a change in surface layer

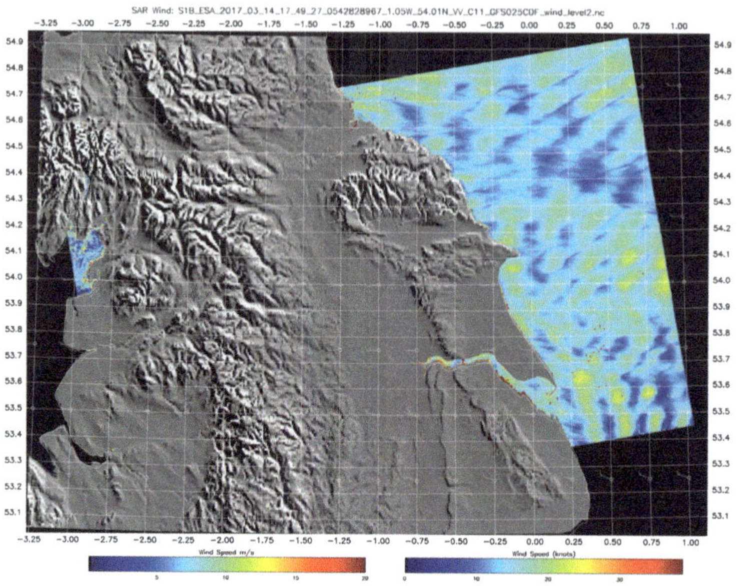

Figure 10.4 A processed SAR image showing 10 m wind speeds indicating gravity waves in the North Sea. Note that radar reflections can be seen off ships and the wind turbines of the Westermost Rough offshore wind farm located at 53.81°N, 0.15°E as well as the Humber Gateway offshore wind farm at 53.64°N, 0.29°E. *Source:* wind fields from synthetic aperture radar provided by the Technical University of Denmark, Department of Wind Energy (DTU Wind Energy). CC BY 4.0.

stability from neutral or unstable conditions to stable conditions (Emeis, 2013) or a reduction in surface roughness such as from land to sea. Because LLJs are often associated with a transition to stable conditions, they tend to occur more frequently at night over land. During the day, when the ABL transitions to a convective boundary layer, the opposite can happen, and the flow can slow. Due to the inertia of the moving air mass, LLJs associated with cycles of surface heating and cooling are referred to as being the result of an *inertial oscillation*.

It can be convenient to model the shape of the jet wind speed profile, and one such engineering model combines a power law profile with a trigonometric jet shape to give a final LLJ profile (Zhang et al., 2019)

$$u_{LLJ}(z) = \left\{ u(z_{ref}) + u_m \left[1 - \tanh^2\left(C_s \frac{z - z_s}{z_s}\right) \right] \right\} \left[\frac{z}{z_{ref}} \right]^{\alpha} \qquad (10.12)$$

where z_{ref} is a reference height, u_m is the maximum speed of the free jet, C_s is the free jet shape factor, α is the shear exponent, and z_s is the height of the free jet maximum. An example LLJ profile using this formulation is shown in Figure 10.5.

Depending on the location and the time of year, LLJs can occur quite frequently, and if the jet height is close to the hub height of a wind turbine, they can have a significant impact on turbine power output. As wind turbines are generally designed to extract wind power from a wind speed profile that increases monotonically with height, LLJs can also have an adverse effect on turbine loads. The height of the jet maximum can vary between tens and hundreds of metres. For example, an analysis of LLJs near the Dutch North Sea coast suggested that jets occurred at a height of between 50 m and 200 m (Kalverla et al., 2019), and data from the Dutch offshore meteorological mast at IJmuiden suggested that LLJs occurred 12% of the time (Kalverla et al., 2017). The height of the LLJ can potentially have a significant impact on wind farm output. Both wind tunnel studies (Doosttalab et al., 2020) and LES simulations (Gadde and Stevens, 2021) have shown how an LLJ with a peak above or below wind turbine hub height can promote re-energisation of the wind speed in an array of wind

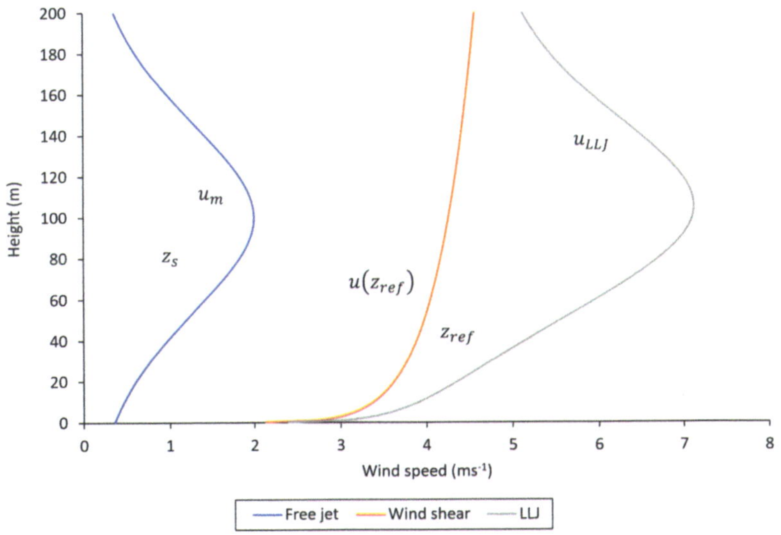

Figure 10.5 Schematic illustration of an LLJ parameterised using an engineering profile. The LLJ profile (right) is formed by combining a free jet (left) and wind shear profile (centre).

turbines, as illustrated in Figure 10.6. When the jet is above hub height, the enhanced positive shear promotes mixing of the free stream flow above the wind farm with the slower-moving air within the wind farm due to the wakes resulting from energy extraction. This mixing-in of faster-moving air increases the wind speed available to downwind turbines and thus increases the output of the array overall compared to a non-LLJ wind profile. An LLJ below hub height creates a similar (though smaller) effect due to the negative shear. The effect is smaller as the free stream flow below the rotors is of a lower magnitude to that above and the strongly stable stratification above the jet in this case restricts the production of turbulence and turbulent mixing. When at hub height, the LLJ presents the first turbine in the row with a higher wind speed and thus results in a higher power output for this machine, but the enhanced wake mixing effect is not seen and the power output of the array is lower overall compared with the first two cases. This effect is yet to be verified from actual observations but points to the fact that the prevalence and characteristics of LLJs may be important when assessing overall wind farm yield.

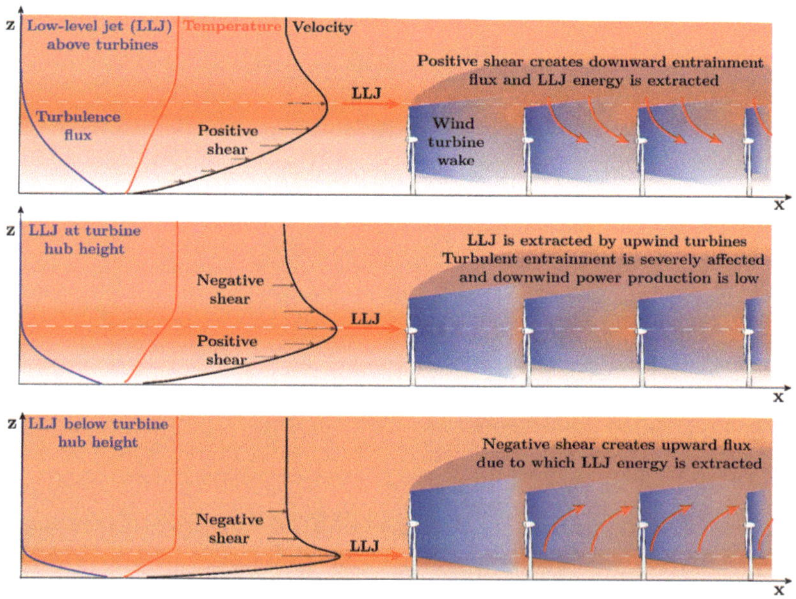

Figure 10.6 Effect of an LLJ on the extraction of wind power from an array of wind turbines (University of Twente, 2021). *Source:* image courtesy of R. Stevens and S. Gadde, University of Twente.

10.4 Convective Cells

Convective cells are formed by patterns of 3D air movement in situations when cold air passes over a warmer surface and can often occur behind a cold front where unstable conditions are created. They can also occur near a subtropical anticyclone (high pressure system). They typically have a spatial scale of a few tens of kilometres and can often be seen over sea areas in satellite images in the form of hexagonal cells as the ascending air causes water droplet condensation and the formation of clouds at the top of the cell. In the case of open cell convection, which tends to be more common, air close to the surface moves outwards from the centre and then, due to surface heating, rises up at the edge of cells to create a ring of cloud surrounding a clear centre. In the case of a closed cell the opposite is seen. Closed cell convection can occur when the temperature difference between the air and the sea is relatively small as the mechanism for the overturning is radiative cooling of the clouds aloft rather than heating from below. Figure 10.7 shows a vertical diagram to illustrate the two types of cell. Open cell convection is frequently seen over areas such as the north-eastern Atlantic, including the North Sea, and is estimated to occur around 20% of the time (Atkinson and Zhang, 1996; Busack et al., 1985). An example of such is shown in Figure 10.8 from a satellite image. Sometimes, roll vortices (described in Section 10.5), which develop as cold air flows from land to sea, may transition into open cell convection as the unstable temperature gradient and the wind shear decrease in magnitude. This may in turn transition to closed cell convection further offshore as the unstable temperature gradient decreases further. The converging or diverging air close to the surface

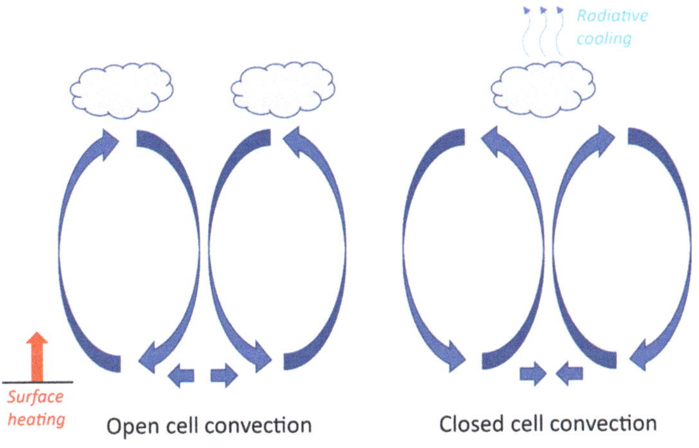

Figure 10.7 Schematic illustration of open cell and closed cell convection.

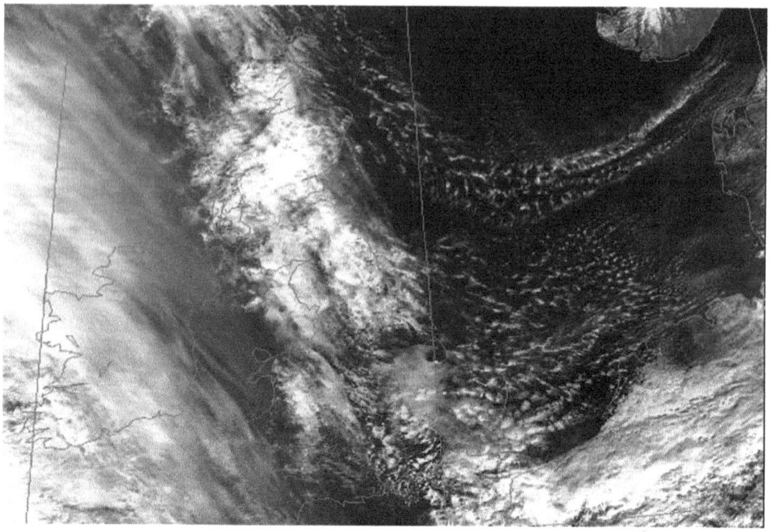

Figure 10.8 An image from the Meteosat 8 HRVIS satellite showing open cell convection over the North Sea on 11 March 2006 at 12:00 UTC. *Source:* EUMeTrain, CC BY-SA 4.0, https://eumetrain.org/satmanu/CMs/OCC/print.html (accessed 5 August 2022).

modulates the mean wind conditions and can cause significant variation in the wind conditions where the cells are present.

This can also have a significant impact on the variability of the output of a wind farm. A study of the Horns Rev I offshore wind farm considered the variability in wind turbine power output comparing a day when open cell convection was present over the wind farm and a day without convective motion (Göçmen et al., 2020). In both cases, the wind speed during the day was less then rated for most of the time. The mean and standard deviation of the wind power are recorded over ten-minute intervals by each turbine. Based on these data, the median standard deviation of the power on the days with and without open cell convection were compared. The results for turbines are shown as contour plots in Figure 10.9. Quite clearly the variability in power output was significantly higher on the day when open cell convection was present. Power fluctuations on the day without convective cells were around 8–10% of rated power, whereas on the day with open cell convection, fluctuations were 10–20% of rated power.

Although open cell convection makes little difference to the overall wind resource available to a wind farm, their prevalence needs to be considered when looking at the variability of the resource at a given location, particularly offshore.

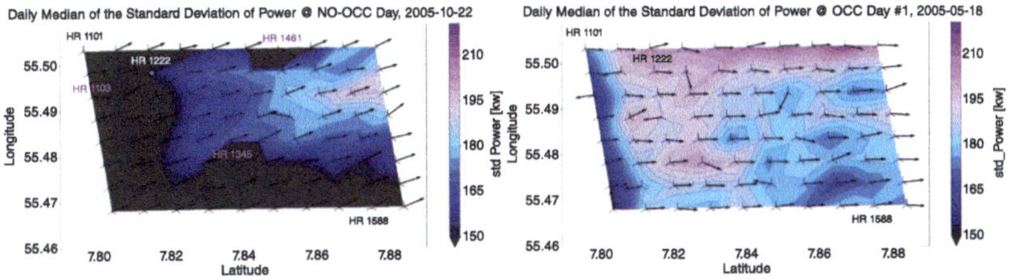

Figure 10.9 A comparison of the median of the standard deviation of the power output of the turbines at the Horns Rev 1 offshore wind farm based on a day where no open cell convection was observed (left) and where open cell convection was present (right). The arrows represent the local wind direction. *Source:* Göçmen et al. (2020) / IOP Publishing / CC BY 3.0.

10.5 Roll Vortices

Roll vortices are related to convective cells but form more organised structures in the ABL. If the ratio of ABL inversion height to Obukhov length, $h_{ABL} / L \gtrsim -5$, roll vertices tend to occur, whereas if $h_{ABL} / L \lesssim -5$, convective cells tend to prevail with some mixture for values in between (Deardorff, 1972; Etling and Brown, 1993; Grossman, 1982). Figure 10.10 shows a satellite SAR image off the northern coast of Denmark where both roll vortices and convective cells are present. Roll vortices are manifested as counter rotating circulations of air flow driven by convection which are organised in long rows aligned at a small angle to the prevailing surface wind direction (see Figure 10.11) which can extend to a distance E of tens up to thousands of kilometres (Etling and Brown, 1993). Their height H typically extends to the top of the ABL (1–2 km above the surface) with a wavelength λ of 2–20 km, giving aspect ratios (wavelength to height) of between 2 and 15. They can persist from a few hours to a few days and can frequently be seen in satellite images over coastal areas during spring at higher latitudes (Svensson et al., 2017). Evidence of roll vortices can be seen in so-called cloud streets, where the vertical rising section of each vortex gives rise to clouds and the sinking section is clear. They can occur at a range of wind speeds, but typical wind speeds are around 15 ms^{-1} (Atkinson and Zhang, 1996). The rolls can create quite significant modulations to the horizontal wind field (Svensson et al., 2017) with typical swings of ±3–4 ms^{-1} relative to the mean wind speed. To date, there has been no study of the impact of convective rolls on wind farm output, though it has been suggested that the relatively slow speed of rotation of the rolls (~minutes) coupled with the additional turbulence created may have an adverse effect on turbine loading (Smedman, 1991, 1992). However, this has not been confirmed.

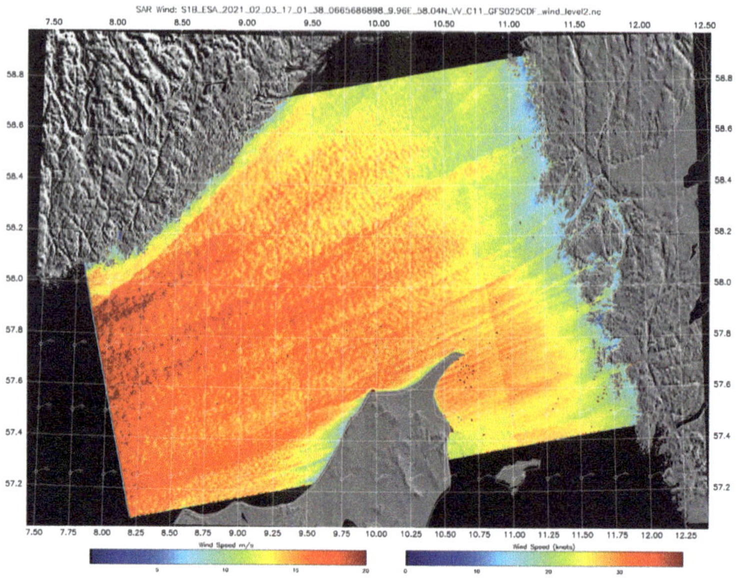

Figure 10.10 SAR image showing the presence of both convective cells (most noticeably at the top of the image) and roll vortices in the Skagerrak and Kattegat north of Denmark. *Source:* wind fields from synthetic aperture radar provided by the Technical University of Denmark, Department of Wind Energy (DTU Wind Energy). CC BY 4.0.

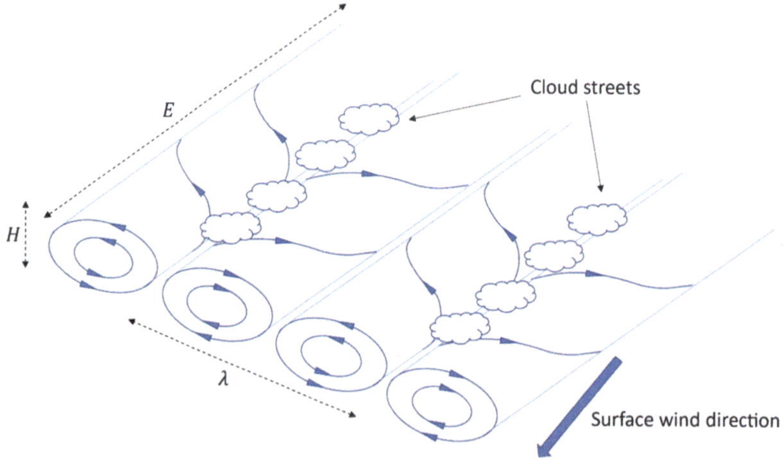

Figure 10.11 Schematic representation of roll vortices showing dimensions and cloud streets where there is rising circulation.

References

Alpers, W., Huang, W., Xilin, G., 2008. Observations of atmospheric gravity waves over the Chinese seas by spaceborne synthetic aperture radar, *Proc. Dragon ESA SP-655*.

Atkinson, B.W., Zhang, J.W., 1996. Mesoscale shallow convection in the atmosphere. *Rev. Geophys.* 34, 403–431. https://doi.org/10.1029/96RG02623.

Busack, B., Bakan, S., Luthardt, H., 1985. Surface conditions during mesoscale cellular convection. *Beiträge Zur Phys. der Atmosphäre* 58, 4–10.

Cheng, C.M., Alpers, W., 2010. Investigation of trapped atmospheric gravity waves over the South China Sea using Envisat synthetic aperture radar images. *Int. J. Remote Sens.* 31, 4725–4742. https://doi.org/10.1080/01431161.2010.485145.

da Silva, J.C.B., Magalhães, J.M., 2009. Satellite observations of large atmospheric gravity waves in the Mozambique Channel. *Int. J. Remote Sens.* 30, 1161–1182. https://doi.org/10.1080/01431160802448943.

Deardorff, J.W., 1972. Numerical investigation of neutral and unstable planetary boundary layers. *J. Atmos. Sci.* 29, 91–115. https://doi.org/10.1175/1520-0469(1972)029<0091:nionau>2.0.co;2.

Doosttalab, A., Siguenza-Alvarado, D., Pulletikurthi, V., Jin, Y., Bocanegra Evans, H., Chamorro, L.P., Castillo, L., 2020. Interaction of low-level jets with wind turbines: on the basic mechanisms for enhanced performance. *J. Renew. Sustain. Energy* 12, 053301. https://doi.org/10.1063/5.0017230.

Emeis, S., 2013. *Wind Energy Meteorology: Atmospheric Physics for Wind Power Generation*. Springer.

Etling, D., Brown, R.A., 1993. Roll vortices in the planetary boundary layer: a review. *Boundary-Layer Meteorol.* https://doi.org/10.1007/BF00705527.

Evdoshenko, M.A., 2020. Imprints of atmospheric waves on the Black Sea surface in data of ocean color scanners. *Oceanologia* 62, 255–266. https://doi.org/10.1016/j.oceano.2020.01.002.

Gadde, S.N., Stevens, R.J.A.M., 2021. Effect of low-level jet height on wind farm performance. *J. Renew. Sustain. Energy* 13, 013305.https://doi.org/10.1063/5.0026232.

Göçmen, T., Larsén, X.G., Imberger, M., 2020. The effects of open cellular convection on wind farm operation and wakes, in: *Journal of Physics: Conference Series*. IOP Publishing Ltd, p. 62014. https://doi.org/10.1088/1742-6596/1618/6/062014.

Goldstein, S., 1931. On the stability of superposed streams of fluids of different densities. *Proc. R. Soc. London. Ser. A* 132, 524–548. https://doi.org/10.1098/rspa.1931.0116.

Grossman, R.L., 1982. An analysis of vertical velocity spectra obtained in the bomex fair-weather, trade-wind boundary layer. *Boundary-Layer Meteorol* 23, 323–357. https://doi.org/10.1007/BF00121120.

Kalverla, P.C., Duncan, J.B., Jr., Steeneveld, G.-J., Holtslag, A.A.M., 2019. Low-level jets over the North Sea based on ERA5 and observations: together they do better. *Wind Energy Sci* 4, 193–209. https://doi.org/10.5194/wes-4-193-2019.

Kalverla, P.C., Steeneveld, G.-J., Ronda, R.J., Holtslag, A.A.M., 2017. An observational climatology of anomalous wind events at offshore meteomast IJmuiden (North Sea). *J. Wind Eng. Ind. Aerodyn.* 165, 86–99. https://doi.org/10.1016/j.jweia.2017.03.008.

Li, X., Zheng, W., Yang, X., Li, Z., Pichel, W.G., 2011. Sea surface imprints of coastal mountain lee waves imaged by synthetic aperture radar. *J. Geophys. Res.* 116, C02014. https://doi.org/10.1029/2010JC006643.

Miller, S.T.K., Keim, B.D., Talbot, R.W., Mao, H., 2003. Sea breeze: structure, forecasting, and impacts. *Rev. Geophys.* 41, 1011. https://doi.org/10.1029/2003RG000124.

Nappo, C.J., 2013. *An Introduction to Atmospheric Gravity Waves*, Second Edition. Academic Press.

Plougonven, R., Zhang, F., 2014. Internal gravity waves from atmospheric jets and fronts. *Rev. Geophys.* 52, 33–76. https://doi.org/10.1002/2012RG000419.

Smedman, A.-S., 1991. Occurrence of roll circulations in a shallow boundary layer. *Boundary-Layer Meteorol* 57, 343–358. https://doi.org/10.1007/BF00120053.

Smedman, A.-S., 1992. Calculations of dynamic wind turbine blade loads from simple meteorological data. *Wind. Eng.* 16, 195–212.

Svensson, N., Sahlée, E., Bergström, H., Nilsson, E., Badger, M., Rutgersson, A., 2017. A case study of offshore advection of boundary layer rolls over a stably stratified sea surface. *Adv. Meteorol.* 2017. https://doi.org/10.1155/2017/9015891.

Taylor, G.I., Research, Y., 1931. Effect of variation in density on the stability of superposed streams of fluid. *Proc. R. Soc. London. Ser. A, Contain. Pap. A Math. Phys. Character* 132, 499–523. https://doi.org/10.1098/rspa.1931.0115.

Thomson, R.E., Vachon, P.W., Borstad, G.A., 1992. Airborne synthetic aperture radar imagery of atmospheric gravity waves. *J. Geophys. Res. Ocean.* 97, 14249–14257. https://doi.org/10.1029/92JC01178.

University of Twente, 2021 Wind farms benefit from strong 'rivers of air' in lower atmosphere. https://www.utwente.nl/en/news/2021/2/978090/wind-farms-benefit-from-strong-rivers-of-air-in-lower-atmosphere (accessed 28 June 2022).

Vachon, P.W., Johannessen, O.M., Johannessen, J.A., 1994. An ERS 1 synthetic aperture radar image of atmospheric lee waves. *J. Geophys. Res.* 99, 22483. https://doi.org/10.1029/94JC01392.

Zhang, X., Yang, C., Li, S., 2019. Influence of low-level jet intensity on aerodynamic loads of horizontal axis wind turbine rotor. *Eng. Appl. Comput. Fluid Mech.* 13, 300–308. https://doi.org/10.1080/19942060.2019.1585953.

Zheng, Q., Yan, X.-H., Klemas, V., Ho, C.-R., Kuo, N.-J., Wang, Z., 1998. Coastal lee waves on ERS-1 SAR images. *J. Geophys. Res. Ocean.* 103, 7979–7993. https://doi.org/10.1029/97JC02176.

11

Long-term Wind Climate Trends

One of the main assumptions made when assessing the long-term wind resource at a potential wind farm site is that the climate is stationary (i.e. no long-term changes in wind speed, or other relevant meteorological variables, are expected). Indeed, methods such as measure–correlate–predict (MCP) use long-term historical climate data as a best guess of what will happen 25–30 years from now. This may or may not be a reasonable assumption under a changing climate. There is very compelling evidence that man-made emissions of greenhouse gases (GHGs) are causing global heating which will have an impact on the future climate (IPCC, 2014). What impact global heating will have on wind speed patterns is still unclear, but this chapter gives an overview of the evidence to date based on analysis of historical data, the models used to predict the future climate and what is projected to happen given current knowledge using a range of possible scenarios.

11.1 The Historical Evidence for Long-term Trends in Wind Speeds

Reliable and widespread measurements of wind speed have only really been available since the middle of the twentieth century. This makes any assessment of long-term changes in the global wind climate challenging. Furthermore, changes in instrumentation, instrument mounting, and meteorological station exposure over the long term make it difficult to extract climate-change-related effects from other factors that can cause systematic changes. Nonetheless, there have been several studies which have analysed wind speed data over multiple decades to assess potential trends. Figure 11.1 shows a map of primarily onshore global surface (~10 m) wind speed trends based on an analysis of 148 studies worldwide (McVicar et al., 2012) using

Handbook of Wind Resource Assessment, First Edition. Simon Watson.

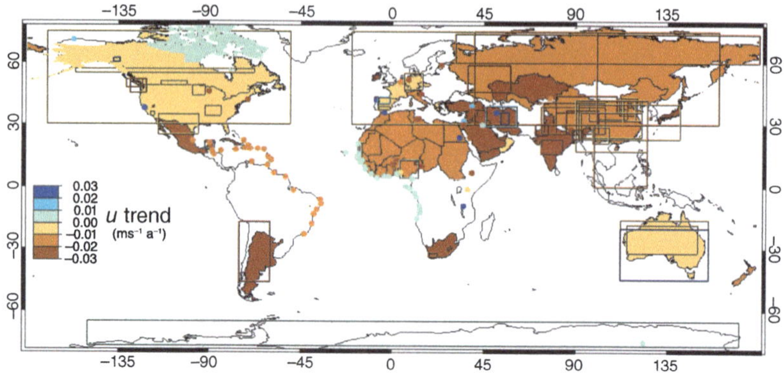

Figure 11.1 Trends in observed onshore surface wind speeds (in ms^{-1} per annum) based on the results of 148 studies covering the late twentieth and early twenty-first centuries. *Source:* McVicar et al. (2012) / with permission of Elsevier.

data of varying lengths with many of the studies over 30 years spanning the latter part of the twentieth century and early twenty-first century. Though there are regional differences, the overall global trend was estimated at −0.014 ms^{-1} per annum based on those sites for which at least 30 years of data were available, suggesting a reduction in global mean wind speeds. This would imply for the lifetime of a windfarm of 25 years a reduction of around 0.5 ms^{-1} at hub height[1], which is not insignificant considering a good onshore site may experience a mean hub height wind speed wind speed of 7–8 ms^{-1}. The reasons for this reduction are not entirely clear, but it has been speculated that some of the change results from a global increase in surface roughness due to human activity and the rest may be due to climate change factors, including, for example, increased heating at the poles compared with the equator, which will tend to reduce temperature gradients and associated wind speeds (Ren, 2010). It should be stressed that this is still speculative and is the subject of ongoing research.

The uncertainty is compounded by a more recent analysis which has suggested that, although there has been a trend for reducing wind speeds worldwide in the late twentieth century and the first decade of the twenty-first century, there is evidence of a recovery in global wind speeds since 2010, as shown in Figure 11.2 (Zeng et al., 2019). Based on an analysis of wind speed data from a large number (>1400) of global surface meteorological stations, this study suggests a wind speed trend of −0.008 ms^{-1} per annum since 2010 and + 0.024 ms^{-1} per annum since. These results would tend to rule out the

1 Assuming a surface roughness of 0.03 m and a turbine hub height of 80 m.

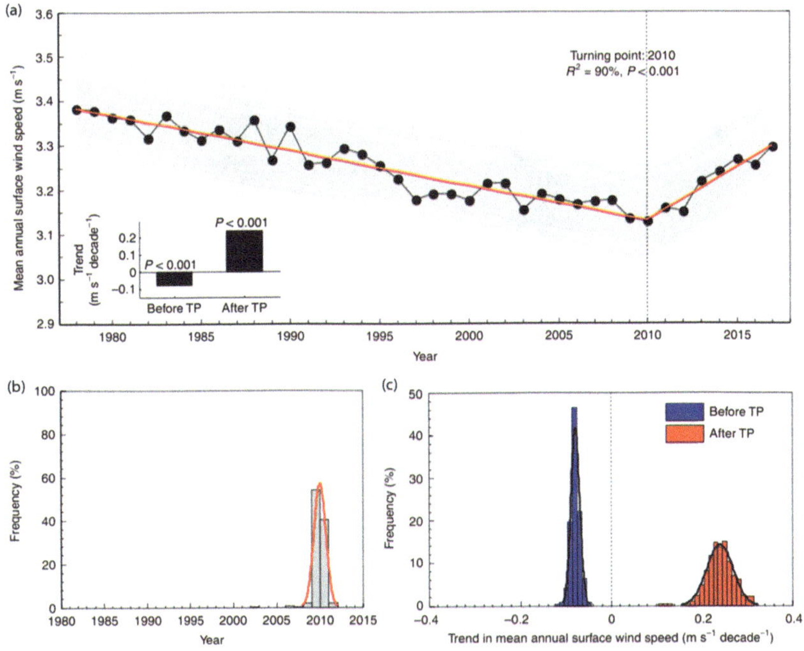

Figure 11.2 Global wind speed trends between 1978 and 2017: (a) Black dots are mean annual wind speeds based on a large number of global meteorological stations; the red line is a regression fit to these values before and after 2010; the grey lines are piecewise linear regression fits to the wind speed data from a random sample of 40% of the stations; the inset is the wind speed trend before and after 2010; (b) the frequency of turning points (TPs) based on the piecewise linear fits showing 2010 to be the median value; (c) frequency of wind speed trends based on the piecewise linear fits before and after each TP. *Source:* Zeng et al., 2019) / with permission of Springer Nature.

hypothesis which relates wind speed changes to increases in surface roughness and more points to changes in global circulation patterns.

By contrast, the evidence from satellite data suggests that offshore wind speeds have, globally averaged, increased slightly in the period 1985–2018 (Young and Ribal, 2019; Young et al., 2011), though there is some spatial variation, as seen in Figure 11.3. The Southern Ocean shows the largest changes (+0.02 ms^{-1} per annum), whereas oceans in the northern hemisphere have experienced smaller changes (0 to +0.01 ms^{-1} per annum). Only the sea areas around South-East Asia show any significant decrease (around −0.02 ms^{-1} per annum). The extreme tail of the wind speed distributions (90th centile) has seen a greater increase with significant positive increases globally and in some areas (again primarily in the Southern Ocean) up to

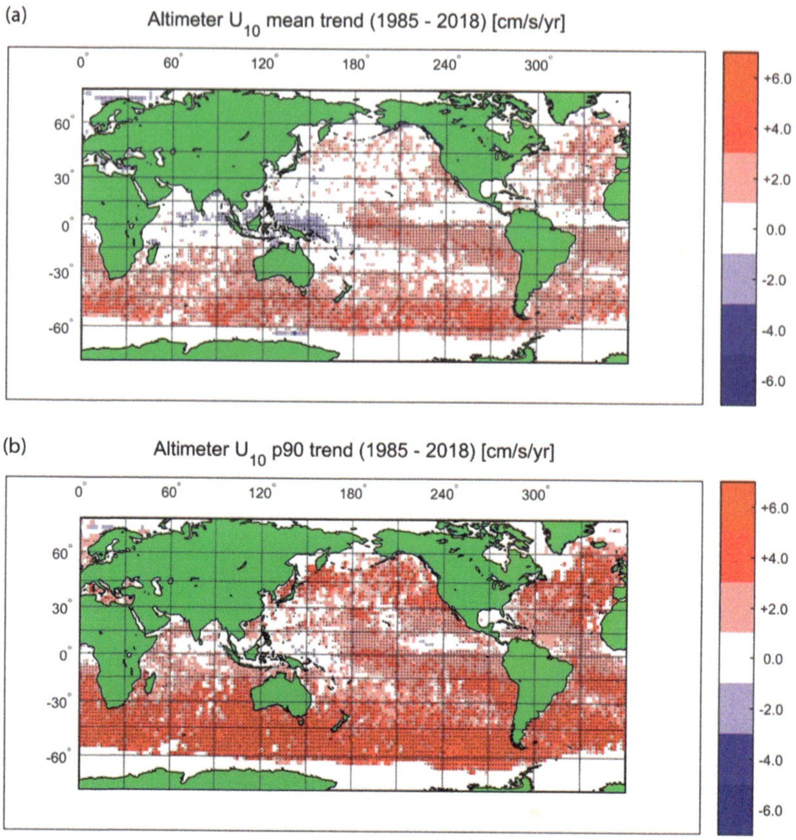

Figure 11.3 Trends in mean and 90th centile 10 m offshore wind speeds over the period 1985–2018 from satellite altimeter data. *Source:* Young and Ribal (2019) / with permission of American Association for the Advancement of Science.

+0.06 ms^{-1} per annum. This suggests a change in the wind speed distributions in the last 30 years with an increase in extreme winds.

Reanalysis datasets provide a potentially more reliable resource to assess spatial and temporal trends in global wind speeds as very localised changes in instrumentation mounting and exposure have a much smaller influence compared with raw measurements. Using data from the three reanalysis datasets ERA-Interim, JRA-55, and MERRA-2, long-term global trends in wind speed are summarised by meteorological season in Figure 11.4 (Torralba et al., 2017). Wind speeds are shown close to the surface (10 m) and at the first model pressure level (850 hPa, which represents a height of typically between 1000 and 1500 m). Although there are some local

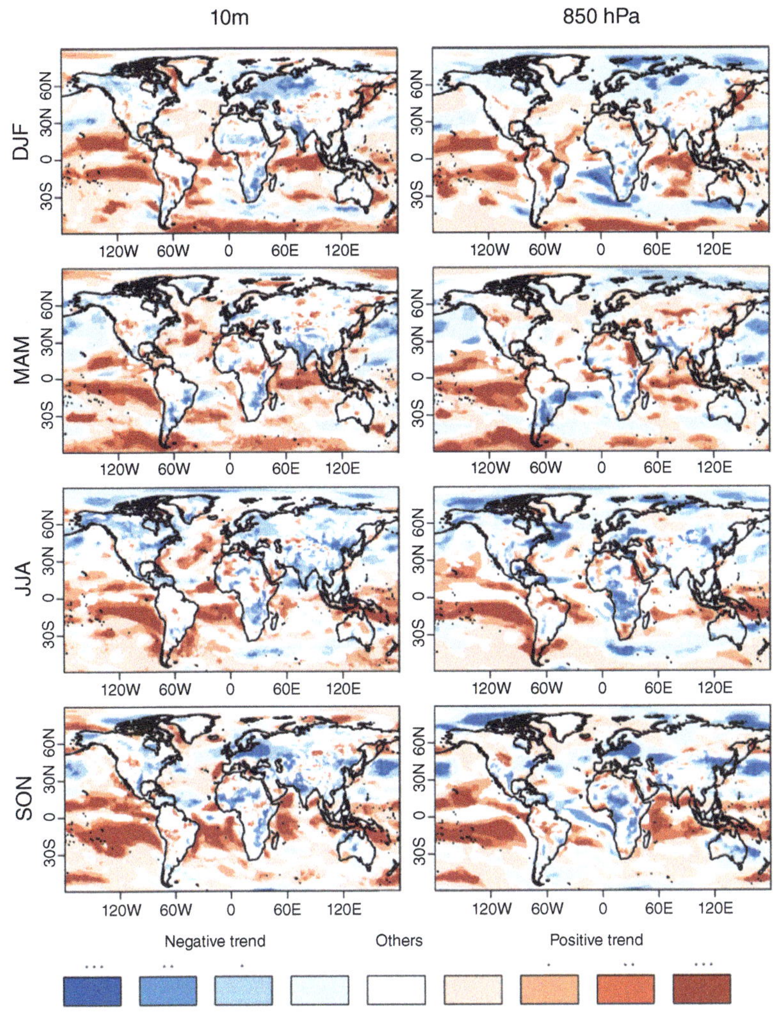

Figure 11.4 Trends in mean wind speed at the surface (10 m) and 850 hPa pressure levels by meteorological season: DJF (December, January, February), MAM (March, April, May), JJA (June, July August) and SON (September, October, November) determined using the ERA-Interim, JRA-55, and MERRA-2 reanalysis datasets over the period 1980–2015. An asterisk in the legend indicates that the trend is significant at the 95% level for one of the reanalysis datasets. The number of asterisks indicates the number of reanalysis datasets which show a significant trend. *Source:* from Torralba et al. (2017) / IOP Publishing / CC BY 3.0.

differences compared with Figures 11.1 and 11.3, the general trend would seem to be confirmed that, on average, wind speeds in the last 30–40 years have decreased on land and increased offshore. Notable are reducing wind speeds over northern Europe, particularly during the autumn and winter periods, when wind speeds are highest, reducing wind speeds over India, and increasing wind speeds over much of the Southern Ocean.

11.2 Climate Models

Climate models include similar physics to those used for weather forecasting and hindcasting to produce the reanalysis datasets described in Chapter 9. Many of these models simulate the coupled response of the principal elements of the earth system which affect the climate, namely: the atmosphere, the oceans, the land, and sea ice. When simulating the climate, attention is paid to the effect of various radiative forcing scenarios, whereby changes to the net-zero balance of incoming radiation from the sun and outgoing radiation are assessed. These changes cause a net-positive energy flux resulting from emissions of GHGs, including those which have an anthropogenic origin, such as (though not exclusively) through the burning of fossil fuels. Until recently, standard climate models were of a class known as *Atmosphere-Ocean General Circulation Models* (AOGCMs). More recently, more complex physics has been incorporated to include biogeochemical cycles such as those involved in the carbon cycle, the sulfur cycle, or ozone. This family of models is known as *Earth System Models* (ESMs) (IPCC, 2013). A large number of AOGCMs and ESMs have been developed in various countries, and their performance has been assessed in a series of large international projects, under the umbrella of the Coupled Model Intercomparison Project (CMIP), the most recent of which is CMIP6 (Eyring et al., 2016). AOGCM and ESM output is generally at a relatively course spatial output (typically ~50–100 km) and so model output is sometimes downscaled to a finer resolution using a statistical process or using a limited area regional climate model (RCM). CMIP model projections are made by assuming different scenarios for future trajectories of GHG concentrations depending on the level of projected global action to mitigate emissions. In the most recent projections, these scenarios are known as *representative concentration pathways* (RCPs), which are labelled by the expected level of radiative forcing which results. That is to say, the net imbalance (incoming vs outgoing radiation from the sun) in terms of radiative flux in $W \cdot m^{-2}$. Four particular scenarios have been extensively used: RCP2.6, 4.5, 6, and 8.5. The lowest value of $2.6\ W \cdot m^{-2}$ represents a very aggressive global effort to limit emissions and $8.5\ W \cdot m^{-2}$ represents a business-as-usual scenario. The remaining two radiative fluxes represent different intermediate levels of effort to limit emissions. It goes without saying that uncertainties in model physics and in expected levels of emissions lead to a significant range of changes to global temperatures, though there is consensus that global average temperatures will increase. Rather less certain is the change to wind speeds which are very dependent on changes to global circulation patterns, including their strength and geographical tracks, of which the ability to predict is still tenuous. Different model projections can give quite different results, sometimes resulting in opposing trends. Nevertheless, attempts have been made to assess localised changes in wind speed during the twenty-first century using ensembles of model predictions.

11.3 Possible Future Changes and Uncertainties

One such model ensemble assessment for Europe based on the best performing of the 21 CMIP5 models (Carvalho et al., 2017) was made by converting wind speed u to theoretical wind power density P_W using the formula

$$P_W = \frac{1}{2}\rho u^3 \tag{11.1}$$

where ρ is the air density assuming a fixed value of $1.225\ \mathrm{kg \cdot m^{-3}}$. Figure 11.5 shows the resulting median absolute deviation of the wind power density from the long-term historical mean based on ERA-Interim over the period 1985–2006. The near-future period shows little change except for Eastern Europe for both the intermediate and high RCP scenarios. Further into the future and into the latter part of the twenty-first century both scenarios indicate lower wind speeds over much of Europe, though the high RCP scenario shows increasing wind speeds over parts of the Baltic region. Noticeably, the intermediate RCP scenario shows significantly decreasing wind speeds over Eastern Europe. These trends have been confirmed by a similar study using EURO-CORDEX regional climate simulations (Tobin et al., 2016), though a more recent study using the CMIP6 models has indicated larger declines in wind speed across Europe, particularly for the RCP8.5 scenario, with an even smaller expected increase in the Baltic region (Carvalho et al., 2021). The findings highlight the nonlinear nature of the climate whereby the strength of change is not uniformly related to the strength of the radiative forcing. This represents a challenge and emphasises the point in Section 11.2 that changes in wind speed will be linked to changes in circulation patterns which are still highly uncertain.

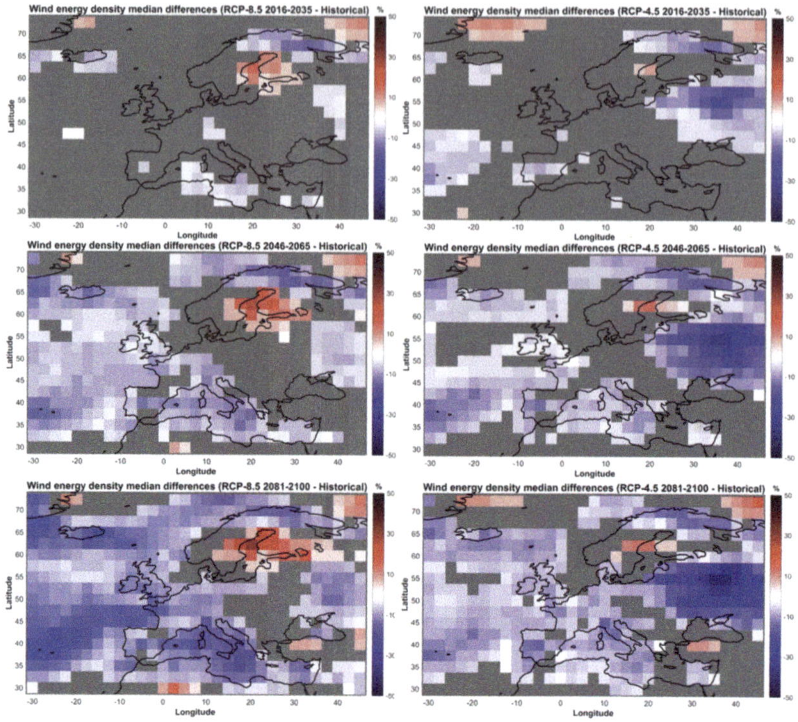

Figure 11.5 Median absolute deviation in wind power density from the long-term historic mean (1985–2006) over Europe based on an ensemble of CMIP5 models for the RCP8.5 and 4.5 scenarios and different future periods. Areas in grey represent no significant change. *Source:* Carvalho et al. (2017) / with permission of Elsevier.

11.4 Impact on Resource Assessment

It can be seen from this chapter that historic and future projected trends in wind speed show a large uncertainty and that trends differ across different geographical areas. Trying to project a consistent trend in any wind resource assessment is thus an almost impossible task at present. Future changes in the strength and trajectory of major weather circulation patterns are far too uncertain to factor into long-term wind speed estimates. The most that can be done is to factor in long-term climatological uncertainty based on historical data close to the site of interest. This is generally the approach taken in an MCP analysis. The question that remains is what length of historical data to use in order to sufficiently represent long-term climate uncertainty.

References

Carvalho, D., Rocha, A., Costoya, X., DeCastro, M., Gómez-Gesteira, M., 2021. Wind energy resource over Europe under CMIP6 future climate projections: what changes from CMIP5 to CMIP6. *Renew. Sustain. Energy Rev.* 151, 111594. https://doi.org/10.1016/J.RSER.2021.111594.

Carvalho, D., Rocha, A., Gómez-Gesteira, M., Silva Santos, C., 2017. Potential impacts of climate change on European wind energy resource under the CMIP5 future climate projections. *Renew. Energy* 101, 29–40. https://doi.org/10.1016/J.RENENE.2016.08.036.

Eyring, V., Bony, S., Meehl, G.A., Senior, C.A., Stevens, B., Stouffer, R.J., Taylor, K.E., 2016. Overview of the Coupled Model Intercomparison Project Phase 6 (CMIP6) experimental design and organization. *Geosci. Model Dev.* 9, 1937–1958. https://doi.org/10.5194/GMD-9-1937-2016.

IPCC, 2013. *Climate Change 2013: The Physical Science Basis: Contribution of Working Group I to the Fifth Assessment Report of the Intergovernmental Panel on Climate Change.* T.F. Stocker, D. Qin, G.-K. Plattner, et al. (eds.). Cambridge University Press.

IPCC, 2014. *Climate Change 2014: Synthesis Report: Contribution of Working Groups I, II and III to the fifth Assessment Report of the Intergovernmental Panel on Climate Change.* Core Writing Team, R.K. Pachauri, L.A. Meyer (eds.) Geneva IPCC.

McVicar, T.R., Roderick, M.L., Donohue, R.J., Li, L.T., Van Niel, T.G., Thomas, A., Grieser, J., Jhajharia, D., Himri, Y., Mahowald, N.M., Mescherskaya, A.V., Kruger, A.C., Rehman, S., Dinpashoh, Y., 2012. Global review and synthesis of trends in observed terrestrial near-surface wind speeds: implications for evaporation. *J. Hydrol.* 416–417, 182–205. https://doi.org/10.1016/J.JHYDROL.2011.10.024.

Ren, D., 2010. Effects of global warming on wind energy availability. *J. Renew. Sustain. Energy* 2, 052301. https://doi.org/10.1063/1.3486072.

Tobin, I., Jerez, S., Vautard, R., Thais, F., Van Meijgaard, E., Prein, A., Déqué, M., Kotlarski, S., Maule, C.F., Nikulin, G., Noël, T., Teichmann, C., 2016. Climate change impacts on the power generation potential of a European mid-century wind farms scenario. *Environ. Res. Lett.* 11, https://doi.org/10.1088/1748-9326/11/3/034013.

Torralba, V., Doblas-Reyes, F.J., Gonzalez-Reviriego, N., 2017. Uncertainty in recent near-surface wind speed trends: a global reanalysis intercomparison. *Environ. Res. Lett.* 12, https://doi.org/10.1088/1748-9326/AA8A58.

Young, I.R., Ribal, A., 2019. Multiplatform evaluation of global trends in wind speed and wave height. *Science* 364, 548–552. https://doi.org/10.1126/SCIENCE.AAV9527.

Young, I.R., Zieger, S., Babanin, A.V., 2011. Global trends in wind speed and wave height. *Science* 332, 451–455. https://doi.org/10.1126/SCIENCE.1197219.

Zeng, Z., Ziegler, A.D., Searchinger, T., Yang, L., Chen, A., Ju, K., Piao, S., Li, L.Z.X., Ciais, P., Chen, D., Liu, J., Azorin-Molina, C., Chappell, A., Medvigy, D., Wood, E.F., 2019. A reversal in global terrestrial stilling and its implications for wind energy production. *Nat. Clim. Chang.* 9, 979–985. https://doi.org/10.1038/S41558-019-0622-6.

Index

a

Adiabatic 17
Adiabatic lapse rate 19
Air density measurement 48
Anabatic wind 265
Analogue ensembles 210–213
Analysis 228
Anemometer
  cup 33, 34, 36, 78
  sonic 45, 78
Anemometer *k* parameter 45
Assimilation 228
Atmosphere-Ocean General Circulation Models (AOGCMs) 286
Atmospheric boundary layer 11
  Ekman layer 11, 27
  height 11, 25, 27, 61
  height measurement 61
  surface layer 11
Azimuth angle 54

b

Background forecast 227
Back-scattering 54, 57, 59, 60
Beat frequency 56
Beaufort scale 2, 4–5
Bessel function 96
Blind test 186
Boom 51
Boussinesq approximation 114, 129
Brunt-Väisälä frequency 26, 146, 266, 267
Buoyancy 16, 17, 114, 264

c

Calms 90, 91
Canopy area index 164
Capacity factor 6
Ceilometer 30, 39, 61, 63
Centrifugal force 22, 23, 24
CFD in wind resource assessment 134
Charnock relationship 12
Climate change 282
Closed cell convection 273
Closure model 121, 127–132
Closure order 139
Closure problem 119
Cloud streets 275
Coastal transition 157
Coefficient of determination 247
Coherence 199

Handbook of Wind Resource Assessment, First Edition. Simon Watson.

Co-kriging 219–220
Cold climates 78–79
Cold to warm surface transition 162
Complex terrain 53, 66, 126, 146, 150, 157–175, 187, 190, 191, 206
Computational fluid dynamics model 66, 127, 134–136
Conservation of a scalar 112
Conservation of mass 111
Continuity 111
Continuity of measurements 76
Convective boundary layer (CBL) 16
Convective cells 273
Conventionally neutral boundary layer (CNBL) 25
Coriolis force 11, 22, 23, 24, 114, 264
Coriolis parameter 24
Correlation 203
Cosine response 41
Cross-isobar angle 24, 28
Cut-off frequency 42

d

Dalton’s law 48
Data logger 50
Data quality 76
Data recording policies 76
Dead band 47
Decision tree 223
Diabatic 14
Diagnostic model 115
Direct numerical simulation (DNS) 112
Dispersion 102
Displacement height 163
Distance constant 38
Distribution 1, 90
  bi-modal Weibull 92
  generalised extreme value (GEV) 102
  Gumbel 102
  Kaimal 100
  Kolmogorov 100
  Rayleigh 6, 91
  singly truncated from below normal Weibull 92
  turbulence 99–101
  two-component mixture Weibull 92
  von Kármán 100
  von Mises 95
  Weibull 90
  wind direction 95–99
Doppler shift 54
Downscaling 137, 144, 223, 251, 252, 286

e

Earth System Models (ESMs) 286
Ekman spiral 131
Elevation angle 54
Equation of state 19, 112
European Wind Atlas 122
Exceedance risk 102
Exposure 89
Extreme value statistics 101–107
Extreme wind speed 34

f

Far wake 170
Floating lidar 69
Flow over hills 65, 119, 123–125, 179–196

Flow separation 126, 127, 129, 148, 157, 179, 183, 184, 186, 189, 193
Forecast 228
Forested canopies 163–169
Free atmosphere 11
Friction velocity 12, 14, 24
Froude number 146

g

Gaussian grid 229
Geophysical model function (GMF) 71
Geostrophic wind 22–27
Global heating 202, 281
Global model 142
Governing equations 111–114
Gradient wind 22–27
Grey zone modelling 141
Grid 134
Guidelines for calculating speed-up over a hill 123–125
Gust 43, 105–107
Gust factor 106

h

Heterodyne 54
Hindcast 228
Homodyne 54
Homodyne lidar 180-degree ambiguity 59
Hybrid levels 232
Hydraulic jump 195
Hydrostatic balance 19
Hyperparameters 223

i

Icing 78
Inertial oscillation 270
Inertial sublayer 11
Inner layer 119
Instantaneous wind speed 34
Instrument mounting 50–53
Intermediate layer 183
Internal boundary layer (IBL) 123, 157
Internal gravity waves 74, 266–269, 270
Inverse distance weighting (IDW) 217
Inversion layer 25, 134
Inviscid 120

k

Katabatic wind 265
Kolmogorov constant 39
Kriging 218–219

l

Land breeze 264
Large eddy simulation (LES) model 133–134
Leaf area density (LAD) 165
Leaf area index (LAI) 164
Lidar 53–64
 continuous wave (CW) 57–58
 pulsed 57–58
Lidars *vs.* sodars 61, 79
Lidars, sodars and radars in non-uniform terrain 66
Linear regression 203–209
Linearised model 119–125
Line-of-sight wind speed 56
Local closure 140
Local oscillator 56
Logarithmic profile 14, 157, 163
Long-term trends 281–286

Long-term variation 88–90
Low-level jet 143, 269–272

m
Machine learning (ML) 222–224
Mass consistent model 115–119
Mast shading 52
Matrix MCP 205
Mean wind speed 34
Measure-Correlate-Predict (MCP) 202–210
Measuring stability 18
Measuring temperature profiles 63
Measuring the wind 33–80
  above the atmospheric boundary layer 70, 72–73
Mesh 134
Mesoscale 148, 190, 191, 263–276
Mesoscale model 136–142, 143
Meteorological mast 8, 50, 51, 78, 250
Meteorological station 75–78
  geographical spread 77–78
Method of independent storms 103
Microscale 114
Microscale model 114–136, 143, 186
Microwave radiometer (MWR) 63–64
Mixing length 121
Mode 102
Model chain 142–146
Modified pressure 132
Momentum 14, 15, 130, 132, 133, 165
Mountain breeze 265
Mountain wave 148, 150, 266, 267
Mountain wind 265

n
Navier Stokes equations 112
Navier Stokes model 127
Need for wind resource assessment 6–7
Nested grid 137
Neutral boundary layer 25
Nonhydrostatic 136
Nonlinear regression 209–210
Nonlocal closure 140
Nudging 137
Numerical modelling 111–150
Numerical weather prediction model 89, 251

o
Obstacle model 170
Obukhov length 14, 17
Offshore 157–175
One in *T* year wind speed 102
One-way coupling 138
Open cell convection 273
Orography 126, 146, 150, 157, 179–196, 263, 266
Outer layer 119
Overspeeding 36

p
Peak-over-threshold 103
Pearson correlation coefficient 201, 205, 246
Planetary boundary layer (PBL) scheme 138
Point measurements 64–66
Potential flow region 120

Power law profile 13
Power spectrum 85–87
Prandtl layer 28
Pressure gradient 23
Pressure sensor
  capacitive 49
  resistance 49
Probe volume 58

r

Radar profiler 64
Radiative forcing 286
Radio acoustic sounding system (RASS) 60–61
Radiosonde 70–71
Rank order 102
Rayleigh length 69
Reanalysis 89, 227–229, 233, 244, 250
  comparison 244–250
Reanalysis generations 229
Reference site 201
Regional climate model (RCM) 286
Relative humidity sensor 50
Remote sensing device 53
Representative concentration pathways (RCPs) 286
Return period 102
Reversal height 94
Reynolds averaged Navier-Stokes (RANS) equations 114
Reynolds number 130
Reynolds' decomposition 112
Richardson number
  bulk 18
  gradient 18
Robust coefficient of variation (RCoV) 214–216
Roll vortices 269, 273, 275–276
Rossby number 94
Rotor inertia 37, 42
Roughness classification scheme 12
Roughness length 11, 172
Roughness sublayer 11
Rough-to-smooth surface transition 124
Ruggedness Index (RIX) 125, 252, 253

s

Sampling interval 42
Satellite 71
  altimeter 72
  lidar 72
  scatterometer 71
  synthetic aperture radar (SAR) 72, 270
Scan pattern
  Doppler beam swinging (DBS) 54
  plan position indicator (PPI) 53
  range height indicator (RHI) 54
  velocity azimuth display (VAD) 54
Scorer parameter 268
Sea breeze 136, 143, 263–264
Sea surface roughness 12
Shear exponent factor 12, 14
Shear stress 14
Shelter model 172
Sigma coordinates 117
Site exposure 76
Smooth-to-rough surface transition 124

Sodar 60–61
Source terms 112, 114
Spatially-averaged measurements 64–66
Spectral gap 87–88
Spectral model 229
Spectral truncation 229
Speed-up factor 124
Spin-up 251
Splines 217–218
Stability 146–150
  neutral 14, 16, 146, 149, 150
  stable 16, 146, 149, 150
  unstable 16
Stability dependent functions 16, 39
Stability distributions 21
Stability parameter 24
Stair step terrain 117
Steep terrain 125–126
Sub-filter-scale (SFS) parameterisation scheme 141
Sub-grid-scale (SGS) model 134
Synoptic 86

t

Tall wind profiles 27
Target site 201
Temperature
  absolute 15
  potential 15
  virtual absolute 15
  virtual potential 15
Temperature sensor
  platinum resistance thermometer (PRT) 49
  resistance temperature detector (RTD) 49
Terra incognita 141
Terrain following coordinates 117
Test case 179
Tetens' formula 48
Thermally driven winds 188, 196, 263–266
Trend surface analysis (TSA) 217
Turbulence 14, 25, 34, 67, 68, 86, 88, 99, 119, 122, 129, 133, 139, 140, 141, 165, 166
  distributions 99–101
  longitudinal 35, 200
  measuring 67–70
  transverse 40
  vertical 40
Turbulence intensity 42
  longitudinal 35
  transverse 40
  vertical 40
Turbulent stress transport 114
Two-way coupling 138

u

Unsteady Reynolds averaged Navier Stokes (URANS) model 132–133
Urban areas 169–175

v

Valley breeze 268
Valley wind 136, 143, 268
Van der Hoven spectrum 85
Variance ratio method 205
Viscosity
  dynamic 14
  eddy 14, 129
  effective 133
  kinematic 128, 130
von Kármán constant 14

W

Warm to cold surface transition 162
Weather balloons 70
Weathervane 1, 47
Weibull scale parameter 90
Weibull shape parameter 90
Wind atlas 7, 104, 118, 144, 250–254
  validation 252, 253
Wind index 89, 213–214
Wind rose 97, 208
Wind shear 11, 16, 17, 146
Wind speed profile 13–22, 158, 172, 270
Wind speed turbulent fluctuation 34
Wind speed variability 85–107
Wind vane 34, 47–48
  potentiometer 47
  reed switch 47

Printed in the USA/Agawam, MA
September 16, 2026

910661.180